Lecture Notes in Mobility

Series Editor

Gereon Meyer

AF785836

For further volumes:
http://www.springer.com/series/11573

Editor
Institute for Mobility Research (ifmo)
A Research Facility of the BMW Group
Munich
Germany

ISBN 978-3-642-43705-2 ISBN 978-3-642-34735-1 (eBook)
DOI 10.1007/978-3-642-34735-1
Springer Heidelberg New York Dordrecht London

© Springer-Verlag Berlin Heidelberg 2013
Softcover reprint of the hardcover 1st edition 2013
This work is subject to copyright. All rights are reserved by the Publisher, whether the whole or part of the material is concerned, specifically the rights of translation, reprinting, reuse of illustrations, recitation, broadcasting, reproduction on microfilms or in any other physical way, and transmission or information storage and retrieval, electronic adaptation, computer software, or by similar or dissimilar methodology now known or hereafter developed. Exempted from this legal reservation are brief excerpts in connection with reviews or scholarly analysis or material supplied specifically for the purpose of being entered and executed on a computer system, for exclusive use by the purchaser of the work. Duplication of this publication or parts thereof is permitted only under the provisions of the Copyright Law of the Publisher's location, in its current version, and permission for use must always be obtained from Springer. Permissions for use may be obtained through RightsLink at the Copyright Clearance Center. Violations are liable to prosecution under the respective Copyright Law.
The use of general descriptive names, registered names, trademarks, service marks, etc. in this publication does not imply, even in the absence of a specific statement, that such names are exempt from the relevant protective laws and regulations and therefore free for general use.
While the advice and information in this book are believed to be true and accurate at the date of publication, neither the authors nor the editors nor the publisher can accept any legal responsibility for any errors or omissions that may be made. The publisher makes no warranty, express or implied, with respect to the material contained herein.

Printed on acid-free paper

Springer is part of Springer Science+Business Media (www.springer.com)

Foreword

What explains the enormous diversity of mobility patterns found in megacities, the variety in the means of transport, and the multiplicity of aspirations concerning the future of urban transport, as so impressively presented in this collection of narratives on mobility cultures around the world? The stories suggest that we tend to underestimate this diversity. It is important to grasp what forces lie behind the differences in mobility cultures if we are to improve the liveability of cities, to increase the quality of transport services and to craft a transport sector that supports economic development whilst minimising social costs.

Major differences in mobility reflect the different stages of urban development. In their early stages, many large cities are characterised by high densities of settlement, with residential locations and jobs in close proximity. The travel distances involved in commuting, shipping intermediate goods, and shopping are all short. To a large extent, people get around by walking and cycling, and even the transporting of goods is for the most part independent of motorisation.

As development continues, and the mass production of goods comes to dominate economic activity—with a vast array of services springing up to support this shift—an increasing demand for land by the producers in emerging sectors often brings about a spatial separation between jobs and residential locations. This separation arises from demand by businesses for locations close to other businesses, whereby they gain productivity benefits from the joint use of infrastructures and the varieties of services and inputs. Meanwhile, poorer households seek to escape the rapidly increasing land rents, and wealthier residents try to avoid the disadvantages of inner cities that are part and parcel of economic progress, such as noise and air pollution and the beginnings of inner-city congestion. These processes lead to a rapidly increasing demand for urban transport. The way in which transport and land-use policies respond to the challenges of pollution and congestion plays a major part in determining which mobility cultures emerge.

A focus on facilitating residential relocation to the outskirts of cities will lead to a strong dependence on individual mobility, reflecting the difficulty of serving dispersed settlements by public transport. The larger the sprawl, the higher will be the dependence on fossil fuels, the health costs arising from local air pollution, and the time loss due to congestion. Allowing cities to sprawl and individual motorisation to surge will make longer-term increases in transport needs and a shift to a higher energy intensity in transport inevitable.

Failure to guide the often dramatic increase in motorisation that takes place when a country joins the ranks of middle-income nations reduces the liveability of cities, and also their long-term wealth. Positive long-term growth rates depend on the production and use of knowledge, a resource which is not exhausted by being used. Despite all the progress made in information and communication technologies, post-industrial cities still depend on face-to-face communication between the producers of knowledge. Extreme levels of congestion in some of the world's megacities run the risk of strangling the interaction between consumers and service producers by, effectively, fragmenting them into disconnected parts. Instead of realising the productivity potential of the local diversity and scale, large cities thus tend to run into stagnation.

Three types of urban settlement and mobility have emerged in high-income countries, which can be summarised in terms of energy intensity and the modal structure of transport. Differences between their large cities translate into differences in national energy use in the transport sector. North American countries have a high degree of sprawl, and consume four to six times as much energy per head as European countries, which mark a middle ground. Asian countries such as Japan, Korea, Singapore, and Hong Kong are benchmarks of energy efficiency in transport. These three differing groups of nations teach us the lesson that countries have a choice. Containing emissions in transport and maintaining relatively compact cities has not proved to be a barrier to high growth rates, as has often been feared. In fact, these Asian countries are considered to be 'development miracles' of recent decades.

The narratives of the mobility cultures in the megacities portrayed in this book provide precious insights into the details of the interaction between development, available and affordable technologies, and the preferences of citizens; moreover, they constitute a valuable source of advice on how urban mobility patterns shape growth and welfare even at the national level.

Dr. Andreas Kopp
Lead Transport Economist, The World Bank Group
Member of the Board of Trustees of the
Institute for Mobility Research

Preface

The global increase of the urban population reflects the high expectations and great opportunities associated with urban life. These opportunities, however, can only be exploited if accessibility and mobility are assured—meaning that mobility is a key factor to the welfare of urban societies. In turn, the issue of urban mobility presents both challenges and opportunities for all those who provide for mobility infrastructure, services and products. However, mobility as specifically applied to megacities often appears to be characterised by an impenetrable complexity which is evident in both the physical characteristics of urban transport systems, and in the underlying forces—economic, cultural, and so forth—which drive and shape them. Combined, they constitute what we call in this book 'mobility culture'. The complexity of urban mobility culture makes the long-term planning and provision of mobility in modern agglomerations a herculean task for all stakeholders—including policymakers, administrators, planners, businesses and entrepreneurs.

Against this background, the Institute for Mobility Research (ifmo) initiated in 2010 a two-year research programme on megacity mobility culture. The ifmo is a research facility of the BMW Group, established in 1998. It deals with future developments and challenges relating to mobility across all modes of transport, with automobility being only one aspect among many. Taking on an international perspective, ifmo's activities focus on social science and sociopolitical, economic and ecological issues, but also extend to cultural questions related to the key challenges facing the future of mobility.

This book is a result of the programme. In this preface, ifmo, as the editor, presents the genesis of its contents throughout the history of the project. We also gratefully acknowledge the relevant parties involved in both generating the insights documented in it and providing guidance throughout the research programme.

This 'Mobility Culture in Megacities' research initiative has aimed to gain a better understanding of the complex and dynamic issues associated with megacity mobility culture, and to provide information that will be of use in supporting decisions in this area. The programme included a pre-study, a post-doctoral fellowship programme and expert workshops dedicated to different domains of megacity mobility culture. This book compiles relevant results from this research and includes contributions by internationally renowned experts on megacities and urban mobility.

The first phase of the ifmo research into megacity mobility culture comprised a pre-study which serves to categorise cities. Every city, and even more so every megacity, is unique in many respects. This fact, however, brings with it the danger of getting lost in the complexity and heterogeneity of megacity mobility culture. The pre-study was therefore established to create a comparative categorisation of cities based on quantifiable mobility characteristics and thus provide some basic orientation for navigating the great diversity of global cities. But this idea is feasible only if harmonised data are available. The Millennium Cities Database for Sustainable Transport, dating from 2001 and containing 1995 data, is the only comprehensive database that brings together a very wide range of comparable mobility-related characteristics of urban areas around the globe. 40 cities out of the Millennium Cities Database were selected as the subjects of a cluster analysis based on almost 60 data items.

In this pre-study a research team, drawing members from Technische Universität München and Goethe-Universität Frankfurt (both in Germany), developed a taxonomy comprising six relevant clusters of cities (plus one outlier, Manila) for urban agglomerations. With regard to the framework conditions for mobility, these clusters differ from each other fundamentally—for example, in urban density and in their level of economic development. These different framework conditions, as well as cultural issues, lead to very different quantifiable characteristics of the transport system, and diverging mobility behaviour. Two of the six clusters cover cities with a relatively low level of development: (1) *Non-Motorised Cities*, which are characterised by high density that enables non-motorised transport to dominate; (2) *Paratransit Cities*, which have generated a need and a market for inexpensive informal public transport to satisfy mobility demand in low-density environments. Another cluster of cities is characterised by a higher level of development, and correspondingly increased demand for motorised transport, but also by an inadequate supply of transport infrastructure, leading to (3) *Traffic-Saturated Cities*. The remaining three clusters each comprise relatively wealthy types of agglomerations: (4) *Auto Cities* have low densities, associated with an absolute dominance of the car; (5) *Hybrid Cities* exhibit dense public transport-orientated urban cores with a significant modal share of non-motorised means of transport, surrounded by low-density car-orientated suburban areas; finally, in the high density (6) *Transit Cities*, public transport dominates urban mobility culture.

The chapter presenting the findings of this cluster study evidently takes a key position in this book. First, it provides a categorisation of cities which helps to supply orientation in the overwhelming diversity of cities worldwide. Second—as the subsequent 'Reader's Guide to Mobility Culture' illustrates—it also provides a framework for mapping change in the urban transport system. The discussions in this Guide make it clear that the key to understanding mobility culture is understanding *change* and understanding the driving forces for changes.

The next stage of the ifmo megacities research programme was dedicated to exploring in more detail the urban mobility cultures in these city clusters, and the associated challenges and prospects. Moreover, the previous cluster study was based only on a cross-sectional snapshot of quantifiable characteristics of the transport system and local mobility, taken at the time that the data were surveyed. Mobility culture, however, also encompasses the barely observable underlying cultural driving forces that shape the transport system. Moreover, mobility culture is in a constant state of flux. It is important to understand the processes that have created a given situation before attempting to draw conclusions about likely future developments. For this reason, the next stage of the research programme aimed to also understand the longitudinal aspects of mobility culture from a long-term historical perspective. To accomplish this, case studies on selected cities and selected aspects of megacity mobility culture were initiated. This requires in-depth knowledge of the cities under scrutiny, and professional networks able to access specific information and data.

Consequently, ifmo launched a post-doctoral fellowship programme on mobility culture in megacities following the pre-study. Throughout 2011, this fellowship programme brought mobility scholars from different disciplines, each representing a city, together in Munich. The fellows and cities were selected by an expert committee so as to cover a wide range of urban agglomerations from different geographic areas across the world. With eight cities from five continents, the sample studied in the fellowship programme covers a broad variety of agglomerations.

Despite the fact that not all city clusters were represented in the sample of study cities, the cities selected are suitable for representing relevant mobility culture issues that have been identified for all the city clusters listed above. Atlanta, for example, which was technically categorised as a Hybrid City because of some public transport supply in its urban core, can serve very well to illustrate car dependency, thus shedding light on Auto Cities. Likewise, public transport as a backbone for urban transport, and issues of traffic saturation, represent central aspects of mobility culture across various clusters. This illustrates that the assignment of cities to clusters is based on the dominant local mobility culture, whereas various mobility cultures usually coexist in one city. These gain or lose weight in the specific urban context as a city moves along evolutionary paths through different stages of development.

Under the supervision of an academic team from the two universities, the fellows advanced the research on mobility culture in numerous ways. In their individual research, they combined their interdisciplinary perspectives on mobility culture with specific research questions pertaining to their disciplines, such as travel mode choice modelling and analysing public discourses. The results of this work do not form part of this book, as they are of interest to specific scientific communities, and are published suitably elsewhere.

Moreover, the fellows were able to carve out important dimensions of the identity of the study cities' mobility culture which have not been described with such clarity before. The fellowship brought together scholars from different cultural, urban and academic backgrounds. Their areas of expertise included history, geography, public policy and architecture, as well as industrial and civil engineering. The distinct multinational and multidisciplinary character of the fellowship enabled a systematic comparative look at the study cities from these different

What is Mobility Culture?

When attempting to define the term 'mobility culture', first, it is necessary to agree to a common understanding of the term 'mobility'. In this context, we understand mobility as the '*ability to travel from one point to another*' and '*actual physical travel*'. Second, the term 'culture' can be defined as '*the set of values, conventions, or social practices associated with a particular field, activity, or societal characteristic*'.[1] Combined, '*mobility culture*' is, then, '*the set of values, conventions, or social practices associated with the ability to travel from one point to another, and with actual physical travel.*'

This appears to be a straightforward definition. However, it turns out to be more blurred once we acknowledge that in modern societies the ability to travel is not purely the result of the traveller's individual situation, decisions and activities. Our ability to travel is almost always subject to numerous external preconditions, ranging from the existence of infrastructure and the availability of technology or energy, to the presence of someone to operate the vehicles in which we travel. In short, our ability to travel is a societal achievement that has been shaped, and continues to be shaped, by countless actors, other activities and framework conditions. Hence, mobility culture is not only intrinsic to our individual ability to travel and the way we travel, but also to how the preconditions for individual travel have come about.

Against this background, it is evident how all-embracing any description of the phenomenon 'mobility culture' has to be. Accordingly, Götz and Deffner define 'mobility culture' as "the entire set of symbolic and physical social practices that relate to the ability to travel. This includes the design of infrastructure and public spaces, transport policy role models and associated public discourses, as well as individual travel behaviour, including the mobility and lifestyle orientations which shape it" (Götz and Deffner 2009). Hence, mobility culture is shaped by many dimensions of our lives, ranging from psychological and social, across demographic and economic, to infrastructural and technological aspects. In reciprocal fashion, mobility culture contributes significantly to shaping these same fields.

In order to pin down the mechanisms that form mobility culture, let us identify the various relevant domains which interact in shaping society's ability to travel:

First, there are *actors*. These are individuals, or groups of persons, whose values, conventions and social practices constitute mobility culture. Different relevant groups of actors can be defined: there are travellers who make individual decisions about their ability to travel, for example choices about vehicle ownership or about actual travel. There are public decision-makers, typically policymakers or planners, who consciously make decisions for larger communities. Finally, there are initiators of innovation which has the potential to shape mobility culture. These innovators can come from very diverse spheres, such as research or administration, but they are generally entrepreneurs who develop technologies, products and services. The activities of innovators also have the potential to influence the mobility

[1] Merriam-Webster, 2011.

of large groups, but often without the innovators themselves being aware of the public impact which their activities have.

Second, there are *actions* by these actors. These actions are part of social practices, and generate observable manifestations of mobility culture which are—at least in part—shaped by the underlying values and conventions of the actors. The myriad actions which enable travel include innovation, communication, decision-making, planning, and designing and building the physical environment and infrastructure. Most of these actions can be purely individual in that they are intended to shape one person's mobility. The actions can also be collective in that they are intended to shape others' ability to travel, or that of larger groups, i.e. they influence the framework conditions for travel.

Third, there are *framework conditions* for the actors and their actions. These framework conditions represent physical or social constraints for mobility—that is to say, they influence the values, conventions and social practices associated with mobility. There are framework conditions relevant to mobility culture which are entirely external, such as topology or climate. However, many of these constraints have been created, or at least adapted, by human actors in a process that is intrinsically influenced by mobility culture: there is the historically shaped geography, including the infrastructure. There are the capacities of decision-makers and innovators, which, for example, include their budgets, knowledge and political power. There is the existence, availability and affordability of technology. There are fundamental societal characteristics, such as religion or segregation of the society by caste, class and race. And there are the socioeconomic conditions, psychological factors and lifestyles of individuals and larger groups.

Moreover, urban transport systems and travel behaviour are never stable, but are constantly changing. On the one hand, this is due to fundamental external influences such as the development of economic and demographic conditions. On the other hand, constant change is deliberately caused by the actions of the actors listed above. These driving forces create processes and events which underlie the development and transformation of urban transport systems. In the process of undergoing this change, mobility culture makes a difference in determining the future of a city's transport system: first, because the actions of actors themselves are a driving force—being influenced by their values, conventions and social practices, which is to say their own mobility culture; second, because mobility culture largely determines how the actors react on external driving forces for change.

Hence, to gain an understanding of mobility culture we need to take on a longitudinal perspective, observing and analysing the change of urban transport systems over time. In this book we will see many examples of how the actors, their actions and the relevant framework conditions all constantly shape and reshape the mobility culture of specific places over a period of time. This leads us to another prerequisite for being able to analyse mobility culture: the choice of a suitable geographic study area. It is immediately evident that mobility culture differs from locale to locale. Specifically, for analysing everyday mobility, an adequate geographic coverage of more or less self-contained entities needs to be selected. For this book, we have chosen a…

Focus on Megacities

The term 'megacity' is associated with metropolises with several million inhabitants. It is the pressing challenges in many domains of urban life that have moved megacities, especially those in the developing world, to the centre of attention. These challenges—with transport being only one of them—are presented in the opening chapters of this book. Megacities also represent suitable, although challenging, study areas in which to understand mobility culture. Each city is truly unique, and the diversity between cities is overwhelming, as we shall see in "The Diversity of Megacities Worldwide: Challenges for the Future of Mobility" in this book. This diversity helps us to understand that each city features its own unique mobility culture, which not only differs from that of other cities, but also stands out from the rest of the country in which that city is located.

This book is a resource for the study of megacity mobility culture in the sense that it is intended to provide help in understanding the particularly complex and multifaceted world of mobility in megacities. However, many aspects of urban mobility culture which are relevant for understanding megacities also exist outside these huge urban masses, and can be illustrated using the example of cities of other sizes. Exemplary cities which do not technically qualify as megacities often epitomise certain challenges posed by, or perspectives on, urban mobility culture, and can thus serve well as a point of reference. For this reason, this book ventures beyond what are technically considered to be megacities, and also includes some cities of a smaller scale.

With the strong recent focus on urban mobility, specifically that in megacities, numerous publications tailored to the perspective of public decision-makers and planners have emerged. We will see in the book, however, that the fate of technical and infrastructural solutions is strongly conditioned by a wide range of ambient factors, ranging from topology to psychology. On the one hand, this requires a multidisciplinary perspective. On the other, this also makes the book relevant to various different groups of readers who are stakeholders in the field of urban mobility culture. This comprehensive approach means that the book is unique among numerous recent publications on urban or megacity mobility and transport, leading us to the…

Purpose of the Book

In the light of the pressing challenges with which many megacities are confronted, windows of opportunities have to be made use of wisely if continuing mobility is to be assured in the future. The need to make political or planning decisions, for example about infrastructure investments, or innovations in the field of mobility services and products, present these kinds of opportunities. This wide spectrum of possible approaches for addressing the challenges facing urban mobility illustrates once more the broad variety of actors in this field, which range from local administrations to international suppliers of mobility products.

This book is aimed at the entire spectrum of stakeholders in the field of urban mobility—including but not limited to planners, politicians, businesspeople and entrepreneurs. All of these actors need to understand local mobility and how it is embedded in a specific environment. However, not only do the perspectives of the various stakeholders differ, but their informational needs also vary: while many planners, specifically those in local administrations, *are already familiar* with the local context, international actors, by contrast, *have to familiarise themselves* with local conditions adequately before being able to make sensible decisions.

The process of getting to know local conditions requires help when it comes to navigating through the complex world of mobility and its interaction with other areas of life. This book provides for such navigation. Its purpose is to provide useful insights for various stakeholders, enabling them to understand the principles and interactions which constitute mobility culture, and, ultimately, supporting them as they make decisions. Last but not least, the book also speaks to researchers, providing food for thought which, it is hoped, will stimulate further research into the issues of mobility culture, with the aim of encouraging them to take a fresh and comprehensive look at transport.

Structure of the Book

The book is laid out in three parts, starting with "**Setting the Context**", in which the first chapter, by Philip Rode, introduces us to the world of megacities. Titled "Trends and Challenges: Global Urbanisation and Urban Mobility", it presents facts, figures and current trends connected with urbanisation. Rode introduces the pressing problems that the cities face, and how these are connected to some of the fundamental challenges facing mankind today. This sets the context for understanding the environment in which the development of transport takes place in the modern world. The chapter also gives a flavour of the huge heterogeneity of existing cities, and hence of mobility cultures.

The second chapter of Part I, "The Diversity of Megacities Worldwide: Challenges for the Future of Mobility", also starts off by emphasising that every city is unique, illustrating this point with numerous statistics on transport. However, it does not leave the reader all at sea in the great diversity of mobility cultures. Instead it provides help in navigating the complexity of global urban mobility culture by categorising cities into clusters. The study presented in the chapter develops six prototypical megacity clusters based on data relating to 40 cities, derived from the 1995 Millennium Cities Database.

Since 1995, time has moved on, and some cities have, naturally, changed significantly, even to the extent that they might now belong in a different cluster. This, however, does not affect the usefulness or relevance of the clusters themselves, which provide for a general taxonomy and also represent stages of development in the life cycle of cities. This is subject of "The Reader's Guide to Mobility Culture", which leads from the quantitative framework provided by the preceding cluster analysis to a more comprehensive analysis of mobility culture. This reader's guide to mobility culture discusses four approaches to understanding it which assist the reader to discover various forms of mobility culture in the megacity stories of the subsequent section of the book.

In Part II of the book we visit eight cities of the world in our "**Stories from the Megacity**". In these chapters, the fellows of the 'Mobility Culture in Megacities' fellowship programme present stories about conflicts and tensions, the inertia of urban mobility systems, the windows of opportunity, and great challenges which sometimes have unexpected solutions. These stories present facets of mobility culture which are first of all important to the specific city, but are also likely to have significant application to many other metropolises worldwide.

Part II begins with Wilson's contribution on **Ahmedabad**. Of all the study cities, this Indian city is probably the one where ancient types of transport are still most prevalent. However, with Ahmedabad's Bus Rapid Transit system, the take-off to modernisation has been initiated, and modern and medieval mobility cultures are in collision. **Beijing**, the city under focus in the next chapter by Song, has already gone a long way: classified as a Non-Motorised City in the mid-1990s, it has undergone a most remarkable development since then, nearly completing a transition to a Transit City. Song's presentation of the forces and measures which shaped this development is a lecture on the dynamics of mobility culture.

Next, we turn to two cities which by 1995 were already positioned on an intermediate level in terms of economic development. Joubert describes how the capital region of South Africa, **Gauteng**, struggles with the mobility culture heritage of apartheid. During decades of race segregation, settlement patterns were enforced which even today make adequate operation of formal public transport close to impossible. Private entrepreneurs have filled this gap by establishing a paratransit system which has become a backbone of transport in Gauteng. Like Gauteng, **São Paulo**, the Brazilian megacity presented by Costa, is characterised by contrasting mobility cultures defining distinct worlds within a single metropolis. Here, social, economic, cultural and political conditions have shaped a mobility situation which imposes stress on nearly all travellers—offering, however, a hugely unequal level of service to the haves and the have-nots.

The next two study cities represent wealthy and car-orientated North American locales. Despite its overall wealth, and largely because of its car-orientation, **Atlanta** is also characterised, by Paget-Seekins, as a city of unequal mobility. Racism in the not-too-distant past, poor planning and fast economic growth have worked together to create a low-density urban fabric with a transport network that limits accessibility, specifically for those without a car. The case of Atlanta illustrates how local mobility culture can be linked to a history of race and class segregation. Another eminently influential factor determining the prevailing mobility culture in the USA is the subject of He's chapter on **Los Angeles**: she discusses the implications of urban density on the Los Angeles mobility culture, and, moreover, points out hidden potential, built into that city's urban structure, for a renaissance of public transport.

The last two megacity stories focus on highly developed European cities which are seeking to further improve their already respected transport systems.

Wengraf's study of **London** shows that the heritage of the built environment limits the possibilities for improving the transport system through building new infrastructure. However, he illustrates how a set of less tangible factors can interpose and create fertile ground for rapid change—in this case the uptake of the G-Wiz electric vehicle. Finally, **Berlin** represents an example of a post-growth metropolis. Heinickel illustrates how—after a history of ambitious planning—Berlin even today banks on transport and urban planning to initiate virtuous circles that will, planners hope, cause the city to make a break from its present-day 'dynamic stagnation'.

These megacity stories contain lessons about urban mobility culture that can be generalised, despite the fact that each of them presents a very particular case. The stories exemplify how a currently prevailing urban mobility culture is deeply rooted in the history and the resulting urban, societal and institutional fabric of a city. This illustrates that any attempt to change and advance urban mobility culture, for example by introducing a new form of mobility, has to consider this legacy and be sensitive to the existing context.

Having discussed the history and status quo of the mobility culture prevailing in selected cities, we turn to the future, looking for "**Perspectives for Megacities on the Move**" in the third part of the book. After the two preceding sections, it is clear that each city's mobility culture is unique, and that the challenges that lie before the cities are many and varied, and differ from one locality to another. Hence, it would seem that there is no easy one-size-fits-all solution to the pressing challenges that urban transport systems around the world are faced with. Nevertheless, there are cities which are perceived as pioneering, and these often serve as role models—they have shown the ability to avoid some of the transport-related problems which other cities are struggling with, because they have taken a path which diverges from the mainstream.

For the beginning of the third part of the book, we have selected one such pioneering city, **Singapore**. Barter's chapter on this city illustrates very clearly how successful it has been in financially disincentivising car ownership and use, in its quest to achieve its transport policy objectives. However, Barter also questions the long-term viability of this concept, mainly because the focus of the Singaporean road infrastructure is on the speed of car travel—neglecting other needs, specifically accessibility for non-motorised modes. This shows that even role-model cities cannot rest on their laurels—they are challenged to develop further and to continue to make preparation for the future.

But what generalised lessons can be learnt from the city case studies in the book? To answer this question, the second chapter of Part III summarises the findings of workshops which were held throughout the fellowship programme, and presents "**Perspectives on Mobility Culture in Megacities**" more generally. First, this chapter reviews the enormous challenges that megacity transport systems are faced with today. Second, it puts forward a list of recommendations that should be considered when developing a local strategy to cope with the future challenges in the transport system. Finally, this chapter promotes a continuance of the discussion of the concept of mobility culture in the research realm, in order to enhance a comprehensive understanding of transport systems.

Stimulating a continued discussion on mobility culture is also our intention in presenting the concluding epilogue to this book. In this epilogue, the fellows of the programme and the editors from ifmo present some food for thought about what we have labelled "**The Seven Mobility Culture Temperaments of Cities**". When studying the cities that are the subject of this book, we found recurrent themes which characterise how urban transport systems move forward. Across all the hugely diverse study cities, we discovered generic mechanisms of interaction between the relevant domains of mobility culture. We labelled these 'temperaments', because, depending on how pronounced any single one of them is in an individual city, they constitute to a greater or lesser degree the mobility culture character of cities.

Given the background of the necessarily all-embracing nature of the concept of mobility culture, we do not claim that this book presents global urban mobility culture in a comprehensive fashion. However, we believe that it brings together a range of stories and analyses so diverse that it has the effect of lifting the curtain and enabling a behind-the-scenes look at the intangible dimensions of transport systems that operate in cities. So then, at the beginning of this journey into the world of megacity mobility, we encourage the reader to see beyond the countless facts, figures and anecdotes in this book and to…

…discover mobility culture!

Dr. Tobias Kuhnimhof
Transportation Engineer
Institute for Mobility Research

Dr. Irene Feige
Head, Institute for Mobility Research

Reference

Götz K and Deffner J (2009). Eine neue Mobilitätskultur in der Stadt—praktische Schritte zur Veränderung. In: Bundesministerium für Verkehr (ed), Urbane Mobilität. Verkehrsforschung des Bundes für die kommunale Praxis,. direkt 65: 39–52. Bonn. (Author's translation)

Part I
Setting the Context

Chapter 1
Trends and Challenges: Global Urbanisation and Urban Mobility

Philipp Rode

Abstract Throughout the world, urban policymakers continue to struggle balancing the ever-increasing levels of activity in cities against the need for more-sustainable forms of urban development. City-making challenges are exacerbated by the global environmental crisis and concerns about climate change, resource depletion, increased levels of pollution, and the loss of biodiversity. These concerns are directly related to current unsustainable patterns of urbanisation. This chapter first focuses on the overall context of global urbanisation and its connections to the challenge of urban transport. With reference to the current growth of cities and the ongoing urbanisation evident in many parts of the world, it details three main components of this growth: people, urban land, and transport. The chapter moves on to highlight the problems, challenges and risks that cities are facing today by looking at the central environmental, economic and social concerns that have emerged in relation to the systemic, spatial component of cities and its linkages to urban transport and access to the city. The chapter concludes by highlighting the importance of working productively with the interrelatedness of urban systems, particularly with regard to land use and urban transport.

Introduction

Throughout the world, urban policymakers continue to struggle balancing the ever-increasing levels of activity in cities against the need for more-sustainable forms of urban development. Questions regarding the size, speed, shape, and spatial distributions of densities, land uses and morphologically differentiated areas of the city and their relationship to transport infrastructure have become increasingly complex and politicised. Related city-making challenges are further exacerbated by the global environmental crisis and concerns about climate change, resource

P. Rode (✉)
London School of Economics, London, UK
e-mail: P.Rode@lse.ac.uk

Institute for Mobility Research (ifmo) (ed.), *Megacity Mobility Culture*,
Lecture Notes in Mobility, DOI: 10.1007/978-3-642-34735-1_1,
© Springer-Verlag Berlin Heidelberg 2013

depletion, increased levels of pollution, and the loss of biodiversity. These concerns are directly related to current unsustainable patterns of urbanisation, which is in turn associated with industrialisation and globalisation.

On the positive side, well-designed cities today find themselves in a unique position to complement national and supra-national agendas in confronting the environmental challenge. With urban areas already encompassing more than 50 % of the world's population, urban territory serves as a proxy for offering answers to the most pressing global challenges [13]. And, cities, with their structural capacity to significantly reduce ecological footprints per capita [70], have good reason to regard themselves as central to the solutions that are being sought for the environmental crisis. Some also see cities as progressive environments, with urban residents tending to be more open to individual behaviour change and therefore providing a better setting for implementing and testing new approaches [4]. Finally, cities—and, more generally, local governments—are closer to 'lives on the ground', potentially allowing for a better understanding of daily routines than state or national governments, a factor that may prove critical in the implementation of change.

This essay is structured in two parts. The first part focuses on the overall context of global urbanisation and its connections to the challenge of urban transport. With reference to the current growth of cities and the ongoing urbanisation evident in many parts of the world, it will first detail three main components of this growth: people, urban land, and transport. The chapter moves on to highlight the problems, challenges and risks that cities are facing today by looking at the central environmental, economic and social concerns that have emerged in relation to the systemic, spatial component of cities and its linkages to urban transport and access to the city. The chapter concludes by highlighting the importance of working productively with the interrelatedness of urban systems, particularly with regard to land use and urban transport.

Dimensions of Urban Growth: People, Territory and Transport

Urban Populations

The world is becoming increasingly urban. In 2007, for the first time in history people living in urban areas outnumbered those living in the countryside (see Fig. 1.1). Only a century ago, this figure stood at 13 %, but it is now predicted to reach 69 % by 2050 [67, 69]. In some regions, cities are expanding rapidly, while in others, rural areas are simply becoming more urban. A significant part of this urbanisation is taking place in developing countries as a result of natural growth within cities and large numbers of rural-to-urban migrants in search of jobs and opportunities. Also, at an individual level, urban regions continue to display

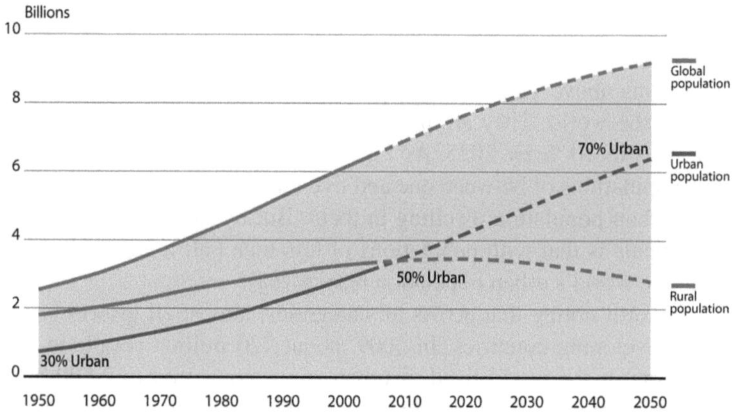

Fig. 1.1 Urban and rural evolution of global population (*Source* LSE Cities at the London School of Economics and Political Science, based on United Nations World Urbanization Prospects, 2007 Revision [67])

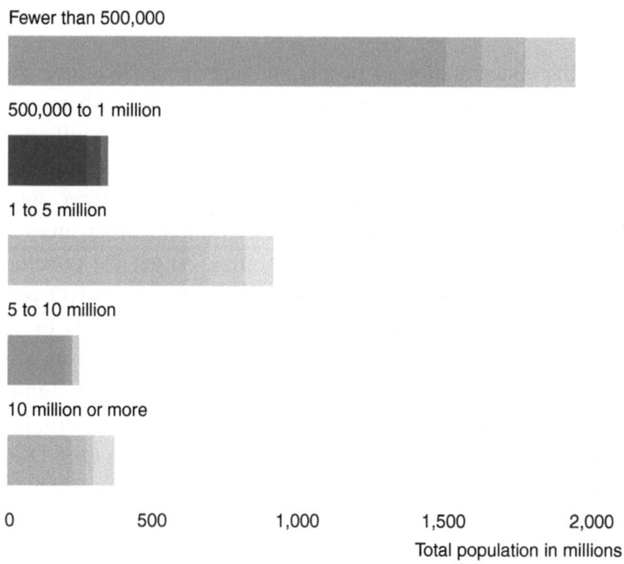

Fig. 1.2 Population by city size class (different shades for 2000, 2005, 2010 and 2015) (*Source* LSE Cities based on the UN World Urbanization Prospects, the 2009 revision [68])

extremely diverse patterns of population development, mostly informed by their specific development stage and their broader socioeconomic contexts. Thus, not all cities grow, and there are in fact an increasing number of cities worldwide that are characterised by population decline.

The UN's differentiation of settlements by city size offers a further perspective on global urbanisation typologies (Fig. 1.2). In 2010, there were 21 megacities with populations above 10 million people, accounting for 8 % of the total urban population in the world. They are predicted to grow modestly to the point where this figure rises to 10 % by 2025. As Fig. 1.2 shows, a more significant share of cities have populations of between one and five million people, with about 22 % of the world's urban population dwelling in them. But by far the largest category of urban settlements is that with populations of less than half a million, where more than half of the world's urban population resides [68].

It is also worth noting that across all categories, almost all growth is expected to occur in developing countries. In 2009, about 920 million people in the more developed parts of the world lived in urban areas, compared to 2,500 million in the developing regions. This ratio contrasts starkly with the balance between the projected figure of 1,010 million in 2025 for the more developed regions and the 3,520 million expected to live in urban areas in the rest of the world [68].

Urban Territory

Cities are about the concentration of people and activities. Therefore, the above figures on absolute urban populations are significantly enriched when considering the land area in which urban populations are living. Globally, the 3.6 billion urban dwellers occupy less than 2 % of the earth's surface [70]. But, rather than increasingly concentrating people, contemporary urbanisation is characterised by a significant trend towards deconcentration, with declining densities within the urban areas themselves.

Looking at a representative sample of 30 cities, Angel [1] concludes that the densities of most of these cities reached their peak more than 100 years ago, and declined on average fourfold from these peaks to average levels of 100 persons/ha (hectare, 0.01 km^2) around the year 2000—the equivalent of an annual rate of decline of 1.5 %. As a result, the increase in urban land has been significant. Since the mid-1950s, urban land area has doubled in OECD (Organisation for Economic Co-operation and Development) countries, while outside the OECD it has grown fivefold [47]. In the US, the total area of the 100 largest urban areas increased by 82 % between 1970 and 1990 [64].

Declining densities are also observable in more recent years. The World Bank has estimated that while urban populations of a representative sample of cities in the developed world grew by only about 5 %, their built-up area increased by 30 % between 1990 and 2000. For a sample of developing world cities, the growth of populations was 20 % as compared to a 40 % increase in urbanised land area [2]. In Angel's study of 120 cities, average built-up density declined during the 1990s from a mean of 144 persons/ha to 112 persons/ha [1]. During that period there were only a few exceptions where densities did not decline, all of them cities in the developing world (Fig. 1.3).

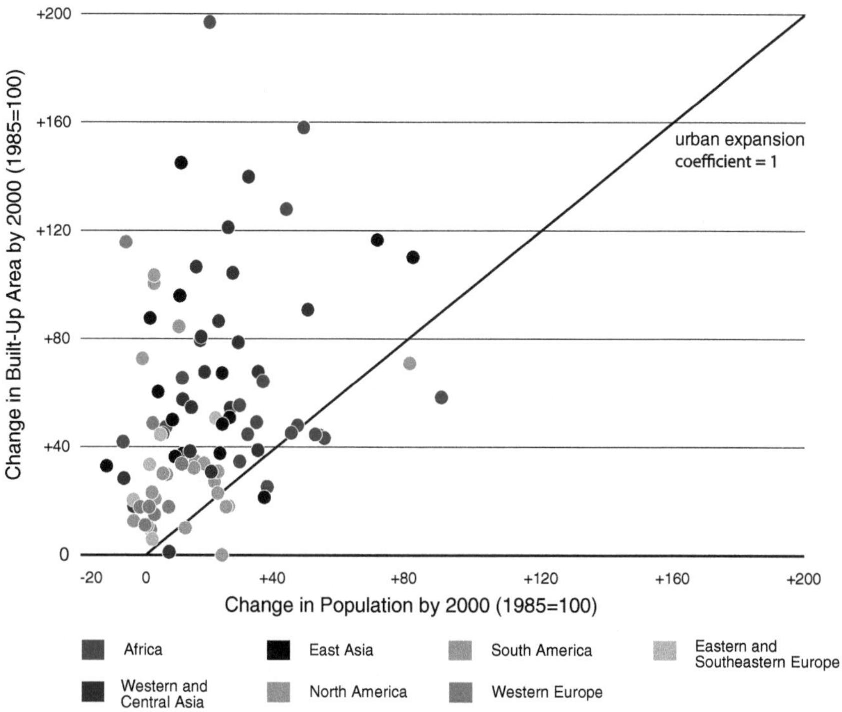

Fig. 1.3 Population change and change of built-up area of selected cities, 1985 to 2000 (*Source* LSE Cities based on Angel [1])

Simultaneously, between 1995 and 2005, suburbs and peripheral development have also grown faster than the urban core in 66 of 78 OECD metro regions [31]. Much of this expansion has occurred with the growth of satellite or dormitory towns and suburban neighbourhoods, triggered by a preference for suburban lifestyles and a combination of many factors: global socioeconomic forces, increasing affluence, prioritisation of personal space over accessibility, cultural traditions, land speculation and land-use planning [17, 24, 65]. At the most basic level, it is of course the complex interplay of land, housing and transport costs—with the introduction of cheap, mechanised transport—that is the most relevant historic factor [1].

On the basis of current trends, the world's urban population may well double in little more than 40 years, but the total extent of urban land in under 20 years. In developing countries, urban population will double by 2030 compared to 2000 levels, and the built-up area is expected to triple over the same period [1]. It is evident that the management of this level of physical urban expansion should be regarded as one of the most central developmental challenges that the world is facing today.

Urban Transport

This growth in urban land area, at a rate that is significantly higher than the growth of urban populations and urban activities, would not have been possible without the development of transport to compensate for the overall decrease in physical proximity. Over the last century, the mechanisation of transport and the related reduction in mobility costs have come in various modes, which have all helped cities to de-densify and expand horizontally. This effect has been well documented following the introduction of streetcars, metro and regional rail systems [26, 28, 32], but has reached an entirely different dimension following mass motorisation with the widespread introduction of privately owned cars [8, 10, 16]. Previously, transit systems allowed for horizontal expansion that both facilitated and required compact, dense urban development and continued to produce human-scale urban environments. Urban designers had to acknowledge the fact that at some point in their journey, all public transport passengers were still pedestrians navigating public urban space.

By contrast, the motor car not only facilitated suburban developments at far lower density levels, but for the first time introduced a transport mode that also required significantly more space (per person) to operate than any previous one. Not surprisingly, many metropolitan regions that once embraced the automobile have become endless cities, with their car fleets demanding more space than ever before. Global urban transport trends further seem to suggest a development cycle whereby cities are affected by the same transformation, related more than anything else to their levels of wealth: initially rising living standards lead to the mechanisation of transport, first facilitated by public transport and then by private vehicles, while higher-income cities are rediscovering walking, cycling and public transport. For cities in emerging economies, the current phase is one of shifting from non-motorised to motorised transport modes, in a manner not unlike the transition of Western cities which was triggered by industrialisation. Other related factors also hint at this analogous development: as travel speeds increase, so do travel distances; and the rate of territorial expansion of cities outstrips population growth, as discussed above.

Given these trends, it is particularly important to more closely address the rapid expansion of car use in cities across the world (Fig. 1.4). This is clearly a defining characteristic of the mid-to-late twentieth-century city, and is set to shape the emerging urban landscape of most expanding metropolitan regions of the world for decades to come. A recent study has forecast that total vehicle stock could increase from 812 million in 2002 to 2 billion by 2030 if motorisation levels across 45 countries (that account for 75 % of the world's population) continue to grow on a 'business-as-usual' basis [20]. Others have predicted that the number of passenger vehicles will reach 2.6 billion by 2050 [76]. Currently, the number of motor vehicles in China increases by more than 20 % every year [49], while car sales in India grow at an annual rate of 16 % [44], with persistent motorisation in emerging economies translating into daily growth rates of about 2,000, 1,000 and 500 new vehicles registered every day in Beijing, São Paulo and Istanbul respectively. But not even these absolute figures adequately reveal the trend towards motorisation: over the last decade,

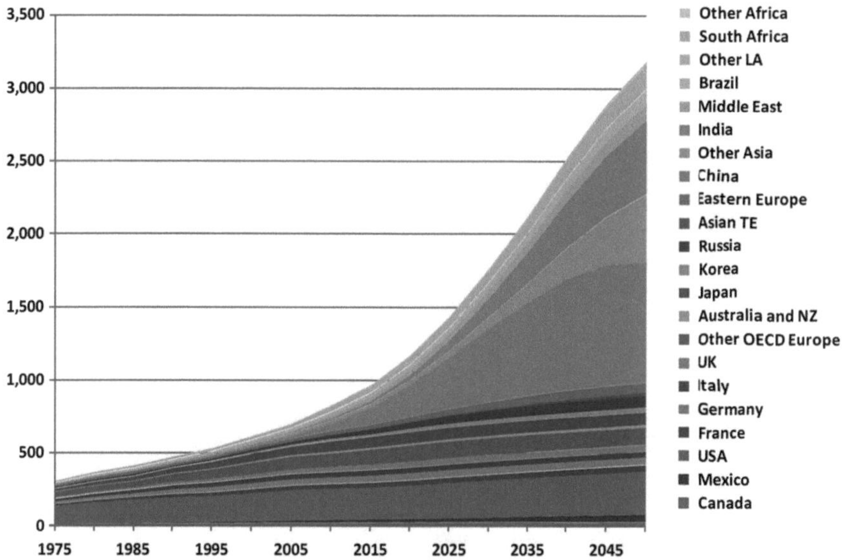

Fig. 1.4 Total stock of motor cars (million vehicles), projection to 2050 (*Source* Fulton/International Energy Agency, Fulton [25])

the growth of motor vehicles in China has been exponential [60], and car ownership nearly tripled over the period between 2002 and 2007, while per-capita incomes increased only twofold [34]. Not surprisingly, in 2010 China became the world's largest car producer and—as with India, Brazil, Turkey and Mexico, and in common with most industrial nations—identified the automotive sector as a 'pillar' industry.

Conclusion

This overview of three key trends related to the strategic spatial development of cities above—urban population, urban territory and urban transport—has already hinted at some of the most relevant interrelationships between them. The following section will identify the most important risks and concerns about the status quo of urban development.

Cities at Risk

Offering any general assessment of contemporary urban development is every bit as impossible as making conclusive remarks on directly related processes, such as industrialisation and globalisation. But if one had to refer generally to global

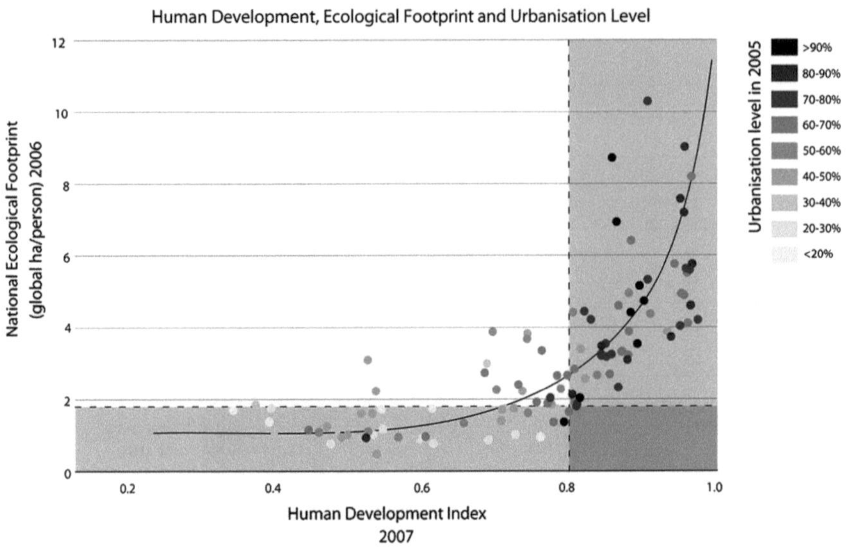

Fig. 1.5 Ecological footprint, human development index and urbanisation by country (*Source* LSE Cities based on multiple sources)

urbanisation in an undifferentiated way, assuming a single universal trend and ignoring variations and local specificities, one observation might be the following: urbanisation has been good for human development, but has compromised environmental sustainability at a global level (see Fig. 1.5). Or, to put it another way, urbanisation has thus far been unable to achieve the decoupling of positive, desirable outcomes from an endless list of negative (side-)effects, with the exception of specific local progress in some wealthier cities, which have relied in large part for this 'progress' on outsourcing their negative externalities to other parts of the world.

Another generalisation, admittedly crude, would be to suggest that the above observation might have been the case to a greater or lesser extent throughout at least the last 200 years of urban development. During the same period, however, the processes of urbanisation have radically changed. Until the early twentieth-century urbanisation could be regarded as synonymous with 'the development of cities', whereas the last 60 years have seen additional urbanisation patterns that have led in part to the destruction of cities and urban life, at least in a classic sense. Suggesting that today's urbanisation 'kills the city' is obviously a normative position, but it does indeed seem difficult to link the dominant urbanisation trends of the last decades—urban sprawl, motorisation, edge cities, gated communities, functional separation, social segregation, loss of public space, and public life to name just a few [7, 33, 77]—to the production of a successful city.

To explore problems, challenges and risks of contemporary urban development more objectively, the discussion below offers an overview of the central environmental, economic and social concerns that have emerged in relation to the systemic, spatial component of cities, and which are strongly related to urban transport and mobility.

Environmental Concerns: Resources, Carbon and Biodiversity

Cities of different wealth levels impact the environment differently. Environmental threats on the local level are most severe in poorer cities, and relate to issues such as fresh water, sewage, health and the degradation of the living environment. As cities become more prosperous, with wider and deeper patterns of consumption and production, their environmental impacts are increasingly felt at the global level (Fig. 1.6). Below there follows an overview on the latter, global aspect; health-related environmental concerns, such as air and noise pollution, will be discussed in their own Sect. 1.3.3.2 , within the overview on social concerns below.

Urban areas in prosperous economies concentrate wealth creation, as well as resource consumption and CO_2 emissions. Globally, with a population share of just above 50 % and occupying less than 2 % of the earth's surface, urban areas concentrate 80 % of economic output, between 60 and 80 % of energy consumption, and roughly an equal share of CO_2 emissions [31, 69]. Most of today's urbanisation goes hand in hand with excessive material consumption, more energy-intensive food supply, and ever-increasing flows of goods and people. This

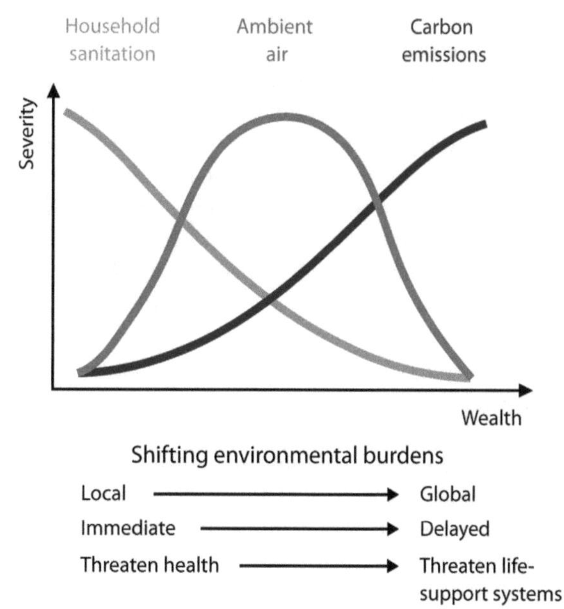

Fig. 1.6 Urban environmental transition (*Source* McGranahan et al. [40])

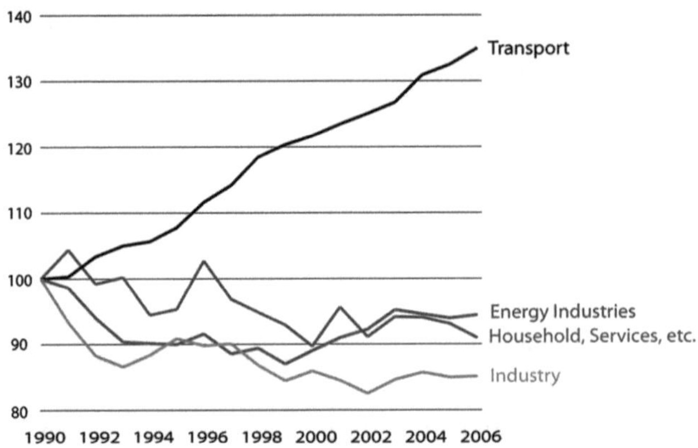

Fig. 1.7 Carbon emissions of the highly urbanised EU-27 by sector and over time (1990 = 100 %) (Source Based on European Commission [23])

pattern is not evenly distributed across the globe, and reflects the concentration of particular activities within individual cities. Buildings, transport, and industry—which are constituent components of cities and urban areas—contribute 25, 22, and 22 %, respectively, of global energy-related greenhouse gas (GHG) emissions [29]. Comparing cities individually, shares of different sectors can vary significantly. For example, transport GHG emissions in Shanghai are 11 %, in New York 23 % and in São Paulo as much as 60 % of all city emissions [66].

Looking at trends within these sectors, transport emissions are particularly worrying. Of the approximately 10 billion trips that are made every day in urban areas around the world—assuming a daily average of three trips per person—a significant and increasing share is with carbon- and energy-intensive private motorised modes. Even within the European Union, a highly urbanised context with ambitious carbon reduction policies in place, transport-related CO_2 emissions increased by a staggering 36 % between 1990 and 2006 (Fig. 1.7), while other key sectors have at least achieved modest emissions reductions [23].

But cities per se are neither drivers of climate change nor the source of ecosystem degradation; for that, certain consumption and production patterns, and particular population groups within cities, are responsible. Moreover, indirect emissions, often not considered, can also have a particularly strong impact: when the emissions from air travel to and from London are taken into account, the city's largest single source of CO_2—at about 34 %—comes from aviation [27]. And considering all consumption-related emissions increases London's carbon footprint to 89.5 m tonnes compared to 44.3 m tonnes of production-related emissions (see Fig. 1.8) [6].

Of course, the environmental impact of cities goes far beyond carbon emissions. London's ecological footprint, a composite measure that translates overall environmental impact into geographic territory, is nearly 300 times the size of Greater London, and twice the size of the UK [5]. The modern city could not function without vast quantities of natural resources—particularly oil, coal and

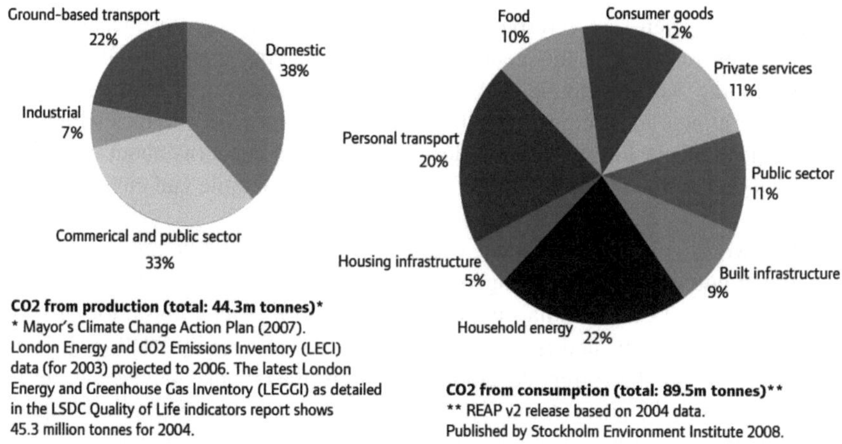

Fig. 1.8 London CO_2 emissions: production and consumption emissions (*Source* BioRegional [6])

gas—for buildings, transport and industrial processes. These resources are not only finite, but will also have to be shared more and more equally among an increasing global population which has just surpassed the seven billion mark.

Even though cities concentrate activities, current urban development has resulted in levels of land-take that are yet another central concern. This is particularly problematic when the land consumed includes valuable farmland or areas providing critical ecosystem services. For example, the current urbanisation patterns in land-scarce China are of great concern. Here, arable land per capita is less than 50 % of the world average, and nearly 1.85 million ha of arable land has been converted to urban use within the last decade [17]. Increasingly, dispersed urban growth consumes green fields and woodlands with, for example, residents of Beijing already having access to only 2.3 m^2 of green space per person [42]. Other ecosystem services have also been compromised by urban land-take, with a significant reduction in biodiversity across city regions worldwide [39]. In addition, unchecked urban expansion has increased the risk exposure of many cities around the world to extreme weather events—for example, as a consequence of local deforestation, or settlements constructed in hazardous locations [61, 63].

Economic Concerns: Infrastructure, Services and Congestion

From an economic perspective, concerns about the lack of infrastructure and service provision in cities have been particularly pronounced in the context of strategic spatial development. In particular, rapid urban growth can overwhelm cities in a variety of regional and political contexts, leading to significant struggles in developing the appropriate infrastructure, mobilising the necessary actors, and managing often scarce resources [18, 41, 46]. But efficiently managing the most relevant resource in this context, land, is something that not only cities in the developing world struggle with. In fact, the trend towards de-densification and sprawl affects most cities, including those in advanced economies which are already at relatively low compaction levels. This raises concerns about economic efficiency and resilience in the face of demographic, economic and environmental change.

Cities worldwide are confronted with demands for infrastructure investments that stretch public and private budgets. The scale of the problem comes into sharp focus in India and China. India's urban population grew from 290 million in 2001 to 340 million in 2008, and is projected to reach 590 million by 2030 [41]. Until then and following the current development models, the country will have to invest US$1.2 trillion in its cities. Annually, it will have to build 700–900 million square metres of residential and commercial space, 350–400 km of subway, and up to 25,000 km of new roads [41]. Similarly, China's urban population is expected to increase from 636 million in 2010 to 905 million by 2030 [68]. It is predicted that by 2050 the country will need to invest 800–900 billion renminbi per year to improve its urban infrastructure, which amounts to about one tenth of China's total GDP (gross domestic product) for 2001 [17].

Naturally, any compromise in the areas of cost-effectiveness and utility maximisation of related expenditures represent serious risks to the economic resilience of urban areas. However, current developments seem to ignore some of the most obvious efficiency drivers, with the prevailing trend of de-densification to levels of sprawling Western metropolitan regions significantly increasing the capital and operating costs of infrastructure. Evidence suggests that linear infrastructure—including streets, railways, water and sewage systems, and other utilities—come at considerably higher cost per unit the lower the urban density [14]. Comparing status-quo dispersed, car-dependent developments with more compact urban development, Todd Litman suggests direct additional costs of anything from US$5,000 to US$75,000 per household unit for building road and utility infrastructure [37]. A recent exercise for Calgary [30] indicates additional costs beyond pure linear infrastructure for schools, fire stations and recreation centres for more dispersed developments, and a study of Tianjin concluded that additional infrastructure cost of more sprawling urbanisation would result in almost double the costs compared to a compact and densely clustered scenario [71].

De-densification and sprawl have also significantly increased operating and maintenance costs of cities, particularly those related to transport. Even with much lower fuel prices than Europe's, sprawling Houston spends about 14 % of its GDP on transport compared to 4 % in relatively compact Copenhagen and a typical 7 % in many Western European cities [35]. Too often, such long-term cost burdens are not taken into account when further increasing the subsidies required for both, private motorised and public transport operations. In the case of New York City, CEO for Cities (a US network of 'urban leaders') [15] estimates that density-related cost savings through reduced expenditure on cars and petrol translates to about US$19 billion annually.

On a daily basis, a critical concern about the concentration of economic activities and the management of urban territory is congestion, which features among the most prominent examples of how economic gains in cities are compromised. Congestion can severely limit the benefits of higher density, as in the case of cities like Shanghai, Bangkok, Manila and Seoul [74]. In many developing world cities, the costs of congestion are enormous: an estimated 3.4 % of GDP in Buenos Aires, 2.6 % in Mexico City and 3.4 % in Dakar [73]. But they are also high in the developed world. Congestion costs in the largely urbanised European Union are estimated at 0.75 % of GDP [73], and in the case of the UK amount to an annual cost of up to £20bn [19]. Less well documented, but nevertheless generating a significant cost burden to urban economies, is the unproductive time that is spent driving cars, and in some cases even chauffeuring others.

Social Concerns: Segregation, Health and Sociability

Segregation

Patterns of contemporary urbanisation in many parts of the world also raise important social concerns. While urbanisation has helped to reduce absolute poverty, the number of people classified as urban poor is on the rise. Between 1993 and 2002, there was an addition of 50 million poor in urban areas, while the number of rural poor declined by 150 million [55]. Urban growth puts pressure on the quality of the local environment in ways that disproportionately affect poorer people, such as lack of adequate access to clean water and sanitation. This results in a huge disease burden that affects livelihoods. Insufficient resources and poor management further compromise electricity supply, waste treatment, transport, and other infrastructure provision.

Typical spatial development patterns further exacerbate the problem, with persistent trends towards greater social segregation. The main components of this pattern include the emergence of socially divisive neighbourhoods, more concentrated poverty, social housing estates, gated communities, shopping centres and business districts. Furthermore, it is characterised by a significant increase in the level of informal development, with large swathes of slum housing without access to basic services, infrastructure and sanitation, and often located in disaster-prone areas which make them more vulnerable to crises. Almost a quarter of the global urban population lives in slums. While in the developing world, the *proportion* of all urban slum dwellers declined from 39 to 32 %, the *absolute* number is on the rise, with 60 million additional slum dwellers in 2010 compared to 2000. UN Habitat estimates that this number will reach a total of 889 million by 2020 [65].

Past and present responses to the social pressures related to housing have been largely unsuccessful. It is widely argued that large concentrated social housing developments, often part of public programmes, as well as other house building programmes aiming to address the lack of adequate housing, had for the most part a negative impact on social dynamics and the economic opportunities of disadvantaged populations in developed world cities [38, 51]. Unfortunately, contemporary initiatives often repeat the housing policy and design mistakes of the past in the developing world [12, 57] and create a new generation of spatially excluded urban dwellers. These are either unable to access critical workplace locations altogether, or are being trapped as 'captive' walkers or public transport users spending unsustainable levels of their time or income on transport.

Health

With regard to the health impacts of contemporary urban development and transport, three main areas of concern are usually referred to: pollution, road safety, and lifestyle-related health impacts. Transport-related air pollution in cities remains a major public health burden, particularly in the developing world. In extreme cases such as Dakar, transport pollution-related health costs are estimated at above 5 % of GDP, while a range between 2 and 3 % is observable for several megacities in Latin America and Asia [73]. In urban areas globally, around 800,000 deaths per year are caused by air pollution [21]. Noise pollution from urban transport is less lethal and tends to be more of a policy concern in developed world cities. However, across a range of cultural contexts, exposure to noise above certain levels severely impacts on human health, increasing the risks of cardiovascular disease, mental disorder and stress. Estimates suggest that more than half of the EU's residents live in areas that do not ensure acoustical comfort [72]. Road traffic is the main factor in transport-related noise, accounting for around 70 % of the total [56].

Furthermore, concerns about road safety in urban as well as rural areas are on the rise, with 1.3 million people being killed annually, and 20–50 million injured in road traffic accidents, which are also the leading cause of death among young people between 15 and 29 years [75]. Per annum, road traffic collisions cost an estimated US$518 billion globally in material, health and other expenditure. For many low- and middle-income countries, the cost of road crashes represents between 1 and 1.5 % of GNP (gross national product), and in some cases exceeds the total amount the countries receive in international development aid [50]. Mohan [43] showed that this is, in fact, an underestimate, and calculated that these costs represent 3.2 % of India's GDP.

There is also a broader set of public health issues connected with healthier lifestyles which receives particular attention in highly urbanised countries. It is estimated that physical inactivity accounts for 3.3 % of all deaths globally, and for 19 million disability-adjusted life-years [11]. Others suggest a death toll of 1.9 million per year caused by inactivity [21]. In the EU, about two thirds of adults are less physically active than the recommended levels, exposing themselves to increased risk of cardiovascular diseases, diabetes and cancer [22]. In the US, these figures are even more dramatic, with about 34 % of adult Americans classified as obese [48].

Sociability: Community Severance

A final key social concern about the contemporary spatial development of cities relates to sociability and community cohesion, which are seen to have come under great pressure as a result of the physical restructuring of cities [58, 59]. At the same time, built environment factors may also directly or indirectly depress social capital, particularly within communities. In relation to transport, Bradbury, Tomlinson and Millington [9] have identified three common types of 'community severance': firstly, physical barriers such as spatial structures limiting interaction, or road traffic causing disruption; secondly, psychological barriers triggered by perceptions related to traffic noise or road safety; and thirdly, long-term social impacts, where communities are disrupted, and which create a more sustained form of being disconnected from certain people and areas close by.

In their study on streets in San Francisco, Appleyard, Gerson and Lintell found that residents on streets with low traffic volumes had three times as many friends and twice as many acquaintances on the same street as those living on streets with high traffic volumes [3]. But it is also the need for physical mobility itself that can have negative impacts. Putnam's research suggests that ten additional minutes spent commuting reduces time spent on community activities by 10 % [53]. A decline in social relationships may have negative impacts not only on physical and mental health but also on economic resilience and productivity [52, 54]. This is particularly the case for disadvantaged people, as community cohesion and social inclusion are linked [36, 45].

Conclusion: From What to How?

To varying degrees, the concerns introduced above are common to the majority of urban areas. They are the direct consequences of a spatial model of urban development based on cheap, carbon-intensive transport (and cheap energy more generally) which dominated most mature cities for decades and is now characteristic of many rapidly urbanising areas. And it is a model that is based on a powerful equilibrium of individual aspirations, personal preferences and business opportunities. However, if one accepts the evidence, arguments and propositions above, this model will prove to be but a temporary equilibrium, one that now requires urgent adjustments and radical transformation—a transformation that will lead to a new mobility culture.

Today, knowledge about the direction of that transformation is reasonably adequate. It is a direction where the production of physical proximity is likely to reemerge as the first principle of city-making. For urban transport, this will further entail a new multimodality based primarily on walking, cycling and public/shared transport, complemented and facilitated by an ever-increasing role for information and communication technology. But urban leaders addressing the required adjustments are struggling to identify the most appropriate tools, procedures and structures for a whole range of policy areas, and are particularly challenged by the requirements for greater policy integration. The question of how to facilitate a coordinated transition is particularly contested when it involves the urban transport sector. Cities and transport each separately receive serious attention as part of development strategies at the local, national and supra-national level. Together, transport and cities also form a complex system that is amongst the most widely debated in the emerging 'urban age'.

The UN Habitat agenda demands that city authorities "support an integrated transport policy approach" and seek to "coordinate land-use and transport planning" [62]. Unfortunately, few cities so far have been in the position to test a significant degree of synchronisation between land use and transport. For cities outside Europe and North America, the innovative practice of Curitiba (in Brazil), Hong Kong and Singapore are particularly exceptional. Elsewhere, transport planning remains an isolated problem-driven policy area with little or no strategic objectives or long-term vision linked to spatial development.

However, new forms of interconnected governance and planning are emerging, and are of particular relevance to cities, with their rapidly increasing complexities and intertwined dependencies. In fact, urban development is often featured as the ultimate testing ground for greater policy integration and has already produced many of the most innovative practices. In recent years, well-documented integrated policy has emerged from cities as diverse as Barcelona, Johannesburg, Bogota, Kolkata, London and Berlin. Furthermore, it is the exceptional interdependence of spatial development and transport that has pushed these two policy areas to the forefront of an agenda for greater integration. Again, it is in cities where this relationship is most pronounced and the need for consistent policy integration most urgent.

Acknowledgments This essay is based on a contribution to the Green Economy Report's cities chapter published by the United Nations Environment Programme [70] and on the author's research on integrated planning, design and transport.

References

1. Angel S (2011) Making room for a planet of cities. Policy Focus Report
2. Angel S, Sheppard SC, Civco DL (2005) The dynamics of global urban expansion. The World Bank, Washington
3. Appleyard DM, Gerson S, Lintell M (eds) (1982) Livable streets. University of California Press, Berkeley
4. Barber B (2012) If Mayors ruled the world: why cities can and should govern globally and how they already do. Yale University Press, New Haven
5. Best Foot Forward (2002) City limits: a resource flow and ecological footprint analysis of greater London. Best Foot Forward, London
6. BioRegional and L. S. D. Commission (2009) Capital consumption: the transition to sustainable consumption and production in London. Chartered Institution of Wastes Management—Environmental Body. London

7. Bookchin M (1992) Urbanization without cities: the rise and decline of citizenship. Black Rose Books, Montreal
8. Bottles SL (1987) Los Angeles and the automobile: the making of the modern city. University of California Press, Berkeley
9. Bradbury A, Tomlinson P, Millington A (2007) Understanding the evolution of community severance and its consequences on mobility and social cohesion over the past century. European transport conference 2007, creating a livable environment seminar. Association for European Transport and Contributors
10. Buchanan CD (1963) Traffic in towns: a study of the long term problems of traffic in urban areas. H.M. Stationery Office, London
11. Bull FC, Armstrong TP, Dixon T, Ham S, Neiman A, Pratt M (eds) (2004) Physical inactivity. Comparative quantification of health risks. World Health Organization, Geneva
12. Burdett R, Rode P (2011) Living in the urban age. In: Burdett R, Sudjic D (eds) Living in the endless city. Phaidon Press, London
13. Burdett R, Sudjic D (2011) The endless city. Phaidon Press, London
14. Carruthers JI, Ulfarsson GF (2003) Urban sprawl and the cost of public services. Environ Plann B: Plann Des 30(4):503–522
15. CEO for Cities (2011) New York City's green dividend. http://www.ceosforcities.org/work/nycs_green_dividend
16. Cervero R (1998) The transit metropolis: a global inquiry. Island Press, Washington
17. Chen H, Jia B, Lau S (2008) Sustainable Urban form for Chinese compact cities: challenges of a rapid urbanized economy. Habitat Int 32(1):28–40
18. Cohen B (2006) Urbanization in developing countries: current trends, future projections, and key challenges for sustainability. Technol Soc 28(1–2):63–80
19. Confederation of British Industry (2004) The UK as a place to do Business: is transport holding the UK Back? CBI Report
20. Dargay J, Gatley D, Sommer M (2007) Vehicle ownership and income growth, Worldwide: 1960–2030. Energy J 28(4):143–170
21. Dora C (2007) Health burden of urban transport: the technical challenge. Sādhanā 32(4):285–292. http://www.ias.ac.in/sadhana/Pdf2007Aug/285.PDF
22. Edwards PTA (2008) A healthy city is an active city: a physical activity planning guide. World Health Organization Regional Office for Europe, Copenhagen
23. European Commission—Directorate-General Energy and Transport (2007) EU 27 CO2 Emissions by Sector, EU-27. http://ec.europa.eu/dgs/energy_transport/figures/pocketbook/doc/2007/2007_environment_en.xls
24. European Environment Agency (ed) (2006) Urban sprawl in Europe: the ignored challenge. EEA Report No. 10
25. Fulton L (2008) CO_2 reductions in transport: what's hot and Not. In: IEA Day, COP-14. International Energy Agency. Poznan
26. Gayda S, Haag G, Besussi E, Lautso K, Noel C, Martino A et al (2005) SCATTER—Sprawling cities and transport. Stratec S.A.
27. Greater London Authority (2007) Action today to protect tomorrow—The Mayor's climate change action plan. Greater London Authority
28. Heinze GW, Kill HH (1991) Chancen des ÖPNV am Ende der autogerechten Stadt. Verkehrspolitische Lehren für einen traditionellen Verkehrsträger im Strukturbruch. Jahrbuch für Regionalwissenschaft 12/13:105–136
29. Herzog T (2009) World greenhouse gas emissions in 2005: WRI working paper. World Resources Institute, Washington
30. IBI Group (2009) The implications of alternative growth patters on infrastructure costs. Plan it Calgary, Calgary
31. Kamal-Chaoui L, Robert A (2009) Competitive cities and climate change: OECD regional development working papers N° 2. OECD Publishing, Paris
32. Knoflacher H, Rode P, Tiwari G (2008) How roads kill cities. In: Burdett R, Sudjic D (eds) The endless city. Phaidon Press, London, pp 394–411

33. Koolhaas R, Boeri S, Kwinter D, Tazi N, Obrist HU (2001) Mutations. ACTAR, Barcelona
34. Kutzbach M (2010) Megacities and megatraffic. ACCESS 37:31–35
35. Laconte P (2005) Urban and transport management—International trends and practices: international symposium. Sustainable Urban transport and city. Tongji University, Nagoya University, Shanghai
36. Litman T (2006) Cities connect: how urbanity helps achieve social inclusion objectives. In: Metropolis conference, Victoria Transport Policy Institute, Toronto
37. Litman T (2011) Understanding smart growth savings,http://www.vtpi.org/sg_save.pdf
38. Massey DS, Denton NA (eds) (1993) American apartheid: segregation and the making of the underclass. Harvard University Press, Cambridge
39. McDonald R, Kareiva P, Forman RTT (2008) The implications of current and future urbanization for global protected areas and biodiversity conservation. Biol Conserv 141:1695–1703
40. McGranahan G, Jacobi P, Songsore J, Surfjadi C, Kjellèn M (2001) The citizens at risk: from urban sanitation to sustainable cities. Earthscan, London
41. McKinsey Global Institute (2010) India's urban awakening: building inclusive cities, Sustaining Economic Growth
42. Melchert L (2004) Transnational urban spaces and urban environmental reforms: analyzing Beijing's environmental restructuring in the light of globalization. Cities 21(4):321–328
43. Mohan D (2002) Social cost of road traffic crashes in India. In: First safe community conference on cost of injury, Viborg, Denmark pp 33–38, http://tripp.iitd.ernet.in/publications/paper/safety/dnmrk01.PDF
44. Mohan D, Tsimhoni O, Sivak M, Flannagan M (eds) (2009) Road safety in India: challenges and opportunities. Transportation Research Institute, The University of Michigan Ann Arbor, Michigan, http://www.ki.se/csp/pdf/Publications/DM_UMTRI_2009_1_1.pdf
45. O'Connor KM, Sauer SJ (2006) Recognizing social capital in social networks: experimental results (Johnson School Research Paper Series). pp 16–18
46. OECD (2006) Competitive cities the global economy (OECD Territorial Reviews). Paris
47. OECD (2009) Cities and climate change: key messages from the OECD
48. Ogden CL, Carroll MD (2010) Prevalence of overweight, obesity, and extreme obesity among adults: United States, Trends 1960–1962 through 2007–2008
49. Pan H (2011) Implementing sustainable travel policies in China: international transport forum. OECD, Leipzig
50. Peden M, Scurfield R, Sleet D, Mohan D, Hyder AA, Jarawan E, Mathers C (2004) World report on road traffic injury prevention. World Health Organization, Geneva
51. Power A (1987) Property before people: the management of twentieth-century council housing. Allen and Unwin, London
52. Putnam R (2004) Education, diversity, social cohesion and 'Social Capital'. 'Raising the quality of learning for all': Meeting of OECD Education Minister Dublin. Dublin
53. Putnam RD (2000) Bowling alone: the collapse and revival of American Community. Simon and Schuster, New York
54. Putnam RD, Leonardi R, Nanetti RY (1994) Making democracy work: civic traditions in modern Italy. Princeton University Press, Princeton
55. Ravallion M, Chen S, Sangraula P (2007) New evidence on the urbanization of global poverty. Population Dev Rev 33(4):667–701
56. Rodrigue J-P, Comtois C, Slack B (2009) The geography of transport systems. Routledge, New York
57. Satterthwaite D (2011) Surviving in an urban age. In: Burdett R, Sudjic D (eds) Living in the endless city. Phaidon Press, London
58. Sennett R (2012) The open city. In: Burdett R, Sudjic D (eds) The endless city. Phaidon Press, London
59. Sennett R (2012) Together; the rituals, pleasures, and politics of cooperation. Yale University Press, New Haven
60. Staniford S (2010) Chinese transportation growth

61. Steinberg F (2007) Jakarta: environmental problems and sustainability. Habitat Int 31(3–4):354–365
62. UN Habitat (2006) The habitat agenda goals and principles, commitments and the global plan of action
63. UN Habitat (2007) Enhancing Urban Safety and Security—Global Report on Human Settlements 2007
64. UN Habitat (2008) State of the World's Cities 2008/2009: Harmonious Cities, Earthscan
65. UN Habitat (2010) State of the World's Cities 2010/2011: Bridging the Urban Divide
66. UN Habitat (2011) Global Report on Human Settlements 2011: Cities and Climate Change. Nairobi
67. UN Population Division (2007) World Urbanisation Prospects 2007: World Urbanization Level—1950–2005. http://esa.un.org/unup/
68. UN Population Division (2009) World Urbanization Prospects—The 2009 Revision
69. UN Population Division (2010) World Population Prospects—The 2010 Revision
70. UNEP (2011) Towards a Green Economy: Pathways to Sustainable Development and Poverty Eradication
71. Webster D, Bertraud A, Jianming C, Zhenshan Y (2010) Toward Efficient Urban Form in China. United Nations University Working Paper, No. 2010/97
72. Wolfram M, Bührmann S, Martino A, Brigati E (2005) Sustainable urban transport plans (SUTP) and urban environment: policies, effects and simulations. Rupprecht Consult, Cologne
73. World Bank (2002) Cities on the move: a world bank urban transport strategy review. World Bank, Washington
74. World Bank (2009) World development report 2009: reshaping economic geography. World Bank, Washington
75. World Health Organization (2011) Road traffic injuries—Fact sheet No 358
76. Wright L, Fulton L (2005) Climate change mitigation and transport in developing nations. Transp Rev 25(6):691–717
77. Zijderveld A (2008) A theory of urbanity: the economic and civic culture of cities. Transaction Publishers, New Brunswick

Chapter 2
The Diversity of Megacities Worldwide: Challenges for the Future of Mobility

Roland Priester, Jeffrey Kenworthy and Gebhard Wulfhorst

Abstract Megacities around the globe present bewildering combinations of transport patterns, transport infrastructure and other factors related to personal mobility. From the sprawling auto-dependent regions such as Los Angeles and Houston through the rail-based transit giants of Tokyo and Osaka, the informal-based public transport systems of Manila and Johannesburg, to the strong non-motorised sectors of Mumbai and the legendary traffic congestion of Bangkok, megacities are as diverse as they are many. But through this complexity and diversity, is it possible to find defining patterns? Can we simplify this perplexing picture? And through this process, is it possible to better understand and distinguish the array of mobility challenges that face such large cities? This chapter explores these questions through a cluster analysis of some 41 megacities across the globe, representing a vast range in wealth and other features. Exploiting *The Millennium Cities Database for Sustainable Transport*, which presents a very large, consistent and reliable set of transport and land use indicators for 100 cities worldwide, the study examines some 59 representative indicators covering land use and wealth characteristics, private, public and non-motorised mode mobility patterns, private and public transport infrastructure, transport investment patterns, transport energy use, transport externalities and other variables. The results suggest that megacities can be classified into six distinct types, which we have called Hybrid Cities, Transit Cities, Auto Cities, Non-Motorised Cities, Traffic-Saturated Cities and Paratransit Cities. The chapter provides a global mapping and description of these city-types, gives an overview on the variables used and the methodology for the cluster analysis and highlights the diversity of mobility patterns across these clusters by selected

R. Priester (✉) · G. Wulfhorst
Technische Universität München, Munich, Germany
e-mail: roland.priester@tum.de

G. Wulfhorst
e-mail: gebhard.wulfhorst@tum.de

J. Kenworthy
Curtin University Sustainability Policy Institute (CUSP), Curtin University, Perth, Western Australia, Australia
e-mail: J.Kenworthy@curtin.edu.au

Institute for Mobility Research (ifmo) (ed.), *Megacity Mobility Culture*,
Lecture Notes in Mobility, DOI: 10.1007/978-3-642-34735-1_2,

© Springer-Verlag Berlin Heidelberg 2013

key variables. It explains how this advances our understanding of mobility in such cities. Finally, it presents the key mobility challenges that characterise each set of cities. The cluster results provided a framework for the selection of cities explored by scholars in the city mobility stories that form the major part of this book.

Introduction

As of 2008, for the first time in the history of the world, more people live in cities and urban environments than in rural areas. Part of the reason for this global urbanisation process is the increasing proportion of the world's population living in so-called megacities [19, 11]. When considering megacities today, the focus has increasingly shifted from the developed to the developing countries, especially in Asia.

While the development of transport in traditional metropolises such as London and New York City began several centuries ago, the maturing megacities in newly industrialising countries and emerging economies are undergoing the same process at an accelerated pace. They face the task in many cases without the benefit of an already well-established and well-functioning public transport system, which is usually the legacy of the foregoing public transport era, which they did not experience. Many of these cities are therefore rapidly transitioning from cities based primarily on non-motorised transport and inferior bus systems, to an increasing reliance on motorcycles and cars.

Take China as an example—in 2001, 42 % of all trips were still being made on foot or by bike; within just three years this proportion had fallen to 36 %, and it has continued to plummet [4]. The upgrading of transport infrastructure—the improvement of public transport systems, in particular—can hardly keep up with demographic and economic growth. The consequences of these dynamics are an often-uncontrolled expansion of megacity boundaries that goes hand-in-hand with increasing consumption of land, accompanied by saturation of the road system and the beginnings of motorcycle and car dependency. To this is added ever-increasing journey lengths, times and costs, and an upsurge in both energy consumption and carbon dioxide emissions; not to mention severe depletion, even destruction, of the public spaces in cities to make way for new roads and car parks.

But this is not all inevitable. We know that the transport systems of Tokyo and Hong Kong also function primarily on public and non-motorised modes, while at the same time being huge cities which are often regarded as embodying some of the best practice in terms of compact city structure and transport efficiency [12]. Moreover, Singapore transformed itself, from the late 1960s onwards, into an outstanding transit metropolis with highly integrated high-density housing and mixed land uses, together with highly effective controls over growth in car ownership and use.

This is but a brief glimpse into the complex world of cities and their mobility characteristics. We know that differences in the history, background, culture, economic strength and governance of megacities around the world lead to a great variation in patterns of mobility. This chapter concerns itself with better understanding megacities as they are today, and in particular the diversity of types of transport and land-use systems found in them. It approaches this by performing a cluster analysis of extensive data from the Millennium Cities Database for Sustainable Transport (compiled by the UITP, 'Union Internationale des Transports Publics', known as the International Association of Public Transport) in order to arrive at clusters of different types of cities—classifications that can be understood in terms of the dominant patterns of transport. The chapter explains the methodology and data used, and reveals the various clusters, going on to discuss their significance (for detailed information about the methodology and technology used, see [14]. The chapter finishes by reporting on some selected transport-related trends from cities worldwide, partly in order to update and contextualise the perspective gained from the cluster analysis.

Background

One of the overarching concerns of the research programme 'Mobility Culture in Megacities', initiated by ifmo, is to improve our understanding of the diverse mobility patterns and cultures found in megacities, in order to be able to identify challenges, and also opportunities, pertaining to the future of urban mobility. Recognising the dynamics of change, both in the transport system and in people's mobility patterns, provides the basis for choosing the core questions for the subsequent research programme. In fact, cities differ from each other in many ways: by spatial structure, the existing transport supply, the population's individual mobility behaviour, and also by their economic, political and cultural framing conditions. In this regard, every megacity manifests its specific set of characteristics and, through analysis, its probable future requirements for improved urban transport and mobility. As a result, many different approaches to the task of adapting to the upcoming challenges arise. These challenges include the ever-growing demand for transport, especially in fast-growing regions, leading to capacity problems for infrastructure; competition in land-use requirements for transport (and other purposes); post-fossil fuel mobility; climate-change impacts; intermodal transport issues; and, last but not least, the evolution of people's mobility patterns and level of demand. As well as the differences, there are also similar patterns of mobility cultures that can be identified in some groups of cities. For this reason, the major research objective of this preliminary study was to analyse what characteristics certain cities have in common, by inspection of which they can also be distinguished from each other.

Methodology

Identifying groups of similar objects, while at the same time noting the differences between the groups, is a process which can be effectively handled by the statistical method of cluster analysis. Chosen as the major scientific approach in this study, cluster analysis was used to reduce the complexity of research objects, in this case the 41 megacities included in the analysis, and to draw the lines between relatively homogenous groups—the megacity clusters. Although there is no single method of clustering, every procedure has a common approach: "...that the objects within a group be similar (or related) to one another and different from (or unrelated to) the objects in other groups. The greater the similarity (or homogeneity) within a group and the greater the difference between groups, the better or more distinct the clustering" [17].

On the basis of the cluster analysis, the study was able to derive research questions tackling the challenges of future urban mobility. The findings from the cluster analysis and the underlying data formed the starting point for the 'Mobility Culture in Megacities' programme, which is described in the main part of this book—the 'city stories'.

Dataset

As the first step, a reliable and comprehensive—but still manageable—dataset for the analysis had to be compiled. For the purposes of comparison, a standardised set of indicators for each of the cities in the analysis was required. This set ultimately included common parameters describing the current transport supply, as well as mobility behaviour indices. Furthermore, spatial and economic framing conditions on mobility were also included.

The most well-known source of such data was found to be the UITP Millennium Cities Database for Sustainable Transport compiled by Jeffrey Kenworthy and Felix Laube [9]. This database contains 69 primary variables, from which 230 standardised indicators were generated for a base year of 1995. Within the scope of this study, the 59 most relevant descriptors of the land-use/transport system of cities have been selected as key indicators for further analysis. These include:

- general city characteristics (e.g. urban density, job density, gross domestic product [GDP] per capita);
- transport supply indicators (e.g. length of road network, length of public transport network, motorisation rate);
- mobility indicators (e.g. number of daily trips (by mode), trip distances, per-capita travel by car and public transport);

- investment in private and public transport (by mode);
- various transport impact indicators (e.g. energy use, emissions, fatalities).

Regional Scope

While the Millennium Cities Database provides data for 100 international cities and urban areas of various sizes, our analysis focuses only on the largest ones, the so-called megacities. Although official definitions tend to indicate a minimum population size of five million inhabitants, some cities below this threshold were also chosen for this study, for several reasons. Firstly, since 1995, when the data were collected, some cities and their suburbs have grown beyond this limit. Secondly, others are on the point of doing so in the coming years, owing to their enormous growth rate. Finally, a number of further cities that are even smaller in size have been added, in order to provide reference values from some regions (such as Dakar).

Additional selection criteria were the need to have a well-balanced set of cities of various population sizes, and to include some older cities alongside the fast-growing newcomers. Furthermore, we tried to include a wide range of countries from every continent. To avoid any confusion, it must be stated that this analysis does not regard a megacity as being defined by single city administrative boundaries bearing that city's name (e.g. the City of New York), but rather in terms of megacity agglomerations (so, in the New York case, the whole Tri-State Metropolitan Area, which has a population in excess of 20 million people). To give another example, Tokyo does not refer only to Tokyo-to, in itself a megacity, but to the Tokyo Metropolitan Area embracing Tokyo-to and Saitama, Chiba and Kanagawa Prefectures (home to over 36 million people today). The complete list of all 41 'cities' or urban regions included in the analysis is shown in Table 2.1, grouped by population size.

Table 2.1 Cities (agglomerations) in the analysis grouped by population size

>10 million			5–10 million		<5 million	
Tokyo	Shanghai	London	Chicago	Taipei	Riyadh	Montreal
Seoul	Osaka	Paris	Bangkok	Madrid	Kuala Lumpur	Cape Town
Mumbai	Guangzhou		Bogotá	Houston	Singapore	Dakar
New York	Jakarta		Washington	Ruhr	Sydney	Harare
São Paulo	Cairo		Johannesburg	Atlanta	Phoenix	
Manila	Beijing		San Francisco	Toronto	Berlin	
Los Angeles	Tehran		Hong Kong	Ho Chi Minh City	Melbourne	

Selection of the Clusters

The 59 key indicators selected as input variables for the cluster analysis were first reduced to 13 factors. This has been done using the standard statistical process of factor analysis, which is able to identify joint or underlying correlations. The significance of the various indicators is represented by the factors created, which are:

1. Urban sprawl and automobile dependence
2. Taxi traffic
3. Public transport vehicle intensity
4. Shared taxi traffic
5. Transport deaths
6. Congestion
7. Scarcity of public transport supply
8. Public transport usage
9. Preconditions for non-motorised transport
10. Motorised private transport trip length
11. Public transport energy consumption
12. Parking charges
13. Parking management/restrictions.

In this study the procedure of 'hierarchical agglomerative clustering' was used for the cluster analysis itself. This is characterised by regarding every object of the analysis as being at the outset a cluster itself. With each successive step, more and more objects are merged, creating new, larger clusters by means of comparing the similarity of the values of the variables. The dendrogram shown in Fig. 2.1 gives a visual representation of the process of clustering. It shows the 'distances' at which cities are combined to one common cluster. To read the dendrogram, one starts at the left edge, where every city is a cluster of one—itself; that is to say, no merging has been done. With every successive clustering step, more and more clusters are created. By the end of this process, they have all been combined to form one large common cluster—this is shown at the right edge of the dendrogram.

The earlier in the process cities are combined, the more similar it means that these cities are. For instance, after the first iteration the following cities are already merged:

- the Australian cities of Sydney and Melbourne;
- in the USA, three pairs: Atlanta and San Francisco, the long-established metropolises of New York and Chicago, and the southern cities of Phoenix and Houston;
- the Canadian rivals Toronto and Montreal.

It is not much further along the clustering process that the similarities of the Japanese cities, Tokyo and Osaka, are found. When it comes to Europe, it is seen

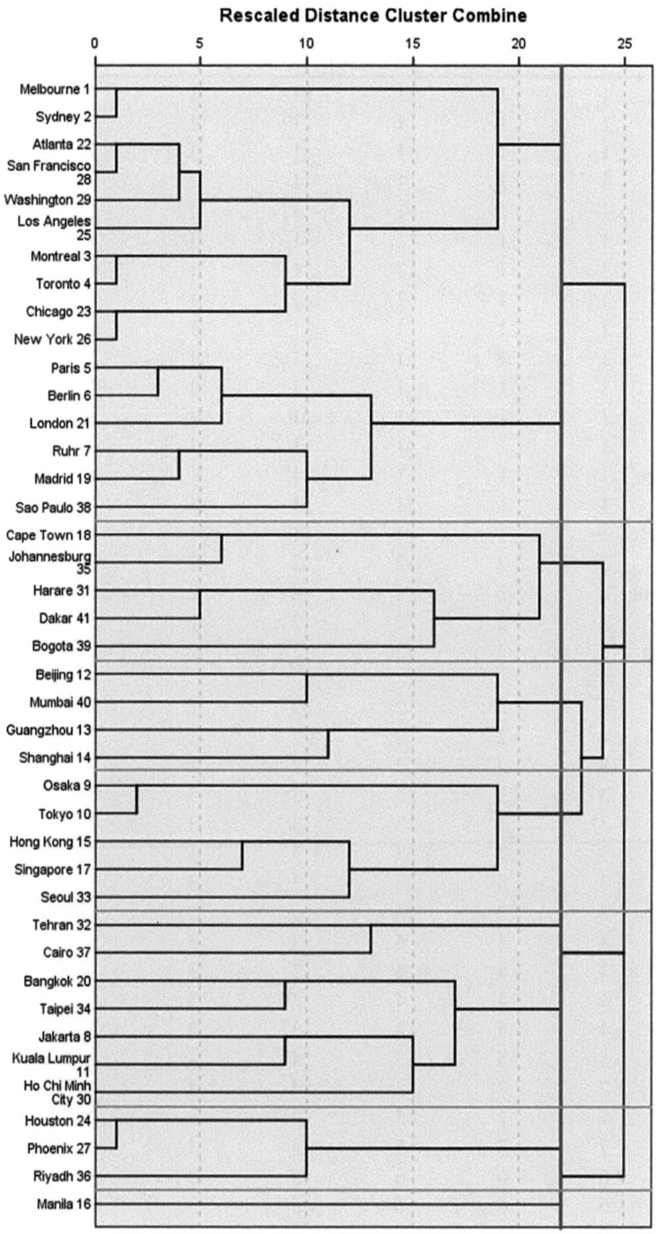

Fig. 2.1 Dendrogram showing each step in the clustering process

Table 2.2 Cluster solutions from 2 up to 8 clusters

Megacities	8 Cluster	7 Cluster	6 Cluster	5 Cluster	4 Cluster	3 Cluster	2 Cluster
Melbourne	1	1	1	1	1	1	1
Sydney	1	1	1	1	1	1	1
Montreal	1	1	1	1	1	1	1
Toronto	1	1	1	1	1	1	1
Paris	1	1	1	1	1	1	1
Berlin	1	1	1	1	1	1	1
Ruhr	1	1	1	1	1	1	1
Madrid	1	1	1	1	1	1	1
London	1	1	1	1	1	1	1
Atlanta	1	1	1	1	1	1	1
Chicago	1	1	1	1	1	1	1
Los Angeles	1	1	1	1	1	1	1
New York	1	1	1	1	1	1	1
San Francisco	1	1	1	1	1	1	1
Washington	1	1	1	1	1	1	1
Sao Paulo	1	1	1	1	1	1	1
Jakarta	2	2	2	2	2	2	2
Kuala Lumpur	2	2	2	2	2	2	2
Bangkok	2	2	2	2	2	2	2
Ho Chi Minh City	2	2	2	2	2	2	2
Tehran	8	2	2	2	2	2	2
Taipei	2	2	2	2	2	2	2
Cairo	8	2	2	2	2	2	2
Osaka	3	3	3	3	3	3	2
Tokyo	3	3	3	3	3	3	2
Hong Kong	3	3	3	3	3	3	2
Singapore	3	3	3	3	3	3	2
Seoul	3	3	3	3	3	3	2
Beijing	4	4	4	3	3	3	2
Guangzhou	4	4	4	3	3	3	2
Shanghai	4	4	4	3	3	3	2
Mumbai	4	4	4	3	3	3	2
Manila	5	5	5	4	4	2	2
Houston	7	7	5	4	4	2	2
Phoenix	7	7	5	4	4	2	2
Riyadh	7	7	5	4	4	2	2
Cape Town	6	6	6	5	3	3	2
Harare	6	6	6	5	3	3	2
Johannesburg	6	6	6	5	3	3	2
Bogota	6	6	6	5	3	3	2
Dakar	6	6	6	5	3	3	2

Source Priester et al. [14]

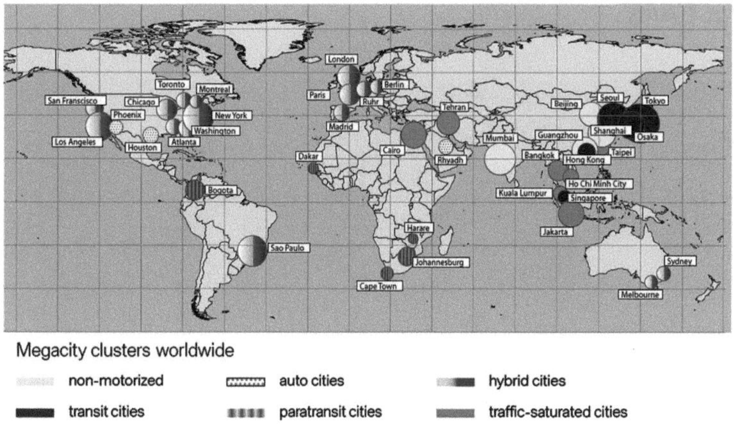

Fig. 2.2 World map of megacity clusters

that the capitals of France and Germany—Paris and Berlin—resemble each other more closely than either does their British counterpart, London. But more significantly, it seems that with respect to mobility culture there is a divide between these three cities and the Spanish capital, Madrid. In fact, the latter seems to have more in common with São Paulo, Brazil's largest city, than it does with its European fellow capitals—a somewhat surprising result. But in the end all the European cities get merged together, and ultimately join with the Australian cities and most of the North American ones to form one large cluster.

Regarding the other cities in the analysis, further iterative steps were needed before similarities could be revealed. Altogether, the dendrogram gives some very useful hints about the internal coherence within the megacity clusters, which were determined in the next step.

In this step, the number of clusters to be selected was determined. To achieve this, a compromise had to be found between the need to distinguish clusters from each other and the risk of going into too much detail by over-separating—a cluster with at least three members was deemed acceptable. In a seven-cluster solution (see Table 2.2) we found what we considered to be a good approach, identifying one fairly comprehensive cluster, five more compact ones, and—an interesting result—Manila as the sole statistical outlier.

Description of the Megacity Clusters

Having selected six megacity clusters (plus the outlier, Manila), the next step consisted in describing their individual characters. An overview of some of the selected indicators is given in the Annex. In Fig. 2.2, all the clusters and their member cities are located on a world map.

Hybrid Cities

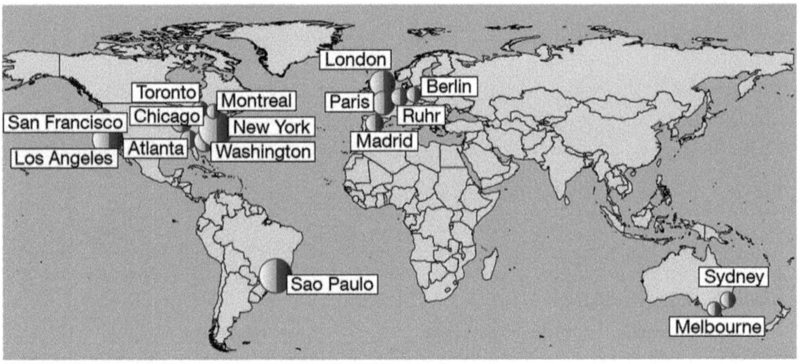

Fig. 2.3 Hybrid cities

Cities in this cluster display a specific character, being best described as of a 'hybrid' nature. This means, on the one hand, that these cities have a consolidated, dense urban core, extending to the inner suburban areas, with good infrastructure and significant usage of public transport and non-motorised modes. On the other hand, it signifies that these urban centres are surrounded by a vast and sprawling suburban area (Fig. 2.3). A typical example is New York. The City of New York itself accommodates some 8.4 million inhabitants in an area of about 800 km^2, with a resulting population density of about 100 persons per ha (hectare, 0.01 km^2). Whereas in Manhattan, the very dense core of the whole region, there were in 2010, 266 persons per ha (US Census, 2010: 1,583,875 persons on 5,950 ha), the whole metropolitan area (known as the Tri-State Metropolitan Area) had, in contrast, an overall density of 19.2 persons per ha, while in some areas the figure falls to around 10 persons per ha or even lower (J. Kenworthy, The Millennium Cities Database for Sustainable Transport, 2012, Unpublished update).

This discrepancy also shows up when the means of transport is looked at. While the network connectivity of the New York public transport system ranks top amongst US cities, it reduces sharply as one reaches the outskirts, with automobile dependence increasing dramatically to levels that are more typical of the vast expanses of other US cities such as Houston and Phoenix. One finds this dichotomy more or less clearly displayed throughout the other members of the Hybrid Cities cluster, not merely in the American cities. For example, Toronto was described in 1990 by the former General Manager of Planning of the Toronto Transit Commission, Mr. Juri Pill, as "Vienna surrounded by Phoenix". In a recent book about Montreal, Kenworthy and Townsend [10] focus exclusively on this dualistic character of the Montreal region.

The hybrid character of this cluster is not only expressed by the difference between urban cores and their surroundings, but also by the way in which transport infrastructure has developed. This could be termed a 'push and push' policy, meaning that for almost every new metre of public transport route constructed, one metre of new road infrastructure is also being built. The Hybrid Cities thus perform strongly in both categories in the analysis. They offer peak values regarding the length of (reserved) public transport routes (see Annex), but also have high figures for freeway length compared to even the really car-dependent 'Auto Cities'. These cities are akin to what Thomson [18] describes in his seminal work *Great Cities and Their Traffic* as 'weak-centred cities', being caught between the competing characteristics of a 'strong-centred city', with its dense and highly centralised core area, and the sprawling characteristics of a low-density 'Automobile City'. These cities are really two city types rolled into one. Even the Los Angeles area falls into this group, although it is at first appearances the least likely member of it. For example, public transport service provision and usage in the City of Los Angeles (e.g. along the Wilshire Corridor through West Los Angeles) attain comparatively high levels. And the population density of the Los Angeles–Long Beach Metropolitan Statistical Area is the highest in the USA, even denser than the comparable New York Urbanized Area. So even Los Angeles has some significant characteristics that push it into the Hybrid City group, rather than placing it in the more extreme Auto City cluster, where you might think it would belong and which is discussed next.

The indicator charts shown in the Annexes to this chapter reveal the reasons for which the European cities are grouped together with most of the American cities and the Australian ones in this cluster. In each of the charts, the difference in indicator value between these 16 cities tends to be much smaller than the difference in value evident between these cities (considered together) and cities of different clusters. Nevertheless, most of the European and Australian cities perform slightly better in terms of public transport, while the American cities rank a little higher regarding road infrastructure and car ownership.

Of course, policies promoting such a high investment in transport infrastructure as demonstrated in the Hybrid Cities do require financial resources. It's therefore not surprising that Hybrid Cities are found among the megacities with the highest metropolitan GDP per capita (see Annex).

Auto Cities

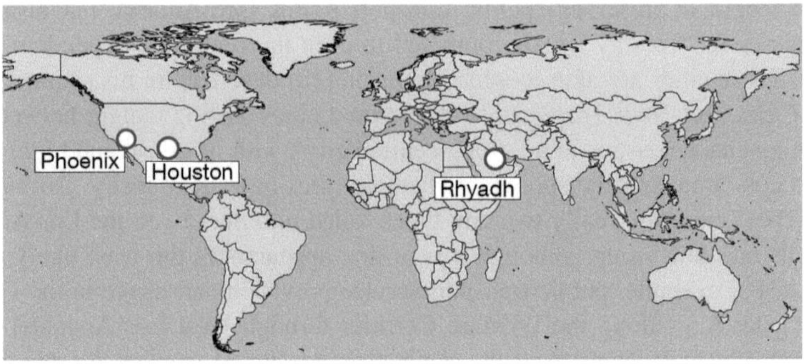

Fig. 2.4 Auto cities

Sprawling, relatively wealthy and completely car-dependent—this is a succinct description of the cluster of so-called Auto Cities (Fig. 2.4).

The cluster members Phoenix and Houston have the lowest urban density (approximately 10 persons per ha) of all cities in the analysis except Hybrid Atlanta. This is in contrast to Hybrid Los Angeles, often used as a symbol for urban sprawl, which ranks—as already mentioned—as the densest US region in this study, though that density is more evenly spread, unlike the New York region, which has an exceptionally peaked and centralised density profile.

When it comes to transport infrastructure, all three cities are at or near the top of the list of length of freeway network per person worldwide, as can be seen in the Annex: Houston ranks number 1, with Phoenix 3rd and Riyadh 7th. Car ownership is, however, a more significant factor in the two US cities than in Riyadh. Regarding public transport, the Auto Cities are grouped together again—but this time at the very bottom. None of these cities has any reserved public transport route. Not surprisingly, their mobility needs are almost exclusively satisfied by the private car (ranging from 94.3 % in Phoenix to 96.5 % in Riyadh).

Of course, this kind of mobility behaviour leads to an enormous figure for energy use per capita. While Phoenix is high enough on the list, having the fourth highest private passenger transport energy consumption at rather over 50,000 MJ per person, Houston uses over 1.5 times as much as even this. The extreme car dependency evident in these cities also causes severe threats to the environment—total emissions in the Auto Cities are very high and transport deaths per 100,000 people in all three are above average.

Transit Cities

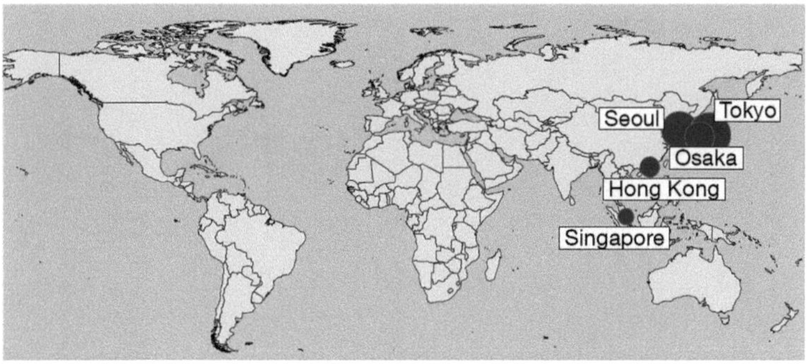

Fig. 2.5 Transit cities

The essence of a Transit City lies in its strong public transport (or transit) profile, in conjunction with the minor role that private cars play in its mobility culture. While car ownership is low or moderate at the utmost (see Annex) and the availability of public transport is high, their public transport modal share puts them near the top of the list. Thus, car kilometres per capita travelled, energy use and particularly emissions are all low or very low (Fig. 2.5).

Furthermore, Transit Cities can be regarded in some sense as the converse of the Auto Cities described above. Although the two groups are similar in affluence—in fact Hybrid, Auto and Transit Cities make up, without exception, the wealthiest half of the cities in the analysis—urban density is revealed to be the criterion that distinguishes them. Transit Cities are usually relatively dense—indeed, two of them are even in the top five in density of all cities analysed. It seems therefore, that depending on its degree of density, a wealthy city can end up more car-dependent (in the case of the less dense) or transit-orientated (if more densely populated).

But cultural and political framing conditions may play a role too, as it is noticeable that all the Transit Cities are located in the wealthier countries of East Asia. Perhaps, however, one of the things that most sharply sets these cities, especially the Japanese ones, apart from the rest is their highly developed rail systems. Although the role of rail does vary between these cities, with very significant contributions to the patronage of public transport taking the form of bus travel, each of them has a world-class rail rapid transit system. In the cases of Singapore, Hong Kong and Seoul, the systems are constantly being extended, upgraded and diversified, especially with the introduction of light rail systems to complement the heavy rail structural backbone of the cities. Rail forms a substantial part of the 'mobility culture' in these cities, and is part of how both residents and visitors map them out in their minds and move around them. Their rail systems are achieving the same level of cultural significance and recognition as the now historic Metro in Paris and Tube in London, both of which are cultural icons, commemorated in everything from board games to placemats on dinner tables.

Non-Motorised Cities

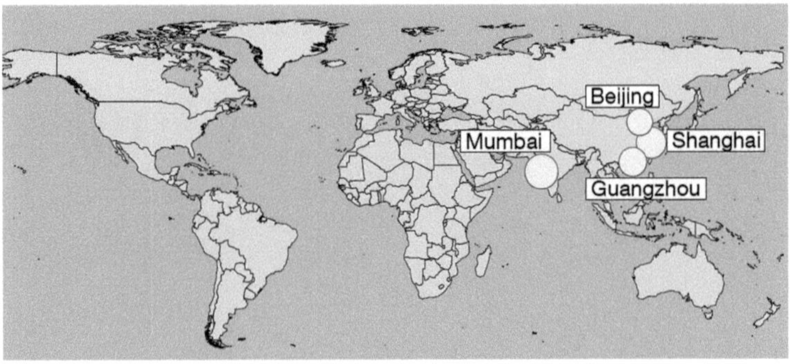

Fig. 2.6 Non-motorised cities

In 1995, the base year of the study data, the as yet relatively undeveloped national economies of the cities in this cluster held them back from both motorisation and investment in public transport as well (see Annex). This accounts for the (then) very low level of transport supply for both private and public transport modes. Some studies, such as Dargay et al. [7], suggest that motorisation starts to grow at incomes over US $3,000 per capita, and in 1995 all of these cities had yet to reach this threshold (Fig. 2.6). For the changes that have occurred in them over recent years, see the section 'Transport Development Trends' below.

This economic background also affects the demand side of transport activity, so that the percentage of trips made by private car puts them at the very bottom of our city ranking (with values from just 7 % in Shanghai up to 24.3 % in Beijing). The popular modes in these cities, accounting for most of the daily trips, are the non-motorised ones, namely foot and bicycle. So it comes as no surprise that formerly bicycle-dominated Chinese cities can be found in this cluster. The proportion of trips accomplished by non-motorised modes in Shanghai, in particular, reached a phenomenal 77.9 % in 1995.

Public transport modes are also used in this cluster, in particular for destinations and activities that are out of the reach of non-motorised transport. In fact, the role of public transport in total daily trips is much higher than one might expect given its poor quality. It accounts for 40.9 % of all trips in Mumbai, for example, which demonstrates the extreme need of the majority of the population for affordable and accessible mobility, notwithstanding its overcrowded and substandard nature. The public transport systems of these cities 'benefit' from vast numbers of captive riders.

Because there is (or was!) relatively little private car traffic in these cities, private passenger transport energy use per capita in the dataset is low. But the same cannot be said of total emissions, because of a comparatively aged fleet of vehicles on their streets, and poor emissions regulations and enforcement. Despite there being a low level of private transport use, these cities still give the impression of being overwhelmed with car traffic. This is because their urban structure is developed around non-motorised transport with very low levels of road supply, which get congested even when rates of car ownership and use are very minimal.

In addition, Non-Motorised Cities are amongst the densest cities in the analysis. Their density, and the associated intensely mixed pattern of land use, ensure that trip distances are relatively short and therefore suited to walking and cycling. The big problems facing these cities are the deleterious effect that motorisation is having on the quality and safety of the public space, and the public policy stance that equates non-motorised modes with 'out-of-date symbols of poverty', at the same time that the developed world is trying desperately to revive the popularity of these modes for the sake of sustainability and the liveability of their cities.

Paratransit Cities

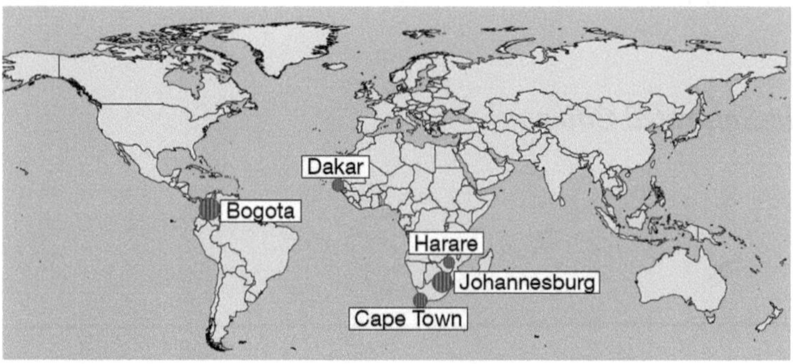

Fig. 2.7 Paratransit cities

In Paratransit Cities, relative poverty and a high level of informality in the practicalities of day-to-day life causes chaotic transport conditions and leads to a very high number of transport deaths (see Annex). In a very real sense, much of the transport system of these cities consists of a semi-organised entrepreneurial response to the general ineffectiveness of government planning and transport infrastructure development (with some notable exceptions, such as Bogotá's TransMilenio bus rapid transit system) (Fig. 2.7).

As in Transit Cities, car ownership and freeway infrastructure development are generally low. But in contrast to that cluster, the reason behind this is not so much density but poverty. Fiscal deficits hinder investment in road infrastructure, and for low-income households it is hard to afford a private car, or in many cases even a motorcycle. The general user cost of private transport as a ratio of GDP per capita is therefore significant.

As a result, private transport modal share is fairly or very low, while use of public transport is, with the exception of the two South African cities, relatively high. But the character of public transport here is very different to the efficient and clean rail-based systems to be found in the Transit Cities in at least two respects.

Firstly, only in Cape Town and Johannesburg are (regional) rail services of a significant scale available. Consequently, 'transit' to a large extent means bus travel, and in this context many of the buses used are poorly maintained, crowded, and often dangerously driven minibuses. Moreover, they spend much of their time stuck in congestion, competing with private vehicles on a poorly developed road infrastructure. Secondly, the informal way of doing things that characterises these cities has invaded all sectors of the economy, including the transport business. So, alongside the few publicly regulated bus services in operation, city streets are crowded with privately operated microbuses. Their drivers are waiting for a full load before setting off, in order to ensure a profit. Of course this means that there are no timetables, and no opportunity for users to plan their travel effectively or rely on any particular travel time. The service is bad, and maintenance of the buses and of the general infrastructure even worse.

Traffic-Saturated Cities

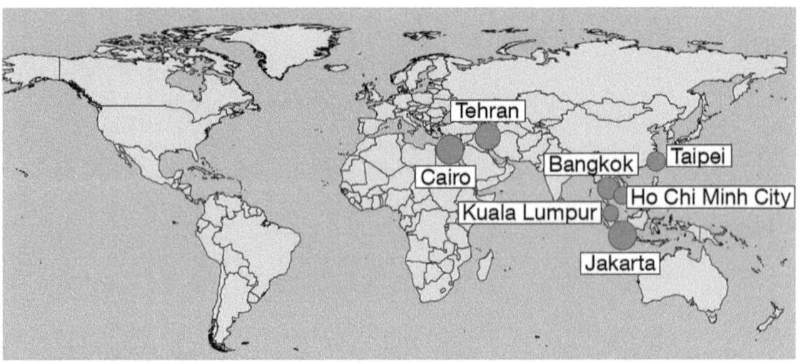

Fig. 2.8 Traffic-saturated cities

This second-largest cluster of megacities is also the most heterogeneous one, characterised by urban regions with intense—in some cases legendary—traffic congestion, Bangkok being an obvious example. In this cluster Taiwan's capital Taipei and the primate cities of Southeast Asian countries are grouped together with the Middle Eastern city of Tehran and the North African city of Cairo. There is thus a broad regional aspect to the cluster (Fig. 2.8). In terms of density, one of the most significant indicators in cluster determination, these cities predominantly rank at the top (see Annex), especially Ho Chi Minh City, which is the densest city in the analysis, but with the exception of Kuala Lumpur, a city of below-average density.

In terms of wealth, this cluster is located in the lower half of the distribution, but still above the Non-Motorised Cities.

What these cities have most in common is a fundamental mismatch between the degree of development of their transport infrastructure—for both private and public transport—and the existing demand for transport. Motorisation is relatively high: the total number of private passenger vehicles per person is for the most part higher than in the wealthy Transit Cities and even, in three of the seven, than Hybrid London. While citizens can increasingly afford motorised transport, typically purchasing first of all a motorcycle and later a car, a phenomenon attributable especially to attractive credit arrangements and few economic disincentives (as is the case in Singapore), public authorities barely have the ability—whether for reasons of limited funding or of bad governance—to provide the requisite transport infrastructure. The result is not surprising: streets saturated with car and motorcycle traffic.

Notwithstanding the above, it does seem that in this group of cities what transport infrastructure dollars there are, although apparently in short supply, are spent on road infrastructure; invariably, however, this expenditure does not keep pace with the fast-growing demand for travel. And without money being invested in attractive and competitive public transport systems, a self-reinforcing mechanism has developed whereby private motorised modes remain at the top of people's transport preferences.

The Special Case of Manila

The city of Manila was isolated from the clusters created when settling on a total of six. Initially, it showed the greatest similarity to the Auto Cities cluster, as can be seen in the dendrogram (Fig. 2.1). But Manila's very distinctive character in terms of mobility culture includes such factors as the dominant role of its 'jeepney' system [1, 2, 16], which yields very high levels of public transport patronage by means of what might be termed saturation levels of service provision, amounting annually to 420 km per capita. Its nearest competitor is Bogotá with 243 km per person. Thus Manila does have some characteristics of a Paratransit City but exceeds the typical Paratransit City by a significant margin when it comes to distance travelled by public transport. It also has an extremely low motorcycle ownership rate compared to most cities, especially its Asian counterparts, and an extremely low car ownership as well—only 90 per 1,000 people. In complete contrast to the Auto Cities, Manila also is exceptionally dense, with over 200 persons per ha; it has minuscule parking provision in its central area (29 spaces per 1,000 CBD [central business district] jobs), and a very low transport death rate of only 8 per 100,000 people. Johannesburg, by contrast, in the Paratransit City cluster, has 26 deaths per 100,000 people, and Bogotá 20. Overall, Manila appears to share similarities with a number of types of cities, but is insufficiently homogenous with respect to a sufficient number of variables to conveniently fall into any one cluster; it thus forms a cluster of its own.

Transport Development Trends

As previously mentioned, this preliminary study is based upon data from the UITP Millennium Cities Database for Sustainable Transport. As this database relies on data from 1995, an update of the data for the year 2005 or later would have been infinitely preferable; however, such an update is not yet, at the time of writing, available. In the intervening years, mobility culture may well have changed substantially in some cities, especially in emerging economies such as India and China, but also in some US cities where rail transit has been reintroduced. This section attempts to summarise a few of these changes.

The cluster findings based on 1995 data are, however, already very interesting, and the age of the data does not negate the general macro comparative differences. In other words, Chinese cities (for example) have not become like US cities, even though car use in them has burgeoned, and lower-density new settlement patterns have emerged on their peripheries. They are still dense cities, and transport patterns are still radically more public-transport-orientated and non-motorised than in US cities, or indeed many other cities in the world.

But motorised private transport in China is on the fast track, as for example in Beijing, where road infrastructure is expanding and car use is rising rapidly. The city is nowadays infamous for its air pollution and massive congestion. But the subway system is radically expanding too, with the total system length now at 323 km [3]. Shanghai displays a similar trend, but here car ownership is more restricted, and the expansion of public transport, especially its huge subway system, keeps pace with the city's growth. Although the Shanghai subway opened as recently as 1995, today it serves the world's largest network. It also already has a Singapore-style 'Certificate of Entitlement' system for permitting car ownership, designed to balance the rate of new car registrations with the attrition rate, thereby achieving a more-or-less stable ownership level.

As nearly everywhere in the cities of the emerging economies, wealth and motorisation are still on the rise, so also are they in the Los Angeles area. But in contrast to the historical backdrop to development, freeway length, as well as car kilometres travelled per person in this city—known for its predominant car culture—is currently stagnating or even declining, and this at the same time that reserved public transport routes and passenger boardings have been increasing (by 28 % between 1995 and 2005). And while improvements to public transport in Houston still represent only a modest effort in this overwhelmingly auto-orientated city, car passenger kilometres and freeway infrastructure per person are in fact decreasing (the latter by 15 % between 1995 and 2005), albeit from extremely high levels in 1995 (figures from the current 2005 update of the cities database, summarised more fully below).

It is more than likely that the city cluster categorisation of this study may change over time—or even that it would be seen to have changed already, if the latest numbers were available as the basis of a fresh analysis. However, the basic cluster definitions are unlikely to have changed—it is merely that some of the membership lists of cities will have altered somewhat. Actually, cities are dynamic systems—they rarely remain static for long, hence their mobility culture is in a constant state of flux. A brief overview of the 1995–2005 data update that follows demonstrates this. Some further final comments about this are made at the end of the chapter to show that in spite of the older data on which it is based, the cluster analysis is still highly relevant and valuable.

From work done to update data on a selection of the cities in the original Millennium Cities Database, it is clear that there are some significant trends emerging between 1995/1996 and 2005/2006, the year of the latest comprehensive update. The 33 metropolitan regions updated comprise 10 in the USA, 5 in Canada, 4 in Australia, 12 in Western Europe, and also Singapore and Hong Kong.

Using the 25 key variables shown in Table 2.3 that have been explored to develop a type of 'sustainability report card' on these 33 cities, it is relatively easy to judge whether a trend supports greater urban transport sustainability or opposes it. The full data on these trends can be found in Kenworthy [8].

We can say that in ten of the 25 variables, or 40 %, the trends are in a consistently positive direction for the groups of cities as a whole—with some intra-group variability, naturally. This means that in each of the groups of cities (American, Canadian, Australian, Western European, and Singapore/Hong Kong considered as a group), on average the variable is moving in a direction that would normally be considered as tending to increased sustainability. Of these ten variables, eight are related to aspects of the public transport system, such as growth in the supply of services and growing patronage of public transport. The details revealed in the data show that the real success story over this period has been urban rail, which has led the public transport renaissance in these cities and shows consistently higher performance than buses. Increasing density of activity (persons plus jobs per ha), and a reduction in transport deaths per 100,000 persons are the other two consistently positive stories to come out of the research so far.

Furthermore, another ten variables achieve a mostly positive trend towards sustainability, with generally only one group of cities diverging from that trend. Six of these mostly positive variables are related to transport infrastructure items, relating to both private and public modes. This means that of the variables examined, 20 out of the 25, or 80 %, demonstrate a generally positive trend with respect to transport sustainability.

Interestingly, only two variables demonstrate a consistently negative trend: car usage in terms of car vehicle kilometres per capita (the actual distance that the cars drive, irrespective of occupancy) and the proportion of jobs located in the CBD of cities. Only one further variable exhibits a mostly negative trend, and that is car ownership. While the 2005 data did not record a reduction per se in car use, they did show a slowing of the growth trend between 1995 and 2005, and this appears to have continued in the post-2005 period [13, 15]. The decline in the proportion of jobs in the CBD does not necessarily imply a decline in the absolute number of jobs located there, and if a good proportion of the non-CBD jobs are shifting to other significant sub-centres serviced by good public transport and accessible by foot and bike, this trend is not necessarily going to cause more car-based travel. On the other hand, if jobs are scattering, the evidence is that this will promote more car use (e.g. [6]).

Table 2.3 A transport sustainability report card on 33 global cities, 1995/1996–2005/2006

Variable	Trend 1995/1996–2005/2006	City group Against trend
Urban form factors		
Urban density (persons per ha)	MGP	Canadian/European
Activity density (persons + jobs per ha)	CP	None
Proportion of jobs in CBD (%)	CN	None
Private transport infrastructure factors		
Length of road per person (metres)	MGP	Canadian
Length of freeway per person (metres)	MGP	Canadian
Parking spaces per 1000 CBD jobs	MGP	European
Cars per 1000 persons	MGN	Canadian
Public transport infrastructure factors		
Total length of reserved public transport route per person (m/1000 persons)	MGP	Australian
Ratio of reserved public transport infrastructure versus freeways	MGP	Australian
Private transport use factors		
Car vehicle kilometres per person	CN	None
Car passenger kilometres per person	GM	US/Australian/European
Public transport service, use and performance		
Total public transport vehicle kilometres per person	CP	None
Total public transport seat kilometres of service per person	CP	None
Total rail seat kilometres per person	CP	None
Total bus seat kilometres per person	GM	Canadian/Australian
Total public transport boardings per person	MGP	Singapore/HK
Total rail boardings per person	CP	None
Total bus boardings per person	MGP	Singapore/HK
Total public transport passenger kilometres per person	CP	None
Total rail passenger kilometres per person	CP	None
Total bus passenger kilometres per person	CP	None
Proportion of total motorised passenger kilometres on public transport (%)	CP	None
Ratio of public transport system speed to road traffic speed	MGP	Canadian

(Continued)

Table 2.3 (Continued)

Variable	Trend 1995/1996–2005/2006	City group Against trend
Energy and externality factors		
Total private transport passenger energy use per person (MJ)	MGP	Australian/European
Total transport deaths per 100,000 people	CP	None
Summary of trends	*Number of cities*	*Percentage %*
Consistently positive (CP)	10	40
Consistently negative (CN)	2	8
Generally mixed (GM)	2	8
Mixed but generally positive (MGP)	10	40
Mixed but generally negative (MGN)	1	4
Total	25	100

Note For the *Generally Mixed* category, the city groups listed against the trend are the ones that go *against* a *positive* sustainability trend

The two genuinely mixed-trend variables are car usage expressed as car passenger kilometres per capita (which does take into account the actual number of people travelling in the cars), where the increases are very small, with two city groups showing declines; and bus service provision in terms of seat kilometres per capita. Of course, a reduction in car passenger kilometres can mean a reduction in car occupancy rather than a reduction in actual kilometres driven by cars.

So, can we say whether the cities in this study, which constitute a significant sample from five regions of the world, are generally moving towards or away from greater transport sustainability? On the basis of these data, we can conclude that most of the trends are in a direction that would tend to support a more sustainable transport system. It is probably true to say that these 33 cities, at least—and possibly many others—find themselves at some sort of tipping point, in that there is evidence of some reversal in factors that have hitherto shown longstanding trends, over many years, in a direction that facilitates and promotes increased car use. Urban transport and land-use policy, and demonstration projects, will be critical in how these trends play out over the next ten years. The important question arises as to why these trends (for example the growth in public transport patronage) are not even stronger than they are, and why car use did not show signs of more pronounced downward movement by 2005/2006. It could be postulated that one missing policy instrument, which if implemented could assist and reinforce the positive trends in physical planning, is economic restraint of car use through pricing signals (as used in Singapore and Shanghai, where prospective car purchasers must bid in an auction system for a Certificate of Entitlement to buy a car; see the chapters on Singapore and Beijing in this book). Governments at all levels—city, state and national—have shown themselves both able and willing to put in place many means of encouraging the use of non-car modes, but seem in general reluctant to reinforce such investment and maximise its effectiveness by constraining the automobile.

The point might be made that it can take some time for investment in public transport, or cutbacks in car infrastructure, to bring about the desired behavioural responses, and that we may therefore expect to observe something of a delay in the desired response in the trend. However, it is difficult to deny that cities generally fail to control congestion, and that most endure systematic daily breakdown in performance of their road system. Moreover, it is hard to imagine that modern cities would tolerate the same kind of daily breakdown in their telephone, sewerage or water systems, or in their electricity supply or Internet provision, simply because of 'congestion', i.e. demand exceeding supply. The transport system thus remains the only essential infrastructure network in most of the world's cities where society is willing to accept such fundamental underperformance, for which until this day no solution has been effectively implemented, whether by overwhelming demand with supply (an approach that has been tried throughout the post-war period, and has failed) or by managing demand so that it matches supply [5].

Final Remarks About the Integrity of the Cluster Analysis

The cluster analysis results come from an analysis of a distinctly aggregate nature, and represent relative differences between cities. The 1995 data were effective for identifying the key city types around the world, which will not have fundamentally changed. Membership of the various clusters will in all likelihood have shifted only slightly in most cases. For example, US cities are still a tremendously long way ahead of all the others in their car dependence, although also still encompassing some interesting Hybrid examples such as New York. European cities are still very orientated to public transport, and are even more walking- and cycling-orientated today than they were in 1995. The Asian cities such as Singapore and Hong Kong are still so far ahead of the others in terms of objective measures of the physical sustainability of their transport systems, that they hardly bear comparison with the rest of the world. The Canadian cities are still sitting quite happily between the US orientation to the car and the European orientation to public transport, and if anything have moved a little closer to the European model. One could further discuss such 'big-picture' perspectives with regard to the more recent data, but fundamentally, in the developed world at least, the perspectives that have been gained from the analysis of 1995 data still hold true for the years that have passed since.

Although motorisation has advanced in the rapidly growing economies, the data so far show that, yes, car ownership has grown, there is more congestion and so on, but the levels of per-capita car use are still very much lower than elsewhere. In Taipei, for example, car vehicle kilometres per capita have increased from 1,813 in 1996 to 2,064 in 2006, which is hardly a huge change (14 % growth in ten years), and certainly keeps the city at the 'very low car use end' of the global spectrum. The general perception is, perhaps, to view the overwhelming congestion in cities that have fundamentally very low road provision, and which are simply not built for the car, as evidence that they have become Auto Cities. However, this is very far from the truth.

What has happened in the case of Taipei, however, is that motorcycle use has nearly tripled from 1,247 motorcycle vehicle km per capita in 1996 to 3,416 in 2006. Even this, though, does not alter the fact that in 1996 Taipei was already fundamentally more of a traffic-saturated 'motorcycle city' than anything else. The same still holds true for 2006—it has merely strengthened. So yes, change has taken place, but not so as to overturn or in any way invalidate the 1995 perspectives in any significant comparative sense. Chinese cities have not become European or American cities.

Conclusions

In summary, this preliminary study, notwithstanding the fact that the data relates to the end of last century, has deepened our understanding of the major archetypes of urban transport in the world's megacities. By understanding the nature of each of the clusters discerned, and the key factors that underpin that cluster, it is possible by further analysis to gain an insight into the key strengths and weaknesses of future mobility in these cities, as well as the opportunities presented to them and the constraints that they are facing. Some key trend data have reinforced the view that cities are in a state of change in their mobility patterns, but at the same time that few cities seem to have fundamentally changed when it comes to the comparative characteristics that place them in the clusters described in this chapter.

References

1. Bacero R (2009) Assessment of jeepney's components, systems and separate technical units for the developments of standards: proceedings of the 17th annual conference of the transportation science society of the Philippines, Manila
2. Bayan JM, Villoria OG, Ieda H (1995) Cost characteristics of bus and jeepney transport systems in metro Manila. National Center for Transportation Studies, University of the Philippines, discussion paper No. 6. Manila
3. Beijing Subway (2012) Company profile. http://www.bjsubway.com/node/76. Accessed: May 24, 2012
4. Bongardt D (2009) Stadtverkehr in Megacities. Bundeszentrale für Politische Bildung
5. Bradley D, Kenworthy J (2012) Congestion offsets: transforming cities by letting buses compete: draft paper for world transport policy and practice 18(4):46–69
6. Cervero R, Landis J (1992) Suburbanisation of jobs and the journey to work: a submarket analysis of commuting in the San Francisco Bay area. J Adv Transp 26(3):275–297
7. Dargay J, Gatley D, Sommer M (2007) Vehicle ownership and income growth, worldwide: 1960–2030. Energy J 28(4):143–170
8. Kenworthy J (2012) Moving to the dark side or will the force be with us?—a sustainability re-port card on trends in transport and urban development in thirty-three global cities, 1995–1996 to 2005–2006: Environmental Research Letters: Special issue: Environmental assessments in the built environment
9. Kenworthy J, Laube F (2001) The millennium cities database for sustainable transport: (CDROM Database). International Union of Public Transport (UITP), Brussels

10. Kenworthy J, Townsend C (2009) Montreal's dualistic transport character: why mon-treal needs upgraded transit and not more high capacity roads. In: Gauthier P (ed) Montreal at the crossroads: superhighways, the turcot and the environment. Black Rose Books, Montreal, pp 29–35
11. Kraas F (ed) (2005) Megacities–our global urban future: earth sciences for society foundation. International Year for Planet Earth, Leiden
12. Newman P, Kenworthy J (1999) Sustainability and cities: overcoming automobile dependence. Island Press, Washington
13. Newman P, Kenworthy J (2011) Peak car use: understanding the demise of automobile dependence. World Transp Policy Pract 17(2):31–42
14. Priester R, Kenworthy J, Wulfhorst G (2010) Mobility cultures in megacities: preliminary study, Technical University of Munich
15. Puentes R, Tomer A (2008) The roadless traveled: an analysis of vehicle miles traveled trends in the U.S. Washington
16. Satre GL (1998) The metro Manila LRT system—a historical perspective (PDF). Japan Railway Transp Rev 16:33–37
17. Tan P-N, Steinbach M, Kumar V (2005) Introduction to data mining. Person Addison-Wesley, Boston
18. Thomson J (1977) Great cities and their traffic. Gollancz, London
19. United Nations: Department of Economic and Social Affairs (UNDESA)–population division (2008) World Urbanization Prospects: The 2007 revision, United Nations

Annex: Selected City Indicators

This Annex shows selected city indicators which were used in the cluster analysis.

The original source of the data for the graphs is: Kenworthy and Laube (2001), The Millennium Cities Database for sustainable Transport.

The Annex uses the following acronyms to label the city clusters:

NM	Non-Motorized Cities
PT	Paratransit Cities
TS	Traffic-Saturated Cities
TR	Transit Cities
HY	Hybrid Cities
AU	Auto Cities

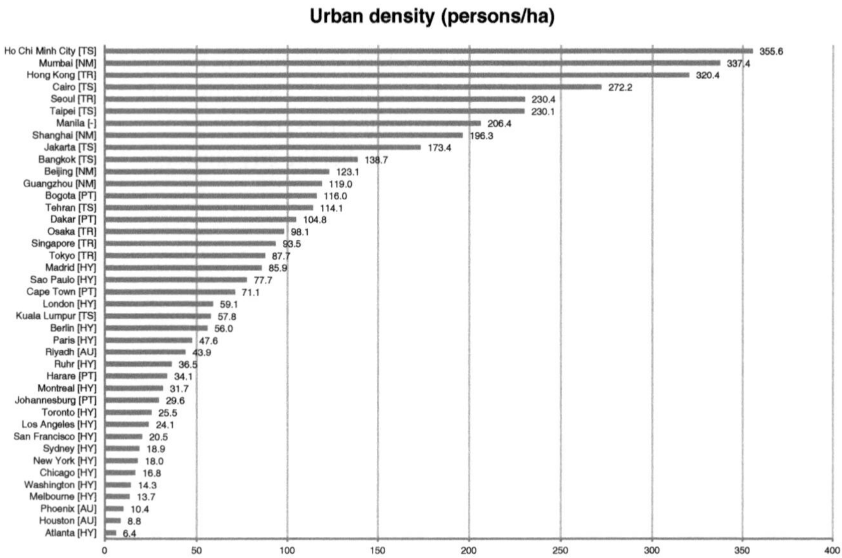
Urban density (persons/ha)
Ho Chi Minh City [TS] 355.6
Mumbai [NM] 337.4
Hong Kong [TR] 320.4
Cairo [TS] 272.2
Seoul [TR] 230.4
Taipei [TS] 230.1
Manila [-] 206.4
Shanghai [NM] 196.3
Jakarta [TS] 173.4
Bangkok [TS] 138.7
Beijing [NM] 123.1
Guangzhou [NM] 119.0
Bogota [PT] 116.0
Tehran [TS] 114.1
Dakar [PT] 104.8
Osaka [TR] 98.1
Singapore [TR] 93.5
Tokyo [TR] 87.7
Madrid [HY] 85.9
Sao Paulo [HY] 77.7
Cape Town [PT] 71.1
London [HY] 59.1
Kuala Lumpur [TS] 57.8
Berlin [HY] 56.0
Paris [HY] 47.6
Riyadh [AU] 43.9
Ruhr [HY] 36.5
Harare [PT] 34.1
Montreal [HY] 31.7
Johannesburg [PT] 29.6
Toronto [HY] 25.5
Los Angeles [HY] 24.1
San Francisco [HY] 20.5
Sydney [HY] 18.9
New York [HY] 18.0
Chicago [HY] 16.8
Washington [HY] 14.3
Melbourne [HY] 13.7
Phoenix [AU] 10.4
Houston [AU] 8.8
Atlanta [HY] 6.4
0
50
100
150
200
250
300
350
400

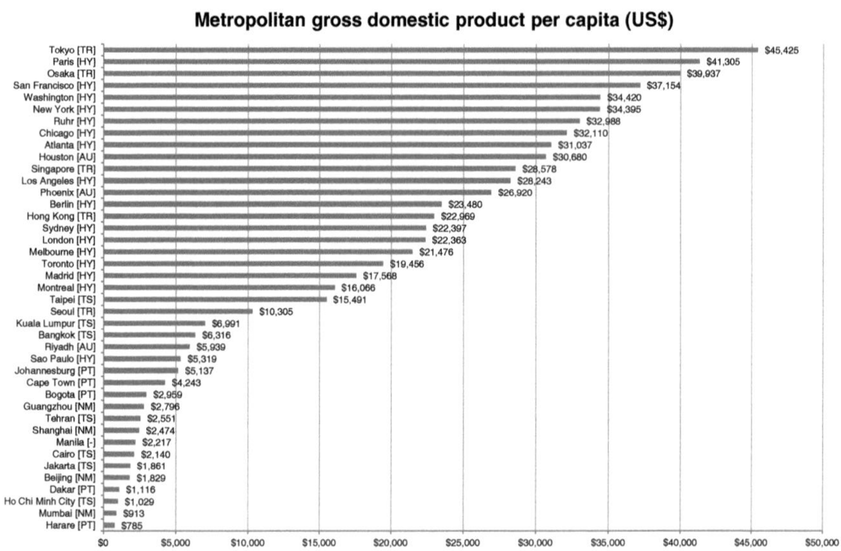
Metropolitan gross domestic product per capita (US$)
Tokyo [TR] $45,425
Paris [HY] $41,305
Osaka [TR] $39,937
San Francisco [HY] $37,154
Washington [HY] $34,420
New York [HY] $34,395
Ruhr [HY] $32,988
Chicago [HY] $32,110
Atlanta [HY] $31,037
Houston [AU] $30,680
Singapore [TR] $28,578
Los Angeles [HY] $28,243
Phoenix [AU] $26,920
Berlin [HY] $23,480
Hong Kong [TR] $22,969
Sydney [HY] $22,397
London [HY] $22,363
Melbourne [HY] $21,476
Toronto [HY] $19,456
Madrid [HY] $17,568
Montreal [HY] $16,066
Taipei [TS] $15,491
Seoul [TR] $10,305
Kuala Lumpur [TS] $6,991
Bangkok [TS] $6,316
Riyadh [AU] $5,939
Sao Paulo [HY] $5,319
Johannesburg [PT] $5,137
Cape Town [PT] $4,243
Bogota [PT] $2,959
Guangzhou [NM] $2,796
Tehran [TS] $2,551
Shanghai [NM] $2,474
Manila [-] $2,217
Cairo [TS] $2,140
Jakarta [TS] $1,861
Beijing [NM] $1,829
Dakar [PT] $1,116
Ho Chi Minh City [TS] $1,029
Mumbai [NM] $913
Harare [PT] $785
$0
$5,000
$10,000
$15,000
$20,000
$25,000
$30,000
$35,000
$40,000
$45,000
$50,000

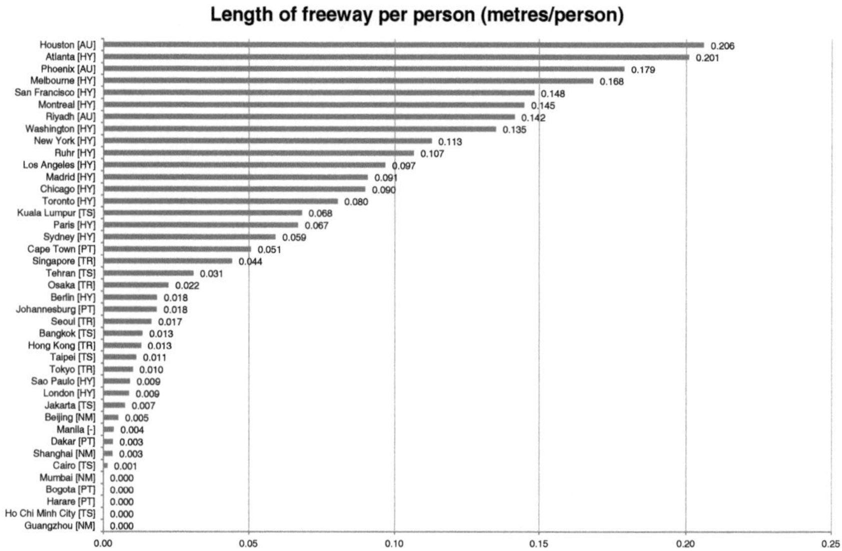
Length of freeway per person (metres/person)
Houston [AU] 0.206
Atlanta [HY] 0.201
Phoenix [AU] 0.179
Melbourne [HY] 0.168
San Francisco [HY] 0.148
Montreal [HY] 0.145
Riyadh [AU] 0.142
Washington [HY] 0.135
New York [HY] 0.113
Ruhr [HY] 0.107
Los Angeles [HY] 0.097
Madrid [HY] 0.091
Chicago [HY] 0.090
Toronto [HY] 0.080
Kuala Lumpur [TS] 0.068
Paris [HY] 0.067
Sydney [HY] 0.059
Cape Town [PT] 0.051
Singapore [TR] 0.044
Tehran [TS] 0.031
Osaka [TR] 0.022
Berlin [HY] 0.018
Johannesburg [PT] 0.018
Seoul [TR] 0.017
Bangkok [TS] 0.013
Hong Kong [TR] 0.013
Taipei [TS] 0.011
Tokyo [TR] 0.010
Sao Paulo [HY] 0.009
London [HY] 0.009
Jakarta [TS] 0.007
Beijing [NM] 0.005
Manila [-] 0.004
Dakar [PT] 0.003
Shanghai [NM] 0.003
Cairo [TS] 0.001
Mumbai [NM] 0.000
Bogota [PT] 0.000
Harare [PT] 0.000
Ho Chi Minh City [TS] 0.000
Guangzhou [NM] 0.000
0.00 0.05 0.10 0.15 0.20 0.25

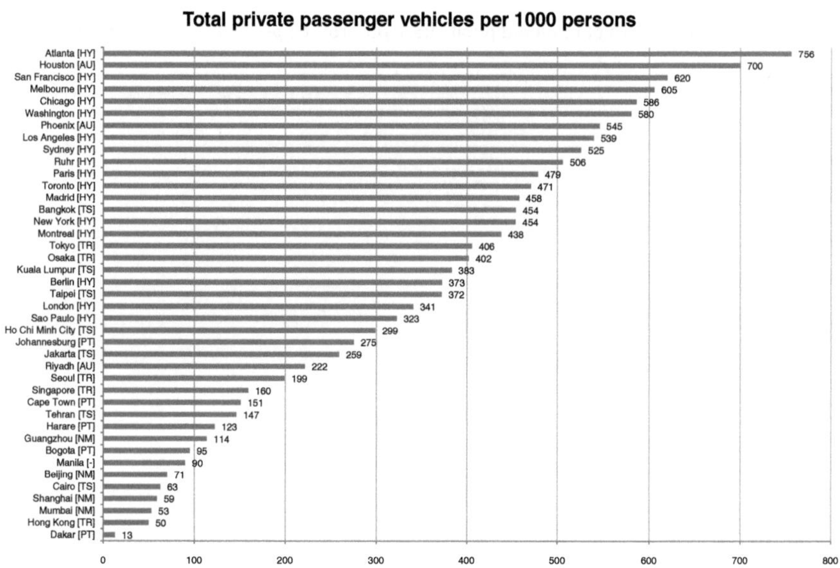
Total private passenger vehicles per 1000 persons
Atlanta [HY] 756
Houston [AU] 700
San Francisco [HY] 620
Melbourne [HY] 605
Chicago [HY] 586
Washington [HY] 580
Phoenix [AU] 545
Los Angeles [HY] 539
Sydney [HY] 525
Ruhr [HY] 506
Paris [HY] 479
Toronto [HY] 471
Madrid [HY] 458
Bangkok [TS] 454
New York [HY] 454
Montreal [HY] 438
Tokyo [TR] 406
Osaka [TR] 402
Kuala Lumpur [TS] 383
Berlin [HY] 373
Taipei [TS] 372
London [HY] 341
Sao Paulo [HY] 323
Ho Chi Minh City [TS] 299
Johannesburg [PT] 275
Jakarta [TS] 259
Riyadh [AU] 222
Seoul [TR] 199
Singapore [TR] 160
Cape Town [PT] 151
Tehran [TS] 147
Harare [PT] 123
Guangzhou [NM] 114
Bogota [PT] 95
Manila [-] 90
Beijing [NM] 71
Cairo [TS] 63
Shanghai [NM] 59
Mumbai [NM] 53
Hong Kong [TR] 50
Dakar [PT] 13
0 100 200 300 400 500 600 700 800

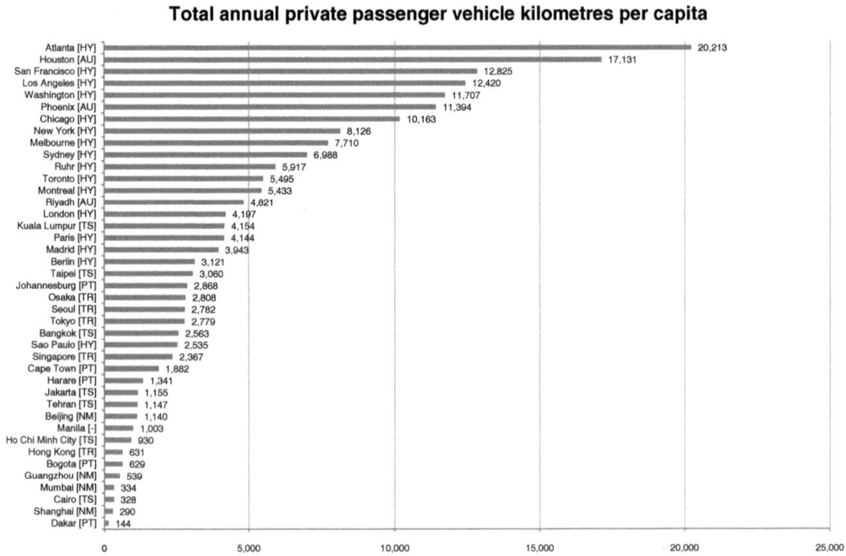
Total annual private passenger vehicle kilometres per capita
Atlanta [HY] 20,213
Houston [AU] 17,131
San Francisco [HY] 12,825
Los Angeles [HY] 12,420
Washington [HY] 11,707
Phoenix [AU] 11,394
Chicago [HY] 10,163
New York [HY] 8,126
Melbourne [HY] 7,710
Sydney [HY] 6,988
Ruhr [HY] 5,917
Toronto [HY] 5,495
Montreal [HY] 5,433
Riyadh [AU] 4,821
London [HY] 4,197
Kuala Lumpur [TS] 4,154
Paris [HY] 4,144
Madrid [HY] 3,943
Berlin [HY] 3,121
Taipei [TS] 3,060
Johannesburg [PT] 2,868
Osaka [TR] 2,808
Seoul [TR] 2,782
Tokyo [TR] 2,779
Bangkok [TS] 2,563
Sao Paulo [HY] 2,535
Singapore [TR] 2,367
Cape Town [PT] 1,882
Harare [PT] 1,341
Jakarta [TS] 1,155
Tehran [TS] 1,147
Beijing [NM] 1,140
Manila [-] 1,003
Ho Chi Minh City [TS] 930
Hong Kong [TR] 631
Bogota [PT] 629
Guangzhou [NM] 539
Mumbai [NM] 334
Cairo [TS] 328
Shanghai [NM] 290
Dakar [PT] 144
0
5,000
10,000
15,000
20,000
25,000

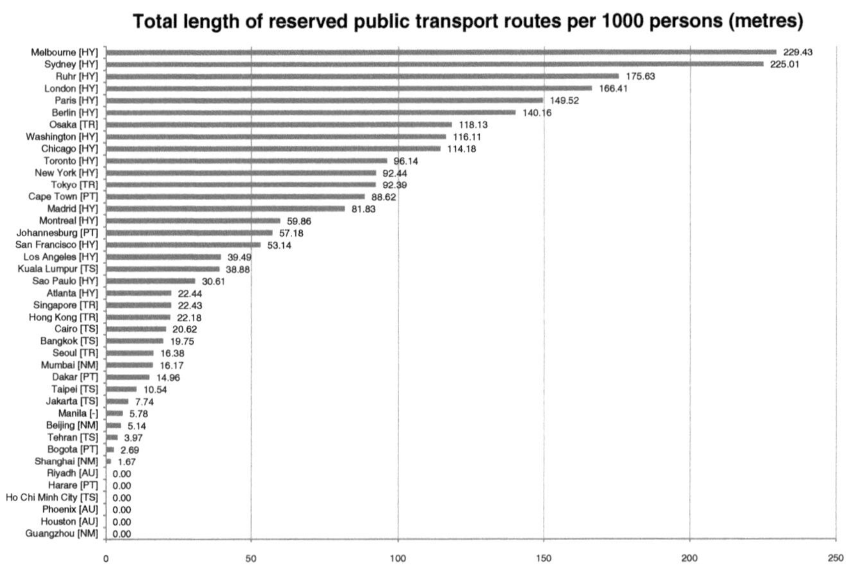
Total length of reserved public transport routes per 1000 persons (metres)
Melbourne [HY] 229.43
Sydney [HY] 225.01
Ruhr [HY] 175.63
London [HY] 166.41
Paris [HY] 149.52
Berlin [HY] 140.16
Osaka [TR] 118.13
Washington [HY] 116.11
Chicago [HY] 114.18
Toronto [HY] 96.14
New York [HY] 92.44
Tokyo [TR] 92.39
Cape Town [PT] 88.62
Madrid [HY] 81.83
Montreal [HY] 59.86
Johannesburg [PT] 57.18
San Francisco [HY] 53.14
Los Angeles [HY] 39.49
Kuala Lumpur [TS] 38.88
Sao Paulo [HY] 30.61
Atlanta [HY] 22.44
Singapore [TR] 22.43
Hong Kong [TR] 22.18
Cairo [TS] 20.62
Bangkok [TS] 19.75
Seoul [TR] 16.38
Mumbai [NM] 16.17
Dakar [PT] 14.96
Taipei [TS] 10.54
Jakarta [TS] 7.74
Manila [-] 5.78
Beijing [NM] 5.14
Tehran [TS] 3.97
Bogota [PT] 2.69
Shanghai [NM] 1.67
Riyadh [AU] 0.00
Harare [PT] 0.00
Ho Chi Minh City [TS] 0.00
Phoenix [AU] 0.00
Houston [AU] 0.00
Guangzhou [NM] 0.00
0
50
100
150
200
250

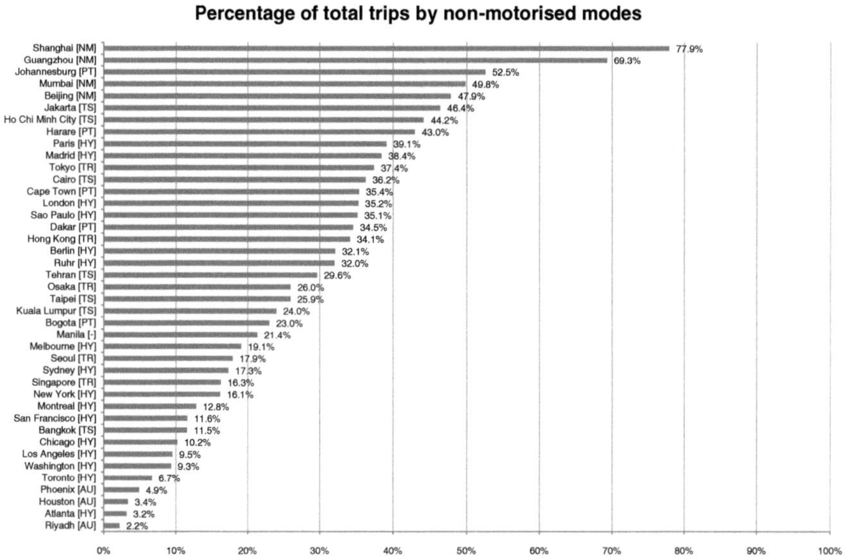
Percentage of total trips by non-motorised modes
Shanghai [NM] 77.9%
Guangzhou [NM] 69.3%
Johannesburg [PT] 52.5%
Mumbai [NM] 49.8%
Beijing [NM] 47.9%
Jakarta [TS] 46.4%
Ho Chi Minh City [TS] 44.2%
Harare [PT] 43.0%
Paris [HY] 39.1%
Madrid [HY] 38.4%
Tokyo [TR] 37.4%
Cairo [TS] 36.2%
Cape Town [PT] 35.4%
London [HY] 35.2%
Sao Paulo [HY] 35.1%
Dakar [PT] 34.5%
Hong Kong [TR] 34.1%
Berlin [HY] 32.1%
Ruhr [HY] 32.0%
Tehran [TS] 29.6%
Osaka [TR] 26.0%
Taipei [TS] 25.9%
Kuala Lumpur [TS] 24.0%
Bogota [PT] 23.0%
Manila [-] 21.4%
Melbourne [HY] 19.1%
Seoul [TR] 17.9%
Sydney [HY] 17.3%
Singapore [TR] 16.3%
New York [HY] 16.1%
Montreal [HY] 12.8%
San Francisco [HY] 11.6%
Bangkok [TS] 11.5%
Chicago [HY] 10.2%
Los Angeles [HY] 9.5%
Washington [HY] 9.3%
Toronto [HY] 6.7%
Phoenix [AU] 4.9%
Houston [AU] 3.4%
Atlanta [HY] 3.2%
Riyadh [AU] 2.2%
0% 10% 20% 30% 40% 50% 60% 70% 80% 90% 100%

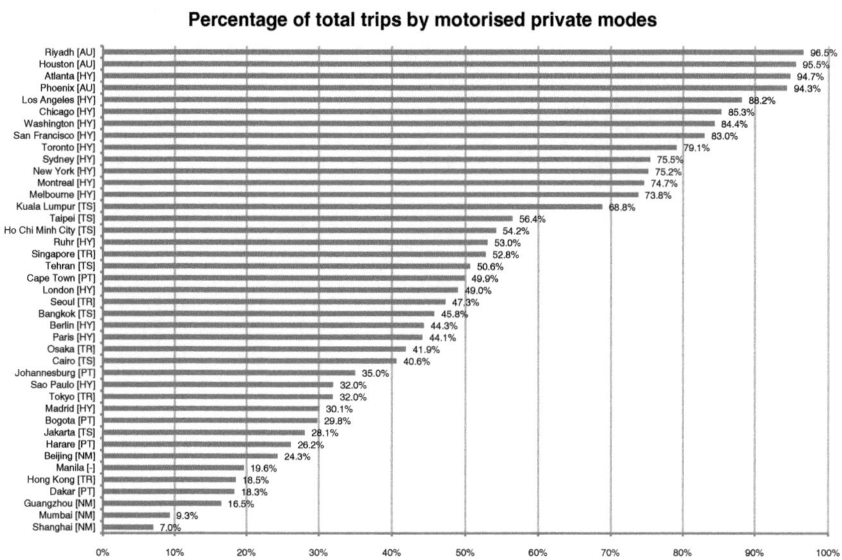
Percentage of total trips by motorised private modes
Riyadh [AU] 96.5%
Houston [AU] 95.5%
Atlanta [HY] 94.7%
Phoenix [AU] 94.3%
Los Angeles [HY] 88.2%
Chicago [HY] 85.3%
Washington [HY] 84.4%
San Francisco [HY] 83.0%
Toronto [HY] 79.1%
Sydney [HY] 75.5%
New York [HY] 75.2%
Montreal [HY] 74.7%
Melbourne [HY] 73.8%
Kuala Lumpur [TS] 68.8%
Taipei [TS] 56.4%
Ho Chi Minh City [TS] 54.2%
Ruhr [HY] 53.0%
Singapore [TR] 52.8%
Tehran [TS] 50.6%
Cape Town [PT] 49.9%
London [HY] 49.0%
Seoul [TR] 47.3%
Bangkok [TS] 45.8%
Berlin [HY] 44.3%
Paris [HY] 44.1%
Osaka [TR] 41.9%
Cairo [TS] 40.6%
Johannesburg [PT] 35.0%
Sao Paulo [HY] 32.0%
Tokyo [TR] 32.0%
Madrid [HY] 30.1%
Bogota [PT] 29.8%
Jakarta [TS] 28.1%
Harare [PT] 26.2%
Beijing [NM] 24.3%
Manila [-] 19.6%
Hong Kong [TR] 18.5%
Dakar [PT] 18.3%
Guangzhou [NM] 16.5%
Mumbai [NM] 9.3%
Shanghai [NM] 7.0%
0% 10% 20% 30% 40% 50% 60% 70% 80% 90% 100%

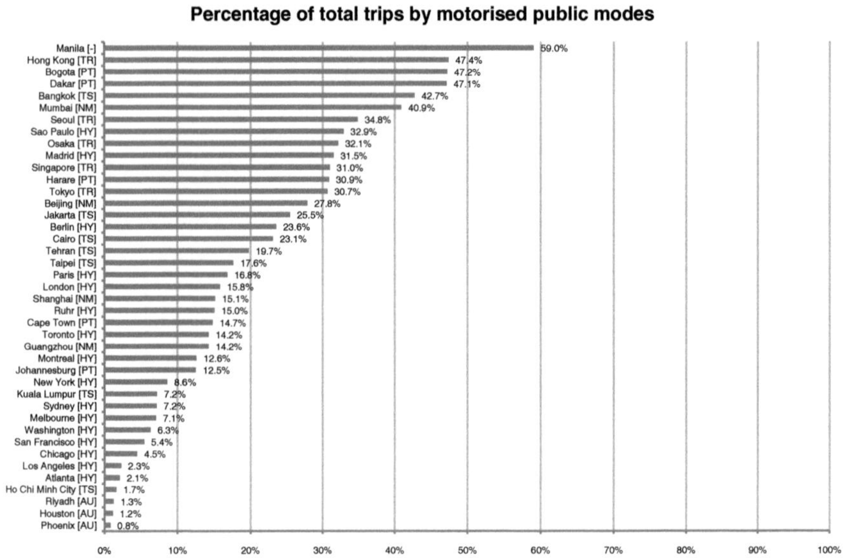
Percentage of total trips by motorised public modes
Manila [-] 59.0%
Hong Kong [TR] 47.4%
Bogota [PT] 47.2%
Dakar [PT] 47.1%
Bangkok [TS] 42.7%
Mumbai [NM] 40.9%
Seoul [TR] 34.8%
Sao Paulo [HY] 32.9%
Osaka [TR] 32.1%
Madrid [HY] 31.5%
Singapore [TR] 31.0%
Harare [PT] 30.9%
Tokyo [TR] 30.7%
Beijing [NM] 27.8%
Jakarta [TS] 25.5%
Berlin [HY] 23.6%
Cairo [TS] 23.1%
Tehran [TS] 19.7%
Taipei [TS] 17.6%
Paris [HY] 16.8%
London [HY] 15.8%
Shanghai [NM] 15.1%
Ruhr [HY] 15.0%
Cape Town [PT] 14.7%
Toronto [HY] 14.2%
Guangzhou [NM] 14.2%
Montreal [HY] 12.6%
Johannesburg [PT] 12.5%
New York [HY] 8.6%
Kuala Lumpur [TS] 7.2%
Sydney [HY] 7.2%
Melbourne [HY] 7.1%
Washington [HY] 6.3%
San Francisco [HY] 5.4%
Chicago [HY] 4.5%
Los Angeles [HY] 2.3%
Atlanta [HY] 2.1%
Ho Chi Minh City [TS] 1.7%
Riyadh [AU] 1.3%
Houston [AU] 1.2%
Phoenix [AU] 0.8%
0% 10% 20% 30% 40% 50% 60% 70% 80% 90% 100%

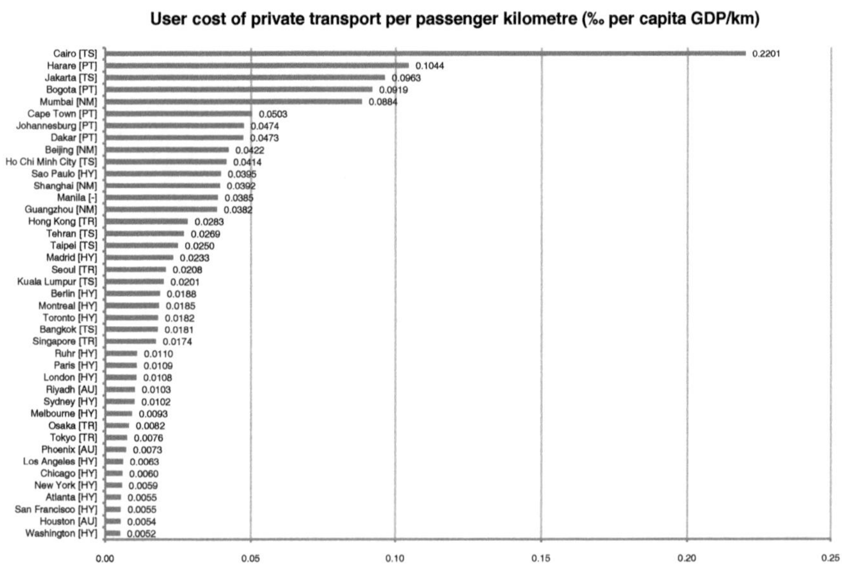
User cost of private transport per passenger kilometre (‰ per capita GDP/km)
Cairo [TS] 0.2201
Harare [PT] 0.1044
Jakarta [TS] 0.0963
Bogota [PT] 0.0919
Mumbai [NM] 0.0884
Cape Town [PT] 0.0503
Johannesburg [PT] 0.0474
Dakar [PT] 0.0473
Beijing [NM] 0.0422
Ho Chi Minh City [TS] 0.0414
Sao Paulo [HY] 0.0395
Shanghai [NM] 0.0392
Manila [-] 0.0385
Guangzhou [NM] 0.0382
Hong Kong [TR] 0.0283
Tehran [TS] 0.0269
Taipei [TS] 0.0250
Madrid [HY] 0.0233
Seoul [TR] 0.0208
Kuala Lumpur [TS] 0.0201
Berlin [HY] 0.0188
Montreal [HY] 0.0185
Toronto [HY] 0.0182
Bangkok [TS] 0.0181
Singapore [TR] 0.0174
Ruhr [HY] 0.0110
Paris [HY] 0.0109
London [HY] 0.0108
Riyadh [AU] 0.0103
Sydney [HY] 0.0102
Melbourne [HY] 0.0093
Osaka [TR] 0.0082
Tokyo [TR] 0.0076
Phoenix [AU] 0.0073
Los Angeles [HY] 0.0063
Chicago [HY] 0.0060
New York [HY] 0.0059
Atlanta [HY] 0.0055
San Francisco [HY] 0.0055
Houston [AU] 0.0054
Washington [HY] 0.0052
0.00 0.05 0.10 0.15 0.20 0.25

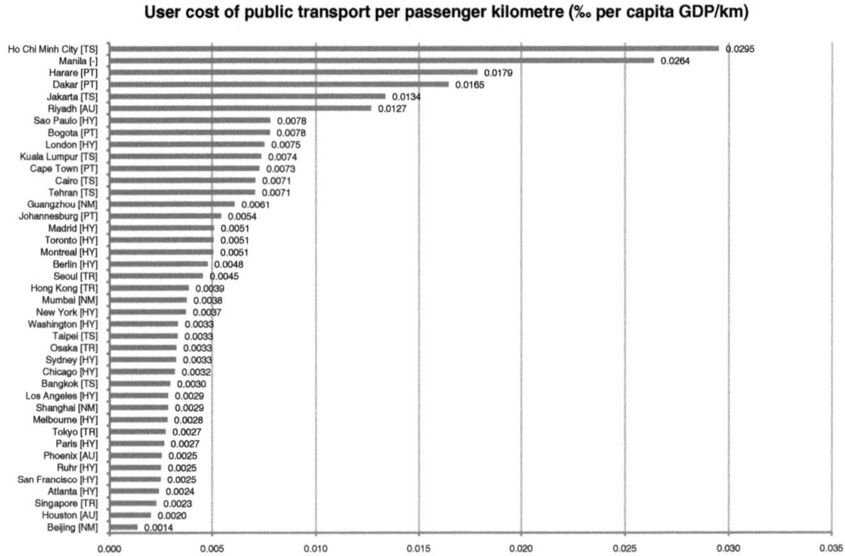
User cost of public transport per passenger kilometre (‰ per capita GDP/km)
Ho Chi Minh City [TS] 0.0295
Manila [-] 0.0264
Harare [PT] 0.0179
Dakar [PT] 0.0165
Jakarta [TS] 0.0134
Riyadh [AU] 0.0127
Sao Paulo [HY] 0.0078
Bogota [PT] 0.0078
London [HY] 0.0075
Kuala Lumpur [TS] 0.0074
Cape Town [PT] 0.0073
Cairo [TS] 0.0071
Tehran [TS] 0.0071
Guangzhou [NM] 0.0061
Johannesburg [PT] 0.0054
Madrid [HY] 0.0051
Toronto [HY] 0.0051
Montreal [HY] 0.0051
Berlin [HY] 0.0048
Seoul [TR] 0.0045
Hong Kong [TR] 0.0039
Mumbai [NM] 0.0038
New York [HY] 0.0037
Washington [HY] 0.0033
Taipei [TS] 0.0033
Osaka [TR] 0.0033
Sydney [HY] 0.0033
Chicago [HY] 0.0032
Bangkok [TS] 0.0030
Los Angeles [HY] 0.0029
Shanghai [NM] 0.0029
Melbourne [HY] 0.0028
Tokyo [TR] 0.0027
Paris [HY] 0.0027
Phoenix [AU] 0.0025
Ruhr [HY] 0.0025
San Francisco [HY] 0.0025
Atlanta [HY] 0.0024
Singapore [TR] 0.0023
Houston [AU] 0.0020
Beijing [NM] 0.0014
0.000
0.005
0.010
0.015
0.020
0.025
0.030
0.035

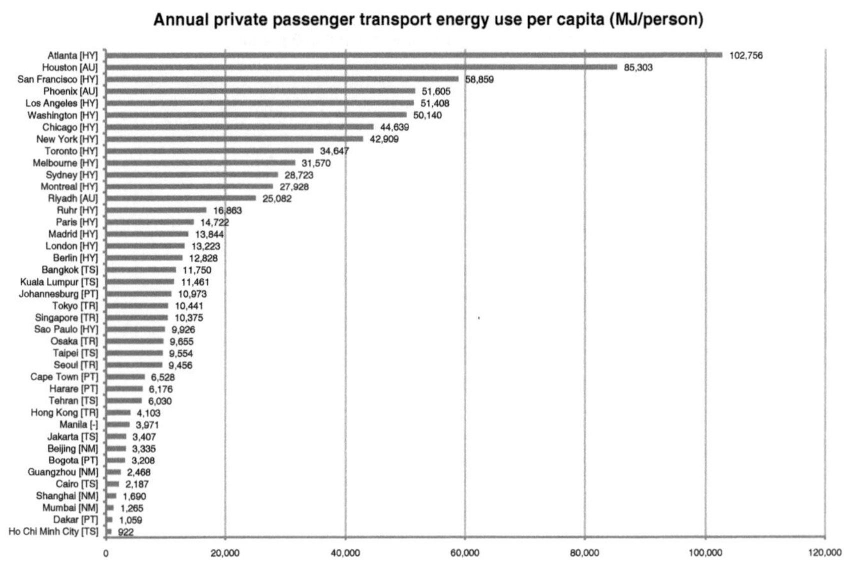
Annual private passenger transport energy use per capita (MJ/person)
Atlanta [HY] 102,756
Houston [AU] 85,303
San Francisco [HY] 58,859
Phoenix [AU] 51,605
Los Angeles [HY] 51,408
Washington [HY] 50,140
Chicago [HY] 44,639
New York [HY] 42,909
Toronto [HY] 34,647
Melbourne [HY] 31,570
Sydney [HY] 28,723
Montreal [HY] 27,928
Riyadh [AU] 25,082
Ruhr [HY] 16,863
Paris [HY] 14,722
Madrid [HY] 13,844
London [HY] 13,223
Berlin [HY] 12,828
Bangkok [TS] 11,750
Kuala Lumpur [TS] 11,461
Johannesburg [PT] 10,973
Tokyo [TR] 10,441
Singapore [TR] 10,375
Sao Paulo [HY] 9,926
Osaka [TR] 9,655
Taipei [TS] 9,554
Seoul [TR] 9,456
Cape Town [PT] 6,528
Harare [PT] 6,176
Tehran [TS] 6,030
Hong Kong [TR] 4,103
Manila [-] 3,971
Jakarta [TS] 3,407
Beijing [NM] 3,335
Bogota [PT] 3,208
Guangzhou [NM] 2,468
Cairo [TS] 2,187
Shanghai [NM] 1,690
Mumbai [NM] 1,265
Dakar [PT] 1,059
Ho Chi Minh City [TS] 922
0
20,000
40,000
60,000
80,000
100,000
120,000

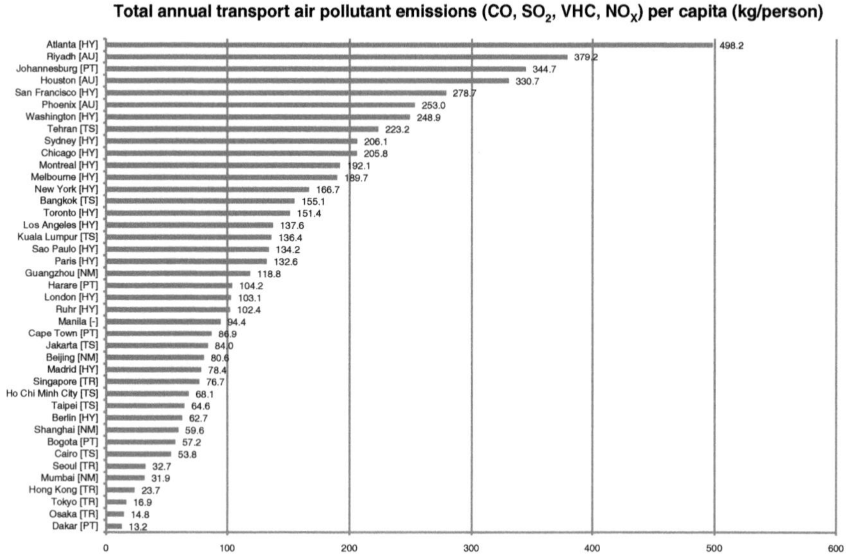
Total annual transport air pollutant emissions (CO, SO2, VHC, NOX) per capita (kg/person)
Atlanta [HY] 498.2
Riyadh [AU] 379.2
Johannesburg [PT] 344.7
Houston [AU] 330.7
San Francisco [HY] 278.7
Phoenix [AU] 253.0
Washington [HY] 248.9
Tehran [TS] 223.2
Sydney [HY] 206.1
Chicago [HY] 205.8
Montreal [HY] 192.1
Melbourne [HY] 189.7
New York [HY] 166.7
Bangkok [TS] 155.1
Toronto [HY] 151.4
Los Angeles [HY] 137.6
Kuala Lumpur [TS] 136.4
Sao Paulo [HY] 134.2
Paris [HY] 132.6
Guangzhou [NM] 118.8
Harare [PT] 104.2
London [HY] 103.1
Ruhr [HY] 102.4
Manila [-] 94.4
Cape Town [PT] 86.9
Jakarta [TS] 84.0
Beijing [NM] 80.6
Madrid [HY] 78.4
Singapore [TR] 76.7
Ho Chi Minh City [TS] 68.1
Taipei [TS] 64.6
Berlin [HY] 62.7
Shanghai [NM] 59.6
Bogota [PT] 57.2
Cairo [TS] 53.8
Seoul [TR] 32.7
Mumbai [NM] 31.9
Hong Kong [TR] 23.7
Tokyo [TR] 16.9
Osaka [TR] 14.8
Dakar [PT] 13.2
0
100
200
300
400
500
600

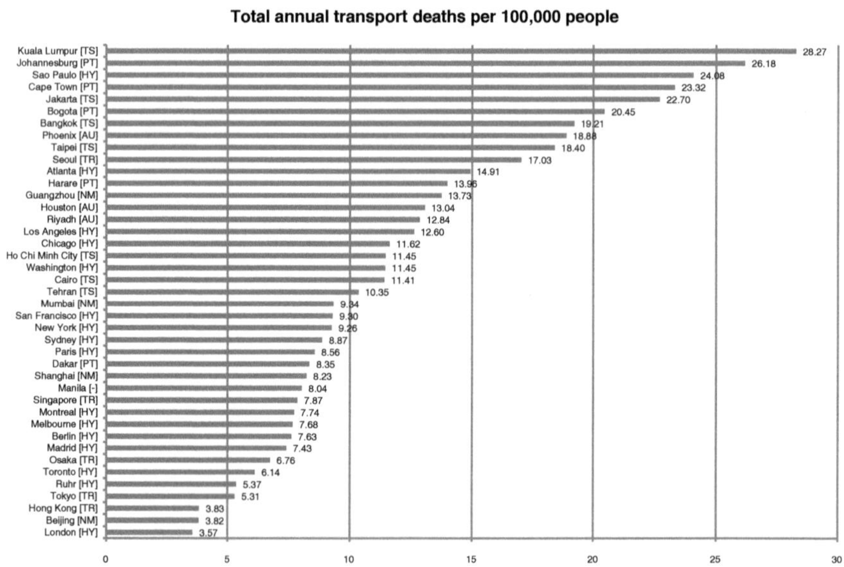
Total annual transport deaths per 100,000 people
Kuala Lumpur [TS] 28.27
Johannesburg [PT] 26.18
Sao Paulo [HY] 24.08
Cape Town [PT] 23.32
Jakarta [TS] 22.70
Bogota [PT] 20.45
Bangkok [TS] 19.21
Phoenix [AU] 18.88
Taipei [TS] 18.40
Seoul [TR] 17.03
Atlanta [HY] 14.91
Harare [PT] 13.96
Guangzhou [NM] 13.73
Houston [AU] 13.04
Riyadh [AU] 12.84
Los Angeles [HY] 12.60
Chicago [HY] 11.62
Ho Chi Minh City [TS] 11.45
Washington [HY] 11.45
Cairo [TS] 11.41
Tehran [TS] 10.35
Mumbai [NM] 9.34
San Francisco [HY] 9.30
New York [HY] 9.26
Sydney [HY] 8.87
Paris [HY] 8.56
Dakar [PT] 8.35
Shanghai [NM] 8.23
Manila [-] 8.04
Singapore [TR] 7.87
Montreal [HY] 7.74
Melbourne [HY] 7.68
Berlin [HY] 7.63
Madrid [HY] 7.43
Osaka [TR] 6.76
Toronto [HY] 6.14
Ruhr [HY] 5.37
Tokyo [TR] 5.31
Hong Kong [TR] 3.83
Beijing [NM] 3.82
London [HY] 3.57
0
5
10
15
20
25
30

Chapter 3
The Reader's Guide to Mobility Culture

Tobias Kuhnimhof and Gebhard Wulfhorst

Abstract In Part One of this book, we have seen ample evidence for the overwhelmingly diverse nature of cities across the world. Despite this diversity, we can distinguish groups of cities that share certain similarities: based on quantifiable characteristics of the urban structure, the transport system, socioeconomic situations and measures of travel behaviour, Priester et al. (see Chap. 2 in this book) have clustered cities into six groups ranging from Non-Motorised Cities through Paratransit and Traffic-Saturated Cities to Auto, Transit and Hybrid Cities.

But—based on this cluster study—are we able to understand and explain how transport systems and travel behaviour develop in cities worldwide? Not yet. The clusters provide a typology of metropolises based on measurable indicators. They don't tell the stories behind these indicators. And the clusters only provide limited insights as regards the character of the cities and the underlying driving forces which shape urban mobility within them. Recalling the terms of our discussion of mobility culture in the introduction, the clusters are based on quantitative figures which are the tangible results of actions taken by the manifold actors in this context. But the clusters don't show why the actors have acted the way they did—they don't reveal the relevant underlying layers of complexity that contribute to mobility culture: the values, perceptions and lifestyle orientations of travellers, as well as the perceptions and preferences of decision-makers within the system of urban mobility.

So, let us move on and look beneath the surface of the quantitative observations. This 'Reader's Guide to Mobility Culture' will assist understanding what shapes the existing mobility system and the prevailing mobility behaviour. Beyond the objective framework conditions, we want to detect the cultural and social driving forces and the ever-evolving public discourses that are at work. Therefore, in this Reader's Guide we suggest four different approaches to identifying and exploring mobility culture in the subsequent chapters of the book.

T. Kuhnimhof (✉)
ifmo, Petuelring 130, 80788 München, Germany
e-mail: tobias.kuhnimhof@ifmo.de

G. Wulfhorst
Technische Universität München, Munich, Germany

Institute for Mobility Research (ifmo) (ed.), *Megacity Mobility Culture*,
Lecture Notes in Mobility, DOI: 10.1007/978-3-642-34735-1_3,
© Springer-Verlag Berlin Heidelberg 2013

In our first approach, we seek to pin down the term 'mobility culture', which is still somewhat vague even after our discussion in the introduction. Therefore, we identify tangible dimensions of mobility culture which play an important role in the following city stories.

Then we acknowledge that mobility culture itself is changing constantly. It is the close scrutiny of these changes that enables us to understand the fundamental principles which constitute mobility culture, and which shape urban transport. Hence, the second approach that we suggest in this Reader's Guide is a framework for discussing change in urban transport systems, based on the cluster analysis of the previous chapter.

The third approach sets forth a number of research questions and hypotheses that have guided the research fellows in their investigations of the selected cities, and have led to the megacity stories in Part Two of this book.

The fourth suggested approach helps to identify basic driving forces which explain the interactions of the different domains of mobility culture by detecting 'temperaments' of the cities.

Approach 1: Getting to Know the Dimensions of Mobility Culture

One of the ultimate objectives of the research on megacity mobility culture in this book is to provide a framework for the successful implementation of appropriate strategies for the future development of urban mobility. Such strategies must be embedded in an interrelated system of the various social, spatial, cultural, political, and economic aspects, i.e. in the local 'mobility culture' [2].

Figure 3.1 visualises the concept of mobility culture as we understand it in this book. To start with, there are factors (shown in boxes) such as urban planning or the socioeconomic situation in the city, which represent largely external entities influencing mobility culture. However, many such external influences not only shape mobility culture, they also form an integral part of it. Hence, the border between external influences *on* and the manifestations *of* mobility culture is often blurred. Manifestations are the observable and tangible—often even measurable—aspects of mobility culture, and include the built environment of cities, transport networks and urban design, which not only manifest but are material representations of mobility culture. There are also less tangible manifestations of mobility culture, e.g. public and political discourses, the image and reputation of the city, and individual travel preferences. These manifestations of mobility culture are complemented by the way in which the populace explores and brings life to the city through travel activities—in other words by the mobility behaviour of the citizens.

We believe three relevant realms of external factors and manifestations of mobility culture can be identified. Together with the mobility behaviour—which usually epitomises the prevailing mobility culture of specific cities better than other indicators—these represent four key dimensions of mobility culture:

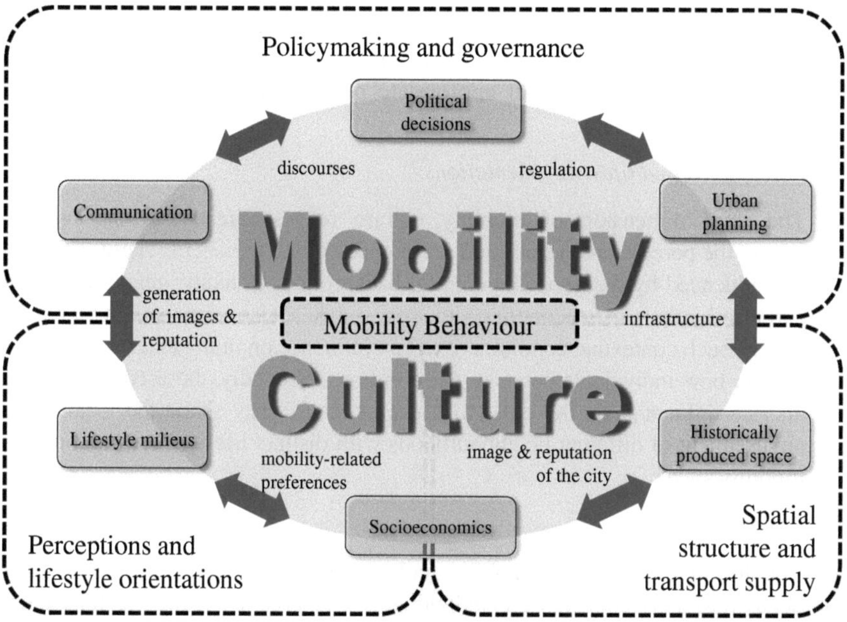

Fig. 3.1 Concept of urban mobility culture (Adapted by the authors from Klinger et al. [3])

(1) *Spatial structure and transport supply*:

Key aspects of this dimension include the characteristics of, and the opportunities and constraints arising from, the fundamental geographical space and its topography, economic and demographic framing conditions, corresponding urban densities, and the transport infrastructure. The quality of the transport supply is characterised by transport networks and service qualities for pedestrians, cyclists, private vehicles and public transport. These properties of urban transport systems can for the most part be described by means of quantitative indicators, such as accessibility measurements. Hence, this dimension of mobility culture was covered in the preceding cluster analysis (see "Chap. 2").

(2) *Policymaking and governance*:

This second dimension of mobility culture covers the societal framework of official and unofficial plans and programmes, and includes: (1) the general political context; (2) the specific political discourse in the urban policy arena; (3) the involvement of the various stakeholders; (4) their quality of governance; (5) participation mechanisms; and (6) the decision-making processes. These aspects of mobility culture strongly influence the local transport system and city structure, and thereby the conditions for everyday mobility behaviour. This influence is mostly exerted by means of the implementation of policy decisions through specific actions and measures, at times conditioned by the availability of appropriate planning instruments and financial budgets.

(3) *Perceptions and lifestyle orientations*:

The third dimension of mobility culture relates directly to the user. It covers the perceptions, values and preferences of the travellers. These are heavily influenced by their specific cultural background, ethnicity, gender and race, and also by their socioeconomic situation, status, social norms and motivations. Firstly, such contextual conditions exert an influence on individual travel options and on how individuals perceive these options. Secondly, these conditions can also be influential on the collective level, because they shape the local social environment of different neighbourhoods with distinct lifestyle orientations and mobility styles.

(4) *Mobility behaviour*:

The fourth dimension of mobility culture encompasses the *actual* mobility behaviour. This concerns the individual social practice regarding long-term, and everyday, mobility decisions. Long-term decisions include the choice of workplace and residential locations (thus influencing urban settlement patterns) and car ownership (thus influencing motorisation trends). Everyday mobility decisions encompass choices about which activities to engage in, trip chaining, destinations, travel distances, mode of travel, departure time, and so on. Since mobility behaviour can be captured—at least partially—with quantitative indicators as well, this dimension is also factored into the cluster analysis.

These four key dimensions of mobility culture will appear in many of the city stories and in the reflections about the perspectives on urban mobility culture (see "Chap. 13" in this book). It are the properties of each of these dimensions, and the way in which these pieces of the puzzle interact, that constitute the unique mobility culture of each city. However, mobility culture is not by any means a stable system, neither does it consist simply in four static building blocks. It is highly dynamic and evolves over time—a property that leads us to our next approach to mobility culture.

Approach 2: Mapping Changes in Urban Transport

It is evident, as was discussed in the previous chapter ("Chap. 2"), that during the last two decades the cities of the cluster study have moved on. Take London and Beijing as examples: simply comparing vehicle ownership development in these two cities exemplifies the heterogeneous developments which have characterised them in recent years. Based on the 1995 data, Beijing was classified as a Non-Motorised City with only about a fifth of the vehicle ownership rate of London. While vehicle ownership in London has decreased slightly, it has grown very significantly in Beijing to the point where today it has a vehicle ownership rate which is roughly two thirds that of London—showing that Beijing has clearly left the non-motorised stage behind.

Does this mean that we can draw only little insight of any long-term value from the cluster analysis, representing as it does a cross-sectional snapshot from 1995? We don't think so. Instead, we believe that the cluster analysis also provides a framework for mapping and interpreting changes in framework conditions for urban transport.

Figure 3.2 plots the cities from the cluster analysis in the previous chapter, grouping them by cluster against two major determinants of urban mobility: the horizontal axis shows the gross domestic product (GDP) per capita (on a log scale) and the vertical axis the population density (also on a log scale). Both of these are key factors influencing the change of urban mobility over time: the economic conditions change—especially in fast-developing megacities—with noticeable speed, which means growth in the case of most cities. In short: over time, most cities move to the right in the space shown in Fig. 3.2.

Urban density, on the other hand, is closely linked to the built environment, and therefore does not change as quickly as the economic situation. Moreover,

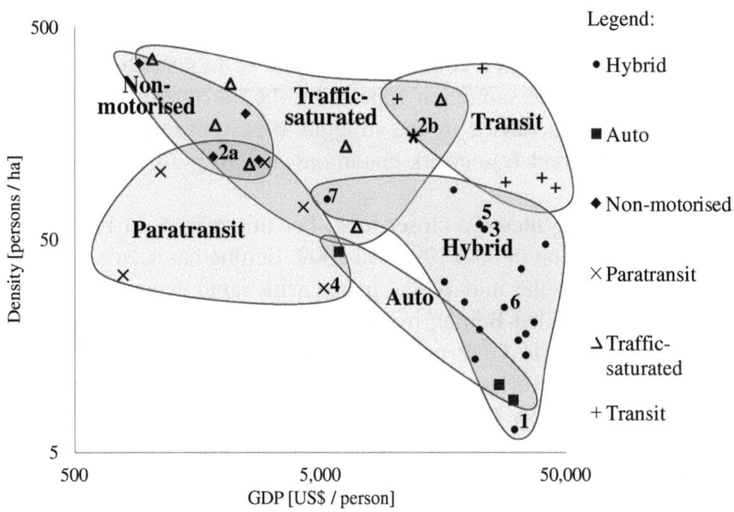

Legend to categorisation and position of study cities:

1: Atlanta 1995 2a: Beijing 1995 2b: Beijing 2009 3: Berlin 1995

4: Johannesburg 1995 5: London 1995 6: Los Angeles 1995 7: São Paulo 1995

Fig. 3.2 Cities and city clusters by population density and GDP per capita (*Source* authors' representation on the basis of "Chap. 2")

the direction of the change is not as clear-cut as in the case of the GDP per capita: while there are cities that have increased their densities over time (see "Chap. 2"), the overall tendency is towards decreasing densities as wealth increases (see "Chap. 1"). Today, many of the rapidly growing megacities still show very high urban densities. As they move on, will they follow the example of some of the relatively rich Hybrid Cities (e.g. Atlanta and Washington in the USA, and the conurbations of the Ruhr region in the west of Germany) and typical Auto Cities (Phoenix, Houston), all of which have very low densities, or will they be able to join the wealthy and densely populated Transit Cities (e.g. Tokyo and Singapore)?

Depicted as in Fig. 3.2, the city clusters represent more than just a cross-sectional snapshot of the state of affairs of urban mobility from the mid-1990s: they allow for mapping the pathways that cities are moving on as they develop. Moreover, thinking in terms of these clusters can assist in identifying upcoming challenges for cities on the move. For example, car ownership tends to grow particularly strongly at levels of GDP per capita between about US$3,000 and 10,000 [1]. Hence, assuming continued economic growth, Fig. 3.2 suggests that some cities are imminently facing a rapid growth in car ownership. If this prospect coincides with high urban densities, the path towards traffic saturation appears to be pre-assigned, with many cities having started down this road already.

Let us turn our attention to pinpointing the individual situations that our eight study cities were facing in the 1990s, and thus set the stage for the megacity stories in the next part of the book.

In the 1990s, Ahmedabad and Beijing were both Non-Motorised Cities[1] characterised by high density, and a continuing low rate of motorisation and relatively low GDP. Because traffic saturation appeared to be inevitable, both cities' recent development was characterised by the struggle to prevent this from happening—albeit with very different framework conditions and very different approaches to planning.

The case of Beijing merits a closer look. For this reason, in Fig. 3.2 we also depict this city's position in both 1995 and 2009. Beijing has seen a slight increase in urban density since the mid-1990s, paired with rapid economic growth. This development has propelled Beijing from being a Non-Motorised City into a position where—according to GDP per capita and population density—the Traffic-Saturated City cluster overlaps with the Transit City cluster (see Fig. 3.2). This raises the question of how Beijing was able to manage this significant development without getting stuck in traffic. In his megacity story about Beijing's transition to a Transit City, Song addresses exactly this issue.

[1] Even though Ahmedabad was not covered in the cluster study, it is evident from the chapter about this city, and the categorisation of neighbouring cities in the cluster study, that it would have been grouped in the Non-Motorized Cities cluster.

In 1995, Johannesburg (which lies in the South African province of Gauteng) and São Paulo represented cities at intermediate stages of economic development. Johannesburg was classified as a Paratransit City, and appears to exhibit too low a density to be immediately threatened with traffic saturation. A key issue for Johannesburg is providing mobility for all parts of the population in a low-density environment. São Paulo represents an outlier of the Hybrid City cluster and—in terms of GDP and urban density—is positioned somewhat like a Traffic-Saturated City. In the case of both São Paulo and Johannesburg, the overwhelming issue is how the transport system is linked to enormous social disparities.

Finally, there are four stories—those of London, Berlin, Atlanta and Los Angeles—which present cities from industrialised countries. In the 1990s these cities already belonged to the wealthiest among those in the cluster study. All of them were assigned to the heterogeneous Hybrid City cluster. Within this cluster, however, they take very different positions—for example, the car-orientated Atlanta exhibits the lowest density in the entire cluster study. For these cities, pressure on the transport system no longer arises from extraordinarily strong growth of car ownership—in the case of some wealthy cities such as London, car ownership has actually decreased in recent years. Nevertheless, the stories of these cities illustrate their attempts to move forward—pushed by external pressures, by their citizens, or by specific stakeholders pursuing various agendas. These cities have already reached a state of development towards which many cities of the world appear to be heading. This makes it particularly interesting to discuss whether these cities are spearheading developments and serving as role models, or whether they are relics of an old world.

We believe that the cluster study and its representation in Fig. 3.2 will assist the reader to more deeply appreciate the rest of the book: it sheds light on the baseline situation for the recent development of transport in our study cities and provides a setting against which to map changes in framework conditions for urban transport, and for identifying imminent challenges. And it leads to the crucial question about the role of mobility culture in determining the fate of cities: do cities run into predictable problems, or do they manage to diverge from an apparently pre-assigned path?

Approach 3: Asking Key Questions About Urban Mobility Culture

We have seen in the previous chapters that the diversity of cities around the world is multidimensional and overwhelming. There is no such thing as a generic pattern of megacity. Instead, there are unique conditions and paths of development. Hence, when analysing mobility culture, we have firstly to acknowledge this strong diversity; secondly, we have to ask the right questions, so that we can understand how this diversity and uniqueness of mobility culture emerges.

In view of all that has been said, we are now in a position to put forward a set of hypotheses which form a framework for acknowledging the diversity of mobility culture:

- There is no general mobility culture found in all megacities. Instead, we find specific and unique combinations of the different properties of urban transport systems which constitute mobility culture.
- The mobility culture of a megacity is the product of individual and institutional practice, and results from the interaction of different stakeholders within given local conditions.
- The specific local mobility culture of a megacity interacts with the social, political and economic framework, at local, regional, national and global levels.

After having acknowledged the singularity of mobility culture in specific places, we can now move on to ask questions which allow us to understand how local mobility culture has evolved, and where it will take us in the future:

(1) *Where have we come from?*

It is self-evident—and all of the subsequent megacity stories illustrate this—that the history of any given setting has shaped the present situation, and that there is a high degree of path-dependency regarding the evolution of mobility culture. This applies to historic conditions pertaining to geography, topography, climate, and politics in general—for example, consider the imprint that the political division of Berlin has left on the city's transport infrastructure. It also applies to very specific decisions which were made in the past regarding the transport system or urban development, as epitomised by London's historic Tube tunnels which today are suffering from capacity limitations. Understanding where we come from helps an accurate appraisal of the present situation, and gives a clearer view of options for the future.

(2) *Where are we today?*
A thorough analysis of local present conditions of the transport system and the identification of a city's position relative to peer cities helps to uncover the strengths and weaknesses of the local situation. After identifying the pressing problems affecting the transport sector, such an analysis must also extend to other realms of urban life, such as public health and urban crime. If this step is omitted, it is not possible to establish the priority of transport problems relative to other issues, and, moreover, potential links between these issues might be missed.

(3) *What does the future hold?*
This in-depth analysis of the past and present situation usually also gives an indication of what opportunities and threats the future might hold. Cities can gain a deeper insight into likely future developments and challenges if they evaluate their own past experiences, and/or examine their present situation,

with reference to the experiences of other cities, and then use this knowledge to generate future scenarios and projections. Furthermore, a city can experience pressing challenges arising from regulations that are imposed by external authorities to which it must submit—a case in point is the need to comply with national clean air standards. By performing this kind of analysis, likely future developments of this kind can be identified and the associated challenges can be addressed.

(4) *What action needs to be taken?*
Having identified likely challenges that the future might hold, these must be evaluated against existing strategies for the future. On this basis, potential solutions can be developed, appraised and implemented. In this process, thinking 'outside of the box' can be the key to success. For example, roadway expansion is by no means the only solution to the problem of congestion—if indeed it is one at all. Moreover, windows of opportunity for initiating change might emerge unexpectedly, even out of situations which are first perceived as a crisis—as these very situations can often bring about the stimulus required to provoke a new beginning.

We will see that this set of four questions guides us through the following megacity stories. By referring to these question, we will become acquainted with their individual mobility cultures—exploring the past, analysing the present, glimpsing the future, and touching on potential strategies and action that may be needed.

Approach 4: Detecting the Mobility Culture Temperaments of Cities

Finally, we want to encourage the reader to search for common denominators in our city stories. True, the cities and their paths of development differ a great deal—but there are nevertheless common threads which run through almost all of the city stories. These threads refer to basic mechanisms of interaction linking all domains and dimensions of mobility culture.

These mechanisms of interaction describe the underlying dynamics and driving forces that shape urban mobility systems. Of course, they are not equally pronounced and operate the same way in all cities. If and how these fundamental mechanisms are at work, is intrinsically determined by the values, conventions and social practices of the local actors in the field of urban transport and city policies. In order to take account of the fact that these mechanisms of interaction are hardly tangible and nevertheless fundamentally constitute the mobility cultural character of a city we label them 'mobility culture temperaments of cities'. Throughout the remaining chapters, this book will show how these mechanisms represent the dynamics that are at work in every single city and constitute the ever-changing entity that is urban mobility culture.

Embarking on the Journey to Mobility Culture in Cities Worldwide

Part One of the book has introduced us to the world of megacities. We got to know the momentous developments and pressing challenges that metropolises worldwide are facing. The cluster study provided an overview over the transport-related issues affecting urban areas of the world. In the next section of the book we will actually visit some of these cities. We will acquaint ourselves with them, the people who live in them, and those who shape their local transport systems. We will get to know the cities' transport histories and future prospects. This Reader's Guide to Mobility Culture has provided the necessary equipment for discerning mobility culture on this international journey. It only remains for us to embark on this exploration of mobility culture across the globe.

References

1. Dargay J, Gatley D, Sommer M (2007) Vehicle ownership and income growth, worldwide: 1960–2030. Energ J 28(4):143–170
2. Götz K, Deffner J (2009) Eine neue Mobilitätskultur in der Stadt—praktische Schritte zur Veränderung. In: Bundesministerium für Verkehr (ed) Urbane Mobilität. Verkehrsforschung des Bundes für die kommunale Praxis. direkt 65. Bundesministerium für Verkehr, Bonn, pp 39–52
3. Klinger T, Kenworthy J, Lanzendorf M (2010) What shapes urban mobility cultures—a comparison of German cities. Paper presented at the European transport conference, Glasgow

Part II
Stories from the Megacity

Chapter 4
Ahmedabad: Leapfrogging from Medieval to Modern Mobility

Swapna Ann Wilson

Abstract Ahmedabad, being a compact city with high density and good mix of land use, is characterised by shorter trip lengths, lower fatality rates and less congestion than is found in many other Indian cities. The region of Greater Ahmedabad has potential for development, and is at present in a phase of growth and expansion. Managing this growth in such a way as to retain its compact nature and at the same time improve its public transport system are the main objectives of the city's various transport and urban development plans. Technologically advanced mass transit modes, such as metro and Bus Rapid Transit, are being designed to improve mobility within the Greater Ahmedabad region. As the city aspires to shift towards modernised modes of public transport, the age-old mobility culture that has evolved from its people's customs and lifestyle cannot be disregarded. Tractors, shared autorickshaws, animal-driven carts and the like cater to the mobility needs of the citizens even today, alongside the newly introduced BRT system. The introduction of modern transport systems and technology in Ahmedabad has to be considered against a backdrop of existing modes and systems that date back centuries, and with which the people have long been acquainted; this local context makes transport reforms more complex to achieve in this city.

Introduction

Every inhabited place—be it a city, a region or an entire country—has aspects of its context that have evolved as a result of its people's customs—and, being deeply rooted in its lifestyle, these can be distinctly seen also in its mobility culture. A cow grazing in a public transport corridor, people squatting on central reservations, or cattle freely resting alongside the edge of the road—or even in the middle—is a common sight in India. The concept of shared space on Indian roads invariably includes room for cattle and stray dogs. A temple under a banyan tree with devotees deeply involved in their daily rituals in the centre of the road,

S. A. Wilson (✉)
Jerry vill, St. Andrews, St. Xaviers College P.O, Trivandrum, Kerala 695586, India
e-mail: swapnaannwilson@gmail.com

Institute for Mobility Research (ifmo) (ed.), *Megacity Mobility Culture*,
Lecture Notes in Mobility, DOI: 10.1007/978-3-642-34735-1_4,
© Springer-Verlag Berlin Heidelberg 2013

vehicles speeding past them, is not a rare sight either. Tractors, pedal rickshaws, camel carts, bullock carts and autorickshaws are the constituents of multimodal travel that are characteristic of Indian cities—and then of course there are the cars, buses, trains and other motor vehicles.

Implementation of public transport systems on Indian roads requires an in-depth understanding of road space in relation to the sociocultural aspects of the place. The introduction of modern systems and technology in an environment where people are acquainted with modes and systems that go back centuries, makes transport reforms more complex to achieve. Transport projects often fall short of their intended functionality because of a lack of cooperation from the public. A lack of technical knowledge on the part of the general public, and their reluctance to abide by new rules and regulations are two of the main problems. Introducing innovative solutions and strategies which help people adapt to new technology and systems represents a greater challenge than the introduction of new transport systems per se. As a response to this complex and varied situation, this chapter discusses the transport initiatives and plans which have been adopted to improve on the existing transport situation in the Greater Ahmedabad region. This region, located in the northern part of the state of Gujarat and comprising Ahmedabad (the commercial capital), Gandhinagar (the political capital), numerous villages and smaller towns, and special economic and investment zones, is rapidly approaching megacity status. It exhibits lifestyle and mobility patterns inherent to its local conditions and socioeconomic factors, amidst rapid urbanisation. New modes of transport such as metro and a Bus Rapid Transit (BRT) system are being introduced in the city alongside age-old animal-driven carts. This chapter explores reform strategies that have been adopted in response to the unique sociocultural systems of the city, and discusses the lessons learnt from successful transport projects such as the BRT system in the city.

The City: An Overview

History

Founded by Sultan Ahmed Shah in 1411 AD, the walled city of Ahmedabad is located on the eastern bank of river Sabarmati. Over the years, the city grew towards the north, the south and the west. The river, being the only prominent physical feature, divides today's city into two parts: the eastern walled city with its peripheral industrial estates, and the contemporary western portion. The city lies in the cotton belt of Gujarat, about 450 km north of Mumbai and about 90 km from the Gulf of Khambat. The history of Ahmedabad dates back to the ancient town of Ashaval about a thousand years ago, and Karnavati in the eleventh century [7]. The walled city was developed as a citadel with the fort, a temple, a mosque, royal tombs, a market and an outer wall to define the city limits. The city within

Fig. 4.1 Aerial view of walled city fabric of Ahmedabad (*Source* author)

this wall is structured into wards, organised by 12 main roads each running out of the city and having an entry controlled by gateways in the outer wall. The urban fabric of the walled city (see Fig. 4.1) is defined by these wards, which are further divided into gated communities referred to as *pols*—each with its own merchants dealing in specialised goods and trade, and often belonging to same caste—and within them further subdivided into streets of ever-diminishing width designed for commuting on foot and use of animal-driven carts. These narrow streets lined by mixed use buildings with balconies overlooking the street, helped develop not only cooler microclimate but also good surveillance that encourages mobility by means of walking.

Merchants, weavers and craftsmen were invited and encouraged by the Shah to settle in the city. The city flourished as a commercial center due to its strategic location to adjacent towns and villages. During the Mughal rule in 1573, the city grew further in density due to its growing merchandise in indigo, cotton and silk [7]. With the advent of British rule in 1780, the city began developing outside of the city wall. The British attempted to shift their domain away from the dense walled city core. The City Municipality was formed in 1858 and the first rail link with Mumbai was established in 1864. This triggered the next phase of development.

The industrial development was confined to the part of the city to the east of the river, mostly outside the walled portion. The British further moved away from the medieval walled core and established more institutional and residential quarters in the western part. This accelerated the construction of bridges across the river, and today the city has 12 river bridges connecting its east and west. The city municipality was given the status of a corporation in 1950. During this time, the population of the walled city declined due to outmigration of the people to the

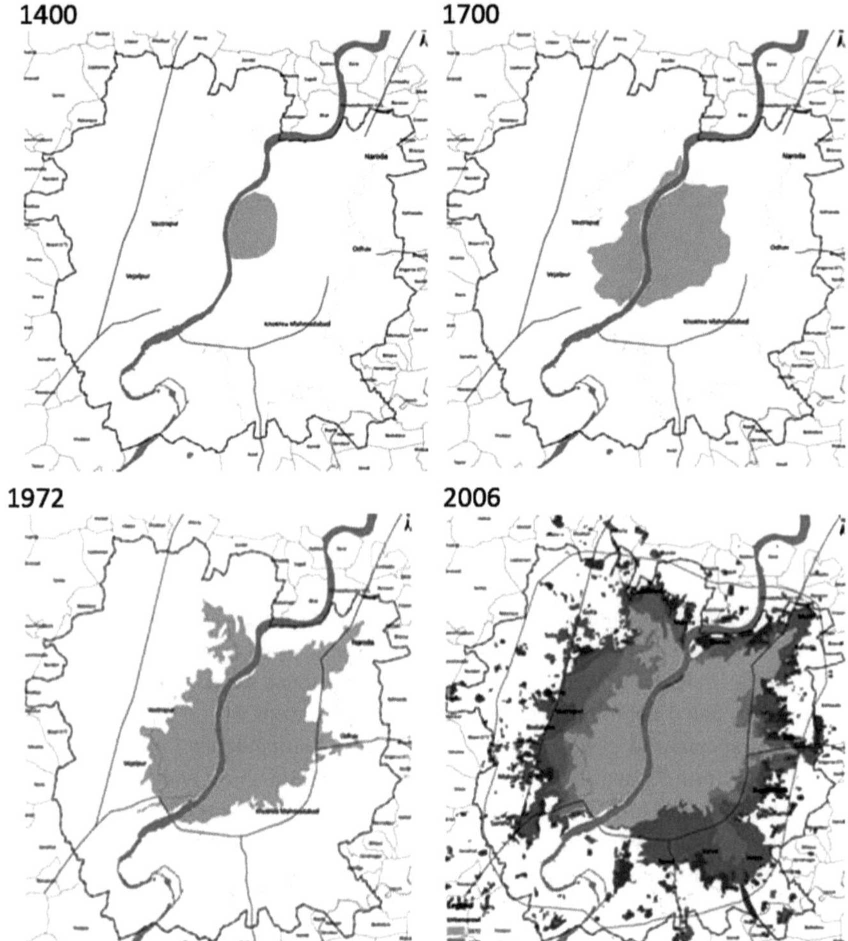

Fig. 4.2 Historical evolution of the city (*Source* Ahmedabad urban development authority)

other developing parts, outside of the walled area. In the latter part of the century, the western part also developed rapidly. Figure 4.2 shows how Ahmedabad has evolved over the centuries. Because of the availability of land in the peripheral areas of the city, low-density development took place at a greater pace. The city's Municipal Corporation now has an area of 466 km^2 within its limits, and an average density of 176 persons per ha (hectare, 0.01 km^2).

Gandhinagar, the capital of the state of Gujarat and located 23 km north of Ahmedabad, is the second largest conurbation of the Greater Ahmedabad Region. It is a planned city divided into 30 sectors around a central government complex, with an urban population of 0.21 million (2011) spread over an urbanised area of 57 km^2, and of lower urban density than Ahmedabad. These two urban centres,

Ahmedabad and Gandhinagar, have good connectivity by means of an expressway, several national and state highways, and railways; they also share an international airport. The Greater Ahmedabad region consists of Ahmedabad, the commercial capital; Gandhinagar, the political capital; numerous villages and smaller towns; and special economic and investment zones within an average radius of 40 km from Ahmedabad.

Growth Trends

Over the years, the city has emerged as a key commercial centre with a strong industrial base of traditional manufacturing industries, especially in textiles, chemicals, plastics, machinery, basic metals and alloys. An important industrial corridor in western India between Delhi and Mumbai passes through the Greater Ahmedabad region. A Dedicated Freight Corridor proposed along the Delhi Mumbai Corridor is expected to trigger huge investment and growth in the near future. Industrial growth is expected as a result of proposals in recent years for growth centres to the north, south, east and west of Ahmedabad. There are 13 industrial estates in the suburbs of the city, well connected by roads and railways. Given these trends, the city is expected to experience rapid growth in population and travel demand.

The city of Ahmedabad, with a population of 5.6 million (2011), has over the years developed to be the fifth largest metropolis in India and the largest in the state. Population projections for the Greater Ahmedabad region show that by 2030 it is expected to grow to 12.5 million, from a figure of 6.5 million in 2010 [2]. The population influx into the city over the past few decades has resulted in growing demand for resources and basic urban infrastructure. The shortcomings that hold Ahmedabad back from being globally competitive, amidst its struggle to accommodate the population influx the resultant growing resource and infrastructure demand, are manifold. Notable amongst them are mobility inadequacies and, until the last decade, a lack of public transport facilities.

Urban Mobility in Ahmedabad

The city of Ahmedabad today faces the issue of rapid motorisation. It developed as a commercial centre, seeing an economic boom and a rise in income levels, especially in lower- and middle-income households. This led to more people being able to afford personalised modes of mobility.

The composition of vehicles in Ahmedabad shows a lower rate of cars/jeeps (13 %) in comparison to 73 % of two-wheelers (2007) in the city [1]. The densely developed commercial capital of the state gave rise to short trips and hence a flexible mode like two-wheeler became popular. Ahmedabad is recorded as one of the

Table 4.1 Percentage of trips in Ahmedabad by mode, 2008

Mode	Percentage of trips
Bicycle	14
Paratransit	6
Private	42
Public	16
Walk	22
Total	100

Source Ministry of urban development [3]

cities with highest rate of increase in the use of two-wheeled and three-wheeled vehicles in the country. Autorickshaw trips are widely dispersed over the whole city and have shorter trip lengths, whereas trips made by car are also widely dispersed, with longer distances travelled per trip. Modal splits of five basic kinds of transport are given in Table 4.1.

The high rate of personalised vehicle ownership in the city is also due to the fact that from 1947 until the last decade, the Ahmedabad Municipal Transport Service was the only public transport system operational in the city. This system, operated by the public sector, failed to meet the demand for travel, and had problems related to insufficient frequency and a deterioration in quality. This led to the increase in the city in privately owned vehicles—mainly two-wheelers—and the emergence of paratransit systems, especially the shared autorickshaw system. The number of privately owned vehicles in Ahmedabad rose 42-fold in five decades: the number of registered vehicles has gone up from 43,000 in 1961 to over 1.8 million in 2009 [2].

Despite the rapid motorisation in the city, use of non-motorised vehicles, constituting walking and cycling, represents a fair share, coming second only to Mumbai (where it is 55 %) in India [5]. This is due to the high-density mixed land-use pattern, which favours shorter trip lengths and encourages cycling and walking. Bicycle trips cover short distances (3 km) and are mainly concentrated largely around the North and central zones of Ahmedabad. A good percentage of trips in Ahmedabad is covered on foot with an average trip length of 2 km [1].

Though trips by pedestrians and cyclists make up a good percentage of those undertaken in the city, the basic infrastructure for these non-motorised modes is poor. The medieval walled city core of Ahmedabad dating from the thirteenth century was designed for pedestrians, and for getting around by animal-driven cart. Nowadays the walled area is struggling with an ever-increasing number of automobiles plying the narrow lanes and alleys that were made for pedestrian movement and community interaction. The increasing traffic gridlock in the city resulting from the boom in car ownership and two-wheeler use has inevitably reduced the effective space available for pedestrians and bicycles to a great extent. The autorickshaws plying these walled city streets (see Fig. 4.3) have proved to be the most effective of the slow-moving modes, in contrast to the speeding two-wheelers.

Fig. 4.3 A typical street in the walled city (*Source* author)

The city of Ahmedabad is compact, with a ring-radial road pattern, displaying a good hierarchy of roads and mixed land use. This has reduced the average trip length to 5.5 km, which is small in comparison to other similar Indian cities such as Hyderabad (where the figure is 8 km) and Bangalore (11 km) [4]. But in recent years the city limits have edged out far beyond their original planned boundaries, thus necessitating longer travel distances and increased vehicular movement.

Ahmedabad has thus faced the challenge of needing to provide efficient mobility and accessibility that is suited to the ever-increasing population and urban sprawl. Traffic congestion, increasing trip lengths, rising road fatality rates and an absence of policies and regulations respecting pedestrian rights all add to the mobility crisis in the city. Over the last decade, the state and local governments have played a major role in initiating improvements in the city's traffic situation. A series of measures to improve urban transport were rigorously implemented.

Transport Initiatives and Plans

The year 2005 was declared as Urban Development Year in the state of Gujarat, heralding efforts to improve the condition of urban areas. Air quality deterioration and noise pollution were the initial focus areas. The city was rated the third most polluted city in 2005, on the basis of the annual air quality monitoring exercise undertaken by India's Central Pollution Control Board. One of the initial steps that the city took towards a cleaner future was to convert the autorickshaws from diesel/petrol to Compressed Natural Gas [3]. Within a time period of two years, many of the city-run buses were also converted, and the bus fleet increased. As a result, in a span of four years, the city fell to 66th in rank in the list of most polluted cities.

The next step was to improve the quality and enhance the connectivity of the mobility network in the city. The BRT system was proposed, and the initial phase of 45 km is now implemented and in operation. Metro links are also proposed, connecting the city with Gandhinagar and establishing connections linking east to west and north to south. At present (2011–2012) the city government is preparing the Development Plan 2031 and the Integrated Mobility Plan (IMP). These plans are prepared so as to be complementary to the other. The IMP focuses mainly on land-use/transport integration and the integrated multimodal public transit system. The public-transport-orientated development envisaged for the city is expected to retain the compact mixed-use nature of the city and effectively also retain the low trip distances of the city [2]. The plan takes into consideration the existing development patterns of and investments in the regional transport network, and sets out to link it with the existing and proposed public transport systems. The concepts discussed in the IMP are new to the Indian context. The proposal for an integrated system of transport in the city is discussed at a conceptual level, and detailed studies will be needed, focused on each specific area. The introduction of concepts of public-transport-oriented development will require detailed understanding of land use that is supportive of public transport in the Indian setting. Infrastructure, when provided for each of the mobility projects in the plan, requires an understanding of its impact on the sociocultural environment of the city. In a city where traditional mobility modes and cultural systems coexist, the challenge to introduce, implement and successfully operate modern public transport system are many and varied.

Adapting to Indian Ways

Many contemporary concepts such as automated doors, escalators, real-time information systems and dedicated central bus corridors along roadways are new to the Indian context. Introducing new concepts to a society that is unaccustomed to modern systems and technology is a challenge for Indian planners. The basic concepts of metro, BRT system and advanced IT systems are new to the developing world. Identification of the major challenges involved in making futuristic designs work in the Indian context is a must. Considering the fact that the target customers for public transport in the country/city come from low-income or middle-income groups, it is important to understand some of the typical Indian customs.

When examined from a distance, Indian cities seem to lack order and discipline. A crowded street with vehicles of all sizes sharing the road with pedestrians, hawkers and street animals is not a rare sight. A *chai wallah* setting up his corner illegally on the roadside and marking out his domain with a few pieces of broken furniture, pavements strewn with spilled cups and plates, street kids and dogs loitering around for a piece of biscuit or bread… these are the components of a 'typical Indian street'. In many sections of the city the distance by which buildings are set back from the edge of the right of way of roads is usually minimal. Hence many roads are in effect multifunctional spaces for household chores such as washing dishes or clothes, or for all manner of household activities. It is not unusual to see people sleeping on streets right outside houses during hot summer nights. Street design in the Indian context thus calls for space which is reserved for household activities. The use of street space changes with time and season, too. The various cultural activities and innumerable festivals of the region require roads to be multifunctional, dynamic features of the city.

Ahmedabad has a fairly high percentage of immigrants from the neighbouring towns and villages. Footpaths often become temporary places of refuge for this section of society. A boiling pot of rice and a baby sleeping inside a cradle hung from a tree on the street along the city footpaths is nothing out of the ordinary. Moreover, the city, as it grew, engulfed a number of small villages which still retain their socioculture and built form. A banyan tree with a temple surrounded by small houses buzzing with cattle and children—this village scene can still be found amidst the dense high-rise structure of the city today. The concept of shared space on roads invariably includes room for people, cattle and stray dogs.

The streets of India have a fair number of informal markets consisting of hawkers lining the streets with goods of various kinds. In certain locations of the city, hawkers occupy defined spaces. The customary system of perpetuating a certain trade in the family includes the inheriting of the space occupied on streets over many generations. This system informally sets out the right of occupancy of specific locations and areas for hawkers. Though developed over the years in a most unplanned manner, there is a logic and an understanding to it. This phenomenon has become an integral part of the streets and of the people's lives –indeed, all these sociocultural and economic aspects of life are built into Indian street design. As the streets of the city are redesigned to cater for any new mass transit, mixed traffic, bicycle lanes, footpaths and parking spaces, the need for 'activity space' is something that cannot be dispensed with.

India is a country whose systems seem chaotic when looked at superficially. But upon close scrutiny, it can be seen that an order does prevail, even in this pandemonium. A kind of supporting mechanism has developed for the very obscure social systems of this region, as a result of constant evolution and adaptation, albeit in an informal setting, over many years. It is important to understand the Indian patterns of behaviour as one attempts to introduce a new transport scheme into

Fig. 4.4 Inside the walled city of Ahmedabad—a sense of shared space on the streets (*Source* author)

Fig. 4.5 Outside the walled city of Ahmedabad—a sense of shared space on the streets (*Source* author)

an ambiguous urban fabric. The disordered Indian streets and transport systems hence exhibit characteristics of immense value. For example, considering the traffic and street systems in the city, the sense of shared space is strong (see Figs. 4.4 and 4.5). Amidst the chaos, an interesting phenomenon that influences traffic on the street is the alluring sense of space and mutual respect amongst road users. A significant proportion of the streets are not grade separated, and all road users still enjoy the freedom of using the full width of streets. Traffic on these streets flows smoothly without traffic lights and other regulations. This is seen in the walled city streets, where the vehicle speed is automatically reduced to 10 km/h as a result of the intensive use of these streets by pedestrians and bicycles. The mutual respect for all road users and modes is characteristic of these roads, and is a model of road development aspired to by many Western planners and designers. The lack of formalised control has resulted in the development of an informal set of rules in the form of hand gestures and eye contact among road users.

Recognising Strengths to Enable Efficient Planning

When new systems are introduced in India, it is important to first understand the order and harmony that prevails beneath the chaotic Indian exterior. For efficient planning of systems, it is important to identify the existing strengths that are inherent in a place. The success of any project lies in the effective integration of these strengths to create cohesive strategies for the development of plans, designs and guidelines. Modern public transport systems in India need to adapt to the Indian mobility culture, or else they run the risk of failing. Past failed attempts to implement BRT system in other cities are examples of this. The failure of BRT system in these cities did not occur because of low passenger numbers, but rather because it could not be operated effectively on Indian streets owing to the prevailing Indian 'traffic' culture and travel patterns. The BRT system in Ahmedabad, however, is widely accepted and is the first successful BRT system in the country. The international quality of the service that it provides, while at the same time being designed and operated in such a way as to blend in with the sociocultural fabric of the city, has made it a 'best practice' model for the country. Planned in 2007, the system began operation in 2009 along a short stretch of 18 km; now, in 2012, a stretch of 45 km has been completed, running 83 buses that cater for nearly 135,000 passengers every day. The project has gained global recognition by winning the Sustainable Transport Award, 2010, and was adjudged the Best Mass Rapid Transit System by the Government of India in 2009 [4]. This section discusses the various challenges and innovations involved in planning and designing such a project, as adapted for the BRT system in Ahmedabad, in such a way as to overcome the complexities involved in introducing it in an Indian context.

Transit Technology Choice

Every city has a prominent mode of transport, from local trains and taxis in Mumbai, through tram and metro in Kolkata, to local buses in Bangalore, Kochi and other cities. In Ahmedabad, local buses run by the Municipal Corporation were, before the advent of the BRT system, already one of the prevalent modes of transport in the city, along with the shared rickshaw services and other paratransit systems. Bus being an accepted mode in the city was one of the factors which favored the choice of a BRT system as the primary mode of mass transit in the city. It would encounter less resistance from the citizens, and paved the way to the introduction of a mass transit system which promoted the idea of enhanced mobility by public transport. Bus allowed more flexibility of design in the planning of a mass transit system on the basis of a ring-radial roadway system and its resultant travel patterns.

Also, for developing countries, where finance is a major concern, the choice of technology and mode plays a vital role. The city could build 88 km of BRT with the allotted funding which would, for the same cost, have enabled the building of a mere 6 km of metro [6]. The Delhi BRT system, designed for a short distance of 6 km, failed to address the need for a city-level mass transit system. In developing countries, it is important to seek transport solutions which expand the public transport coverage, quality and quantity, but ideally without the use of modes that demand high implementation and maintenance costs. In a country like India, where there exists a huge gap between demand for and supply of such travel modes, there is no need to fear any failure of public transport projects occasioned by a lack of demand. The BRT buses were full from day one of their operation in Ahmedabad.

Planning a Complete Network

Although the public transport of the city was poor, the city has for many decades enjoyed a good road structure with a definite hierarchy. The reason that the city developed as a ring-radial might be that this was a derivation—or rather an extension—of the pre-existing street pattern dating back to the days of the historical walled city. The city has a fairly high density of 176 persons per ha, and a very dense network of roads. The ring–radial road structure of Ahmedabad is more complete than those of other cities in the country. The existing road network structure was identified as a potential feature/model on

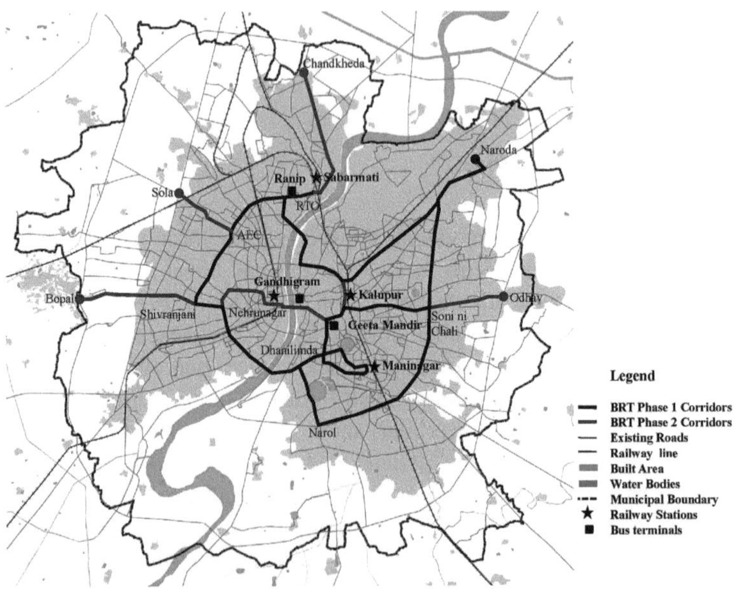

Fig. 4.6 BRT network plan (*Source* CEPT)

which to base the BRT network. The network forms a complete ring encircling the city centre, with corridors radiating out to the city limits. The linear connection in the centre links the east and west sides of the city, passing through its core. Figure 4.6 shows a plan of the BRT system network with the Phase 1 and Phase 2 corridors.

'Network and not just corridors' was one of the basic principles of planning the 90 km long network [6]. The 90 km road network chosen for introducing the BRT system in the city was planned and designed in a comprehensive manner. Provision of bus lanes was not the sole objective of the project. Pedestrian and Bicycle facilities were provided with enhanced utilities like street lighting, pavement improvement etc. Parks, rest areas and activity zones were introduced along the corridor as part of the road way design. The corridor was landscaped and designed to become an active public realm. The network established new links between urban centers and hence new roads were constructed and existing ones widened. Two new river bridges across Sabarmati, various flyovers, rail over bridges and foot over bridges were also constructed as part of the BRT project to enhance connectivity and thereby form a complete network.

'Connect busy places but avoid busy roads' [3] was another of the principles, meaning that priority was given to connecting busy places by choosing the less-congested roads for doing so. The network connects the central city—with its traffic generators such as public transport terminals, markets, industries, various institutions, and the like—to the residential zones in the city.

The Selection of 132 Feet Ring Road as the first stretch for commencement of the BRT construction was crucial. The particular stretch of the BRT network with enough right of way would retain the existing mixed carriage way space in comparison to the narrower parts of the network in the city center. This was a strategy adopted to reduce resistance from the public during implementation and initial operations.

Routes were chosen that ran through low-income neighbourhoods and areas which had previously been less accessible. The corridors were so selected as to help enhance existing links and form new ones, thereby improving accessibility in the city. The selection of the BRT route to run through the central business district of the city was crucial, as this is expected to raise the level of demand for the BRT, and also help to increase public transport patronage in the whole of Ahmedabad. The route connected to a major designed public place, the Kankaria Lake Development project, which also helped to increase patronage. One of the strategies was to synchronise the annual Carnival Fest with the inauguration date of the second stretch of BRT corridor, linking it with the lake. This helped to promote the project and introduce the new link to the city. Kankaria is a lake with multiple entertainment zones around it. Located on the eastern side of the city, it was less accessible from the western part of the city. The idea of linking public spaces such as Kankaria not only helped increase passenger numbers but also improved accessibility [3]. The future public transport lines have been chosen to run along routes linking Sabarmati riverfront and other entertainment zones in the city.

Retaining the City's Compactness

One strength of the city of Ahmedabad is its compact nature and mixed land use, which has evolved over years. The compactness of the city has given rise to short trip lengths. This shorter trip length has, over time, led to a relatively high percentage of pedestrians and cyclists. Average travel times are in the range of 15–20 min [3]. The annual number of road traffic fatalities in the city, at 244 in 2010, is also fairly low when compared to that of other cities with comparable populations such as Bangalore (where it is 840) and Hyderabad (424) [4]. Further, the balanced transport network and the predominance of two-wheelers limits excessive concentration of traffic in any one part, tending towards less congestion. It is important to recognise that short trips and less-congested streets appear to make city streets safe without compromising on mobility [3].

The BRT network was planned to help retain the compactness of the city and to increase the density of its undeveloped/low-density areas. BRT routes were planned to link areas less developed, unused textile mill areas or vacant land in the city center. The introduction of the BRT system is expected to be a catalyst for development and a kind of re-densification of these prime areas. After the first three-year period of operation, newer developments of high-density and mixed land use are already visible along the corridors. There has been a considerable increase in the average height of buildings, and of property values along the BRT corridors, as a result of the positive influence of introducing a mass transit system along them. This is expected to help increase city density within the city limits and prevent the growth boundary from sprawling.

Integrating Multimodes

'No single mode will cater to the mobility needs of the city' [2]. As is typical of Indian cities, Ahmedabad has a mix of modes which are motor- and animal-driven and ply the same road space. A good percentage of the target groups of public transport modes such as BRT consist of people whose access or egress modes are shared autorickshaws, tractors, bullock carts or camel carts. Ahmedabad has a significant proportion of its daily passengers coming from the nearby towns and villages, most of whom are commuting to the city for work. Villagers commute to the city in bullock and camel carts, tractors, shared autorickshaws, bicycles or by foot. But an efficient public transport system should attract other private vehicle users too. For an effective system, there needs to be provision for integrating the various modes of the city, which requires insight into the fundamentals of the access and egress modes of travel. Apart from the major shift of passengers from the local city buses to BRT buses, a mode shift of 25 % from three-wheelers, 20 % from two wheelers and 10 % from cars was observed in the first month of the BRT

operations [3]. The planning of the BRT bus station areas took into consideration formalised parking facilities for two-wheelers, cars, autorickshaws and so on, to help integrate the various modes with the new system; Respecting the high share of non-motorized vehicles in the city, dedicated bicycle lanes and wide footpath were provided along the BRT corridor.

Communication and Branding

The prospective users of BRT, or a metro, lack technical know-how and experience of automated doors, real-time information systems, off-board ticketing systems, and the like. The idea of an underground or overhead high-speed public transport system such as a metro, or dedicated bus lanes for a BRT system, with a bus station in the central reservation, is alien to a vast majority of citizens. Before commencement of operations of any such public transport system in the city, it is important to educate the people about the basic principles of the technology and how to use it. It is important to understand the habits, lifestyles and level of expertise (or lack of it) of the potential public transport passengers, while not shrinking from providing them with technologically advanced solutions.

After the design and implementation of the BRT system, the most important challenge was to bring the system to the people, instill confidence in its reliability, and ensure that it was an acceptable, welcome alternative mode of transport. One of the unique experiments in this process of implementation has been the city government's continuous interaction with the people concerning the system. This has taken the form of various public outreach measures including workshops, stakeholder meetings, free trial runs and so on. User satisfaction surveys were conducted at all BRT bus stations during the trial runs, which helped to improve the overall quality of the system. The surveys are conducted every month, and record levels of user satisfaction in order to improve it. These surveys are in-depth and detailed, for example docking the buses at the stations, bus operator's quality of driving, cleanliness check based on the spit marks made by *paan* (consisting of betel leaf, areca nut and other ingredients, very commonly chewed and spat out throughout India). Surveys of the kind help keep a check on the quality of BRT system operations and maintenance.

The system was operated on a trial-run basis for three months prior to its opening, to encourage more people to use it, operating over a short stretch of 13 km [3]. The whole idea of dedicated routes for buses occupying the centre of the road, automated doors, off-board ticketing, turnstiles and real-time information was new to the people—and to the operators too. Running the bus system on a trial basis was a strategy undertaken by the city government to help communicate the concepts of a bus mass transit to the people. The system was operated free of cost, which encouraged people to try it out. The system's advertising budget was actually used to fund its operation during this trial period, which turned out to be the best advertisement that the project could have.

People from all walks of life were encouraged to use the BRT system for free. Eminent citizens, doctors, industrialists, religious leaders and government employees were invited and accompanied by the Municipal Commissioner to take a ride on the BRT buses and promote the system. Special rides were arranged for school children to help them get accustomed to the new transport system in the city. It was important to educate the citizens about the technicalities of using an unfamiliar system. National and international experts also added refinements to the system. The initial steps towards developing sustainable transport in the city have thus far proved to be effective.

The BRT uses an 'integrated transit management system'. This includes traffic signal management with priority for the BRT buses, smart-card based payment integrating other public transport modes, and, on the buses, a passenger information system and a geographic information system (GIS). The use of an intelligent transport system with automotive doors and dedicated lanes was explained by special instructors at each bus stop and inside the buses during this period. People were appreciative and accepting when given a chance to use the system free of charge, and in this way became accustomed to new methods of mobility.

Another strategy for increasing acceptance was branding the system. The system was named *Janmarg*, meaning 'people's way', by Hon. Chief Minister of Gujarat, Shri Narendra Modi, thus presenting it as a system that would carry all the citizens of Ahmedabad, in safety and comfort, and at relatively high speeds. A new logo, designed by the National Institute of Design, gave it its distinct identity, which was also seen in the special buses with their distinct livery. As a means of branding Janmarg, a catalogue was prepared and made available to the public, outlining the planning, design principles and characteristics involved. Besides this, brochures, cartoon strips, annual newsletters and other literature was also prepared and distributed to help spread the idea of a dedicated bus transit system in Ahmedabad. Models of bus stop and Janmarg-branded pens were made specially and given away as souvenirs. Other initiatives taken included uniform designs for the entire staff [3].

Enforcement of Rules and Regulations

New transport concepts, as they are introduced, tend to encounter general resistance from the public, and especially from the private vehicle owners. BRT system projects have been proposed, and funding sanctioned from central government, in ten Indian cities, and such systems are operational in few of them—at the time of writing (mid-2012) these being Delhi, Pune, Jaipur, Goa and, of course, Ahmedabad. The BRT projects in Delhi and Pune have not been not entirely successful—many issues arose which hindered the full functioning of the systems. These cities experienced difficulties in maintaining dedicated lanes exclusively for BRT buses—they faced the problem of private vehicles entering the BRT lanes owing to a lack of strict regulations. Ahmedabad also faced similar problems.

Fig. 4.7 Ariel view of BRT corridor (*Source* CEPT)

The authorities had a job keeping cattle out of the BRT lanes, and special efforts were made to 'reroute' them. Ahmedabad's BRT was designed as a closed system, which further restricted other vehicles from entering the dedicated lanes through means of barricades and strict monitoring of traffic. Ex-army men were employed to act as BRT traffic control and enforcement. Efforts were made to help keep the BRT lanes as dedicated bus-only lanes and to restrain mixed traffic from entering them. The Ahmedabad BRT is the first full BRT system in India operated as a closed system, and is gradually coming to be seen as a role model for BRT systems in the country; it is also paving the way for other infrastructure planning and development projects in the city.

The BRT system in Ahmedabad is a closed system with median bus stations and dedicated 'BRT bus only' lanes laid in the center of the road (see Fig. 4.7). The median system is cost effective, easy to implement and occupies less space in comparison to curbside lanes. This strictly keeps the BRT buses in the central lane and causes minimum interference from other modes at junctions. The centrally located bus lanes hence do not hinder the general access to shops, houses, office buildings etc. in plots along the BRT corridor by other modes. In a median BRT system, the buses are docked on the driver side of the bus at the centrally located bus stations, unlike that of a curb side BRT system. This helps precise docking of the BRT buses. In Ahmedabad level boarding of bus was a factor given prime importance from the initial operations and trial runs of the system. This helped people accept the new system of boarding the high floor buses at a meter above the ground level, with ease. Level boarding also saves travel time and adds to its efficiency [1].

Lifestyle and Design

Ahmedabad has a dry desert climate, and has developed an interesting city structure, built form and lifestyle—from style of dress to cuisine—to adapt to the climate. The extremes of temperature, ranging from 44 °C in summer to 5 °C in winter, have influenced the city's pattern of life. These factors were also relevant to the design of the BRT system.

The BRT bus stations had to be of closed design, suited to an off-board ticketing system. The station design created by the local architects was more translucent in nature, with steel rope netting covering the length of the station so as to let in enough light, and cross ventilation. The level of transparency offered by the design also helped to ease the claustrophobic effect that can be imparted by a closed bus shelter, making the idea, which was new to the people, more easy to accept. Thus this vital issue was dealt with in a most simple manner, avoiding potentially high panic levels—a particular risk during peak hours. As a public outreach measure, a prototype bus stop was constructed to convey the idea of a closed central bus station.

Another interesting behavioural pattern is the manner in which people crouch alongside heavily trafficked roads, and at times even in the centre of the road. Many people enjoy sitting on the benches alongside roads of all sizes running through their neighbourhood. This habit seems to have arisen from the culture of carrying out all manner of household activities on the verandas and the adjoining streets in the walled city. Furthermore, one of the attractions of crouching on the road amidst heavy traffic, for both people and animals, is the wind generated as vehicles speed by, which is a welcome breeze in the midsummer days with their blistering with heat and still air. The BRT corridor incorporates this roadside culture, so ingrained in the populace, by designing public spaces in the centre of the road, chiefly at the ends of the BRT bus stations and islands of traffic junctions, wherever space is available. These spaces have been made easy to for the public to access by means of pedestrian crossings and station access designs.

The Ahmedabad BRT system is dynamic, and continuously evolving to adapt to its context. As new routes are laid out, new challenges arise, and are resolved in an innovative manner. It was implemented as the first of its kind in the country, and efforts are continuously made to monitor the system, understand hindrances, and rectify them as effectively as possible. Continuous surveys are being conducted by the Centre for Environmental Planning and Technology University, the technical consultant for the project, to obtain and understand people's feedback, and to assess both quality of operations and maintenance of the system. Data obtained includes modal shift to the BRT system from other modes, daily passenger counts, peak-hour boarding and alighting demand at all station locations, and recommendations for improving stations. As a result of these recommendations, the Municipal Corporation has taken steps to alter the system design and operation, to assure improved efficiency. The alterations made to it, based on an extensive set of data collected and analysed every month ever since operations began, have made the Ahmedabad BRT system a dynamically evolving one which attempts to adapt to the city context.

It was observed that people tended to cross the barricaded lanes at certain locations, by jumping across the barricades themselves. During the initial phase of operation, the BRT lanes were opened up at certain locations to enable more pedestrian and traffic crossing. The number of BRT traffic police was increased at critical intersections, and at bus stops where demand was heavy, to promote better management. After two years of operations, the pavement design was changed from asphalt to concrete at BRT bus stops, as settlement of the lanes at these locations had been observed. This change has been incorporated into the rest of the routes under construction, and those that are at the planning stage. Similarly, the absence of overtaking lanes in the Ahmedabad BRT system design has been identified as a major hindrance to its efficient operation; overtaking lanes are therefore now being incorporated into the BRT systems that are being designed in other cities. The manner in which the system adapts to the requirements of the city holds lessons for other cities that aspire to the successful implementation of a BRT system.

People and Government

Ahmedabad, being the commercial centre that it is, is rich in entrepreneurs supportive of ambitious municipal projects. Both the response from the public and the support of the media have been exceptional throughout the planning, execution and operational stages of various projects in the city. One of the recognised strengths of Gujarat is that the general awareness and acceptance of development projects is higher than in the majority of the other states in the country.

One of the important factors which has led to the successful implementation of the system in the city is the presence of strong and decisive leadership at both state and city level. The media response has been positive throughout its implementation and operation, keeping the people involved, giving updates about the project and providing positive criticism. Media coverage over the past few years has played a major role in eliciting public response and heightening awareness of the BRT system in the city. Ahmedabad is populated by entrepreneurs and other individuals who would like to see change for the good. Finally, citizens who see change as positive have been supportive, and have helped by adapting their lifestyles to fit in with the new systems.

Conclusion

Every city has its unique mobility culture temperament and its own patterns of transport that have evolved over time. This is, in the main, a reflection of the lifestyle, employment, commerce and trade, climate, and environment of the region. As developing cities seek to introduce contemporary systems and concepts of transport adopted from developed nations, it is important that this is done only after gaining an understanding of these inherent characteristics, in order that the mix of the modern and the indigenous may be sensitively handled so as to gain the best possible outcome.

As new concepts of city development are proposed, it is vital to grasp the city's strengths, and then work with them to achieve better execution and accomplishment of projects. As discussed in this chapter, Ahmedabad is a city which has successfully initiated transport reforms, implemented the BRT project, and gained the support and cooperation of the people. In doing this, many strategies and methods were adopted by the city government and project consultants. Projects need a detailed level of attention, and appropriate solutions come from spheres as diverse as the political, the scientific and the cultural, as has proved to be the case with the BRT project in Ahmedabad. The city has attained to best practice in the country, and was awarded the 2010 Sustainable Transport Award for visionary achievements in the fields of sustainable transport and urban liveability. But as it prepares for more advanced transport systems and technology, the challenges remain formidable. There is still uncertainty as to how to efficiently integrate the various modes of transport in the city with public transport in such a way that there is greater patronage of the latter. Care needs to be taken to understand the nature of the lively diversity of modes that are indigenous to this city, and to understand the users of them also, if an efficient multimodal system is to be established in it.

The greatest challenge for developing countries is to integrate new technology and systems of transport efficiently in their differing regions in such a way that the people 'take ownership' of it. Strategies for communicating with and educating the masses about innovative systems to help in their adoption are crucial. The success of any project rests upon the details of its integration with the existing systems and ways of life already familiar to those at which it is aimed.

References

1. CEPT (2008) Comprehensive mobility plan and bus rapid transit system plan–Phase 2. Ahmedabad
2. CEPT (2011) Greater Ahmedabad integrated mobility plan. Ahmedabad
3. National Institute for Urban Affairs (2010) Ahmedabad bus rapid transit system, urban transport initiatives in India: best practices in PPP. Ahmedabad. doi:http://www.niua.org/projects/tpt/AHMEDABAD%20BRTS.pdf
4. One World Foundation India (2012) Efficient solution for urban mobility: documentation of best practice. India

5. Sanyal S, Nagrath S, Singla G (2009) The alternative urban futures report–urbanisation and sustainability in India: an interdependent agenda. Thomson Press, India
6. Swamy HMS (2011) Janmarg, Ahmedabad bus rapid transit system, presentation for workshop on Janmarg. Ahmedabad
7. Vivek N (1990) Urban morphology and the concept of 'type', a thematic and comparative study of the urban tissue. Ahmedabad

Chapter 5
Beijing: Transition to a Transit City

Ziqi Song

Abstract Over the past three decades, Beijing has experienced profound changes in mobility culture. The city has changed from being a Non-Motorised City to a nearly Traffic-Saturated City in less than 30 years. Private cars have become the leading power in the motorisation process, and bicycles no longer dominate streets in Beijing. The city has accomplished the building of a surface and underground transport infrastructure system of reasonable quality. In addition, the government transport policy has been modified from encouraging private car purchasing to restricting vehicle ownership and use. No other megacity has ever experienced such rapid and pervasive changes in its transport system within such a relatively short period of time. The encouraging sign is that the city has developed a clear vision for becoming a Transit City, and has been working on promoting public transport for the last couple of years. This chapter investigates the dynamic aspects of the city's mobility culture and describes its evolution path towards a more sustainable future. More specifically, three aspects of the evolving mobility culture are focused, namely travel demand, mobility patterns and transport policy. This chapter concludes with discussions on the strengths of Beijing which will assist it in achieving its long-term vision, and the obstacles to that aim which the city also presents.

Introduction

China is a developing country undergoing fast economic growth and rapid urbanisation. The economic reforms which began over 30 years ago unleashed tremendous energy, which has influenced and changed all aspects of the country. The city of Beijing has a history going back more than 3,000 years, and has served as the capital of China for more than 850 years. It is situated on the

Z. Song (✉)
Department of Civil and Coastal Engineering, University of Florida, 365 Weil Hall, 116580, Gainesville, FL 32611, USA
e-mail: ziqi@ufl.edu

Institute for Mobility Research (ifmo) (ed.), *Megacity Mobility Culture*, Lecture Notes in Mobility, DOI: 10.1007/978-3-642-34735-1_5,
© Springer-Verlag Berlin Heidelberg 2013

north-western fringe of the North China Plain. Beijing is the cultural and political centre of China, and is also one of the four municipalities in China that have provincial status and come under the direct supervision of central government. Since China's era of economic reform, Beijing has entered the fastest growing period of its history.

Beijing municipality currently has an administrative area in excess of 16,000 km^2, containing 14 districts and two counties [2]. The old city, also known as the inner city, lies right in the centre of Beijing. It is the historic urban core. The city has constructed a concentric ring road (beltway) system over the past 30 years. The ring roads were originally designed to serve as urban expressways that circle the urban area. However, as the urban area expands, the old ring roads became urban arterial roads, and new ring roads were subsequently constructed. The first beltway to function with full access control and grade separation is called the Second Ring Road, and was completed to this standard in 1992 (the notional First Ring Road does not function as one; it circumnavigates the Forbidden City, but is not officially defined). The area within it equates roughly to the old city. The latest beltway to be completed is the Sixth Ring Road, which opened to traffic in 2009. The concept of 'urban boundary' is a vague one in Beijing. In some literature, it includes two dense urban districts (Dongcheng and Xicheng), i.e. the old city, and four inner suburban districts (Chaoyang, Haidian, Fengtai and Shijingshan), which are usually referred to as the six urban districts and cover about 8 % of the whole administrative area of the city. In other literature, however, the urban area is defined as the area within the Fifth Ring Road, only accounting for around half of the area of the six urban districts.

The economic reforms that began in 1978 fundamentally transformed Beijing, and indeed the whole of China. Ever since then, the city has been experiencing double-digit gross domestic product (GDP) growth. In 2009, the nominal GDP of Beijing municipality reached 1,215 billion renminbi (about US$191 billion). GDP per capita and average disposable income exceed US$10,000 and 4,000, respectively [2]. The city is also shifting its structure from an industry-based economy towards a service-orientated one. Beijing used to be designated as the 'economic centre' of north China. In the latest Beijing Master Plan 2004–2020 [5], however, this function was dropped. Most companies in heavy and chemical industries were either shut down or moved to the adjacent Hebei Province. By 2009, the tertiary (service) sector was contributing more than 75 % to Beijing's GDP [2].

The historic development of the city's mobility culture in the last half-century can be briefly summarised as follows. From the establishment of the People's Republic of China in 1949 to the late 1970s, China's economy was dominated by the socialist planned economy. Travel demand during this period was low, and trips were made primarily by non-motorised (walking and cycling) modes. These demand characteristics were largely determined by the prevailing social and economic structure, and were also supported by the urban form and infrastructure, which were dedicated to non-motorised modes.

For activities that were out of reach by non-motorised modes, the only viable option was public transport. Nevertheless, due to poor economic conditions, the public transport service was in bad shape. The transition from planned economy to market economy brought a new era of transport development to the city. From the early 1980 to the mid-1990s, the economic reforms began to have a notable influence on every facet of the social and economic structure. The sudden economic prosperity generated huge demand for travel. However, the state-owned public transport services were not ready to handle such large increases in travel demand due to both a lack of funding and bureaucracy. Taxis sprang up to fill this service gap. Meanwhile, minibus was also a popular paratransit mode. As a short-term solution to the problem of limited public transport availability, minibuses were allowed to operate along regular bus routes, providing more frequent services and more flexible stopping locations than regular buses, but generally also charging higher fares. This service was later banned because of its poor safety record, and problems arising from excessive competition, such as overcharging of customers and overloading of vehicles.

Starting in the mid-1990s, the city entered a process of rapid motorisation, which is still going on today. As in many megacities around the world, this explosive motorisation created enormous problems, especially traffic congestion. Traffic conditions have deteriorated dramatically as the traffic volume has approached—and even exceeded—the capacity of the road network in recent years. Despite tremendous efforts by the municipal government to supply more road infrastructure, construction of new road facilities has struggled to keep pace with the increasing travel demand. In 2005, the average speeds of urban express and arterial roads were 46 and 32 km/h respectively during morning peak hours (7–9 a.m.). Within five years, the speeds had dropped by around 25 to 30 %: for morning peak hours in 2010, the average speeds of urban express and arterial roads were as low as 35 and 22 km/h respectively [8, 10]. To tackle the growing traffic gridlock, a strategy of public transport priority was proposed as a sustainable solution in the mid-2000s. Since then, the city has invested heavily in upgrading its public transport system, especially in the building of a comprehensive subway network. The municipal government recently introduced a series of demand-side strategies to further restrict vehicle ownership and use, and has also set out its ambitious goal of building a Transit City as its long-term vision for Beijing [24].

The mobility culture of Beijing is constantly evolving, and is by no means a static concept. This chapter investigates the dynamic aspects of the city's mobility culture and describes its evolution path towards a more sustainable future. The remainder of this chapter is organised as follows: Sect. 2 focuses on three aspects of the evolving mobility culture, namely travel demand, mobility patterns and transport policy; Sect. 3 discusses what should be done to transform the city into a more sustainable one; Sect. 4 concludes this chapter with discussions on the strengths of Beijing which will assist it in achieving its long-term vision, and the obstacles to that aim which the city also presents.

An Evolving Mobility Culture

Unprecedented Travel Demand Growth

Beijing, as other parts of China, has been enjoying huge economic prosperity for more than three decades. Figure 5.1 shows that the city's GDP per capita has been growing at an average annual rate of more than 15 % since 2000. In a similar way to other megacities that are in the process of fast economic growth, Beijing is experiencing a boom in demand for travel. Daily travel demand, including walking trips, increased from 9 million trips per day (in the entire administrative area) in 1986 to almost 37 million per day (within the Sixth Ring Road alone) in 2009, which implies that travel demand has more than quadrupled in less than 25 years [3, 9]. Even though it is not surprising to observe such significant travel demand growth for a city that is undergoing dramatic urban growth, there are still some unique reasons for this that are not shared by other megacities around the world. Rapid urbanisation, motorisation, economic reforms—all these three processes happen almost simultaneously and blend together, generating not only more, but longer trips. These forces eventually trigger an inevitable change in mobility patterns. This section explains the unprecedented travel demand growth from these three interrelated aspects.

Urbanisation and Demographic Change

China is undergoing the most extensive and fastest urbanisation seen in human history. It is predicted that more than 400 million rural residents will relocate to

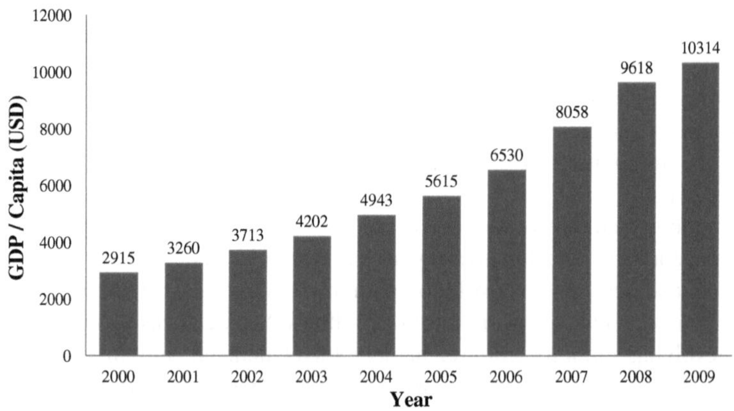

Fig. 5.1 Beijing GDP per capita growth (2000–2009) (*Source* 2001–2010 Beijing statistical yearbooks)

urban areas by 2030, meaning that China's urbanisation rate will reach 65 % by then [15]. Beijing is no exception to this pattern. The population growth in the past 30 years has changed the spatial form of the city, and even well-informed city officials were at times taken by surprise. In the 1983 city master plan, the government planned to control the inhabitants to fewer than ten million (within the administrative area) by 2000. However, this goal had already failed by as early as 1990, ten years ahead of the target planning year. And in 2010, the population in the Beijing administrative area was over 19.6 million, going by the 6th population census [1], which again well surpassed the limit of 18 million by 2020 set out in the latest 2004 city master plan. Table 5.1 summarises the demographic changes of the past three decades in Beijing.

Empirical studies around the world have demonstrated that land-use patterns, such as urban density and degree of centralisation of work, are closely correlated to urban transport characteristics, especially the level of car dependence [20, 21]. To estimate the urban density in Beijing, the population density of the built-up area has been adopted. Although there is no clear urban boundary in Beijing, most of the built-up area coincides with the area of the six urban districts. Therefore, the total urban population of Beijing can be estimated as about 10.8 million, giving an urban density of around 8,000 persons/km^2 in 2009, which is slightly lower than the figure of 95.4 persons/ha (9,540 persons/km^2) that Chen et al. [13] arrived at, based on data in 2000.

A strong trend towards suburbanisation since the 1980s is evident from Table 5.1. The population density in the urban core has been steadily declining, while the suburban districts, especially the inner suburbs, are becoming more densely populated. Although most developed megacities have experienced an era of suburbanisation, the underlying driving forces behind the suburbanisation in China are quite different from those of cities in Western countries [29,

Table 5.1 Population change in Beijing (1982–2009)

Population (thousands)	Old city	Inner suburb	Outer suburb	Administrative area
1982	2,418	2,840	3,972	9,231
1990	2,337	3,989	4,494	10,819
2000	2,115	6,389	5,066	13,569
2009	2,111	8,689	6,750	17,550
Population density (persons per km^2)				
1982	27,763	2,214	256	546
1990	26,826	3,110	289	640
2000	24,278	4,980	326	802
2009	22,849	6,810	449	1,069

Source 1982, 1990, 2000 China population census and 2010 Beijing statistical yearbook

32]. The most significant difference is land ownership. In China, the government is the owner of all urban land. China adopts a unique land leasehold system, which separates land-use rights (LURs) from ownership [30]. Developers can only acquire LURs, not land ownership, from the government—in other words they lease land, for between 30 and 70 years depending on land-use purposes [27] The government plays a decisive role in urban land supply and land use. Hence, Beijing municipal government has substantial control over the suburbanisation process. For example, government-driven urban renewal projects generally relocate residents and polluting factories from the old city to suburban districts. Also, government-subsidised housing for low-income families is usually built in suburban districts to take advantage of the cheap land price. On the other hand, in the last decade or so, there has also been a clear trend of market-orientated suburbanisation [18]. Private developers actively participated in building properties in suburban areas because of the flourishing property market. Residents also started moving to suburban districts to seek better air quality and an improvement in their living conditions. Furthermore, new urban residents to the city, brought in by rapid urbanisation, usually choose to live on the city periphery, where housing prices are much more affordable. That being said, the suburbanisation process is still heavily influenced by the government, which holds a monopoly position when it comes to land supply in China.

Economic Reform and Change in Spatial Patterns

In the past two decades, Beijing has experienced the fastest urban growth in the city's history. The extent of urban built-up area rose from 476 km^2 in 1995 to more than 1,349 km^2 in 2009 [25, 26], a virtual tripling of size in under 15 years. Although the urban area has expanded significantly, and suburban districts have attracted substantial population growth in recent years, the spatial pattern of Beijing remains primarily monocentric despite the many efforts of the municipal planning commission to avert this development pattern. The city planned to build ten scattered clusters on the periphery of its built-up area, according to the 1983 and 1991 city master plans. Unfortunately, the polycentric planning principle was not well implemented due to rapid urbanisation and social preferences. Chen et al. [14] developed an urban growth model and identified three primary reasons for the failure of the polycentric development plan: underestimation of the agglomeration effect; the construction of the ring road system; and a lack of effective means to protect the greenbelts between the built-up area and satellite towns. As a result, most of the jobs and other social resources—such as government agencies, shopping malls, medical centres and the world-famous museums—are still concentrated in the urban area. Around 68 % of all trips occur within the Fifth Ring Road, an area which only accounts for only about 4 % of the total administrative area of Beijing [3].

The monocentric spatial pattern and the trend towards continuing suburbanisation obviously require people to spend more time commuting, and therefore generate more travel demand; another factor that also plays an important role in generating more demand is the reform of the traditional employment system. Before the 1980s, there was literally no private sector due to the political ideology. People worked for either government agencies or state-owned enterprises (SOEs), which are called *danwei* in Chinese, and were generally tied to the same employer for their whole life. All these agencies and SOEs were designed to be self-sufficient. They provided not only jobs but also a comprehensive welfare package, including housing nearby to one's employers. Almost all economic activities, most of the social activities also, were carried out under the auspices of the *danwei*. Therefore, most trips could be completed on foot or by bicycle. When the economic reforms started, the old employment system was challenged. People started moving to private sector jobs that provided better pay. Moreover, foreign companies began to set up joint ventures in China. The reform of the *danwei* system, as a part of the ongoing economic reforms, unties welfare functions, including housing, from people's workplaces, and hence provides employees more freedom to change jobs. Now new employees have to buy their properties on the open market. The pre-reform-era job/housing balance came about as a by-product of the traditional *danwei* system and served as an important reason for the limited travel demand during that period [28, 31]. The reform broke the link between jobs and housing and, predictably, unharnessed a tremendous need for travel.

The Trend Towards Motorisation

Beijing has entered a period of rapid motorisation, led by soaring private vehicle ownership since the mid-1990s. Over the past decade, the city has maintained a double-digit annual growth in automobiles, with the number of private vehicles almost quadrupling in that time span. Figure 5.2 illustrates the increase of vehicle ownership from 2004 to 2009 in Beijing. By the end of 2009, the whole administrative area had more than four million automobiles, of which around three million were private cars [9]. Figure 5.3 shows that, despite the fact that the city is in the process of rapid urbanisation, the rate of vehicle growth has considerably outpaced that of the population growth. The number of motorised vehicles per 1,000 persons has more than doubled in the last decade. Even though the motorisation rate is still relatively low (around 230 vehicles/1,000 persons) compared with North American cities, for a city that used to be designed to accommodate primarily non-motorised traffic, it is already beyond the capacity that the city can properly handle, and has created serious congestion and environmental problems.

Motorisation came earlier and faster to Beijing than to any other Chinese city. It is not a coincidence that Beijing has become the pioneer of motorisation in China. In the literature, income growth has been identified as the primary

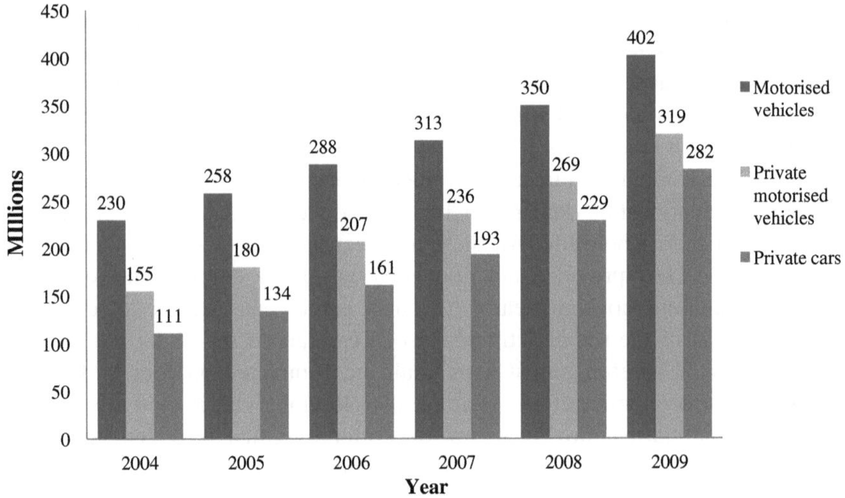

Fig. 5.2 Beijing vehicle ownership evolution (2004–2009) (*Source* 2010 Beijing transportation development annual report)

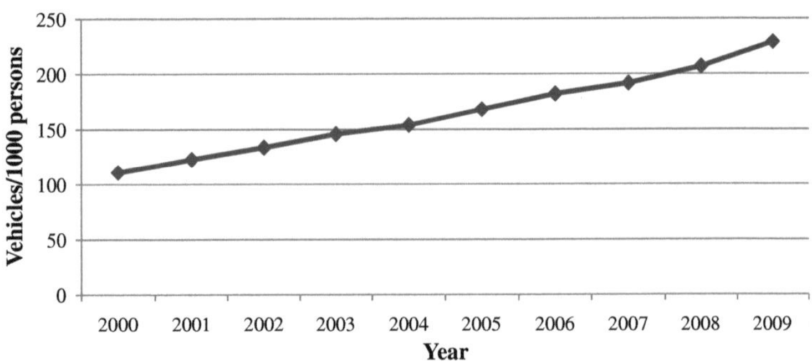

Fig. 5.3 Growth in vehicles per 1,000 persons (2000–2009) (*Source* 2006–2010 Beijing transportation development annual report)

impetus for vehicle ownership (see, for example, [11, 17]). The foremost reason for this fast motorisation is thus that Beijing is a more economically developed city by Chinese standards, and that the economic well-being of its residents makes it possible. Another more practical reason is that people have to travel greater distances to access jobs and other social activities because of the aforementioned three factors, namely the monocentric spatial pattern, suburbanisation, and the reform of the *danwei* system. For instance, the average trip distance to work by car increased from 10 km in 2000 to 14 km in 2005, a 40 % increase [3]. Driving in a private car is evidently the most convenient and comfortable means of travel for achieving the ever-increasing commuting distances.

In addition, being the capital city, Beijing has more monetary and political resources to build a world-class road infrastructure system than do lesser cities, and this to some extent encourages people to own private cars. Since the introduction of its first beltway (the Second Ring Road) in the 1980s, the city has constructed another four beltways (the Third, Fourth, Fifth and Sixth Ring Roads). Moreover, the phenomenon of treating cars as a status symbol has sped up the motorisation process, although it is notable that in recent years, as private cars have become more affordable, the image associated with owning one has begun to change, and owning a car is not considered such a strong indicator of one's prestige. Furthermore, the motorisation of Beijing was also fuelled by the central government's political and economic agenda. In the mid-1990s, the central government wanted to make the automotive industry a 'pillar' of the industrial and economic development. Beijing used to be one of the top three car manufacturing bases in China back in the 1980, and in the 1990s. Even today, car manufacturing still plays an important role in the city's industrial sector.

Rapidly Changing Mobility Patterns

As the level of motorisation increases, mobility patterns in Beijing are undergoing a fast and dramatic change. In 2009, the car modal share increased to 34 %. Two other modes experienced relatively significant changes: subway gained 2 %, while bicycle lost 2.2 % of the total modal share compared to the previous year. The modal split of daily trips in 2009 (excluding walking) is illustrated in Fig. 5.4.

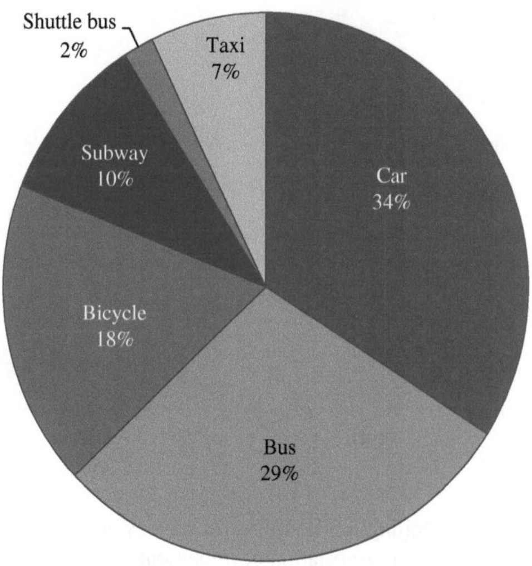

Fig. 5.4 Beijing modal split of daily trips in 2009 (*Source* 2010 Beijing transportation development annual report)

In the 1980s, and even the 1990s, non-motorised modes were the dominant mobility choice in Chinese society at large. It is no wonder that China was nicknamed a bicycle kingdom back then. In Beijing, this situation started to change as the city entered a period of rapid urban growth and motorisation. By 2009, the modal share of bicycle had dropped to 18 %—still an admirable level in the eyes of its peer megacities in the Western world. However, the modal share of bicycle was as high as 63 % in 1986. The modal split evolution in Beijing is illustrated in Fig. 5.5. The bicycle culture has been fundamentally changed in the past two decades. Cyclists are shifting to other modes both voluntarily and involuntarily. In the 1980s, before private cars and the subway were widely available, cycling was almost the only option to commuters, apart from the inefficient and often unreliable bus services. After the economic reforms, people had to travel greater distances to access jobs and other social activities. The relatively short distance that can be travelled in a reasonable timeframe obviously limits the potential of bicycle use. Furthermore, the long, cold Beijing winter is not friendly to cyclists. Therefore, when other modes became more readily available, people started to shift to them. Liu [22] undertook a survey on work-trip mode choice in Shanghai, and found that cycling was perceived as an inferior mode by all income groups. The outcome might well be similar if the survey were conducted in Beijing. On the other hand, many commuters involuntarily switched to other modes simply because of changes in urban form and built environment. Bicycle infrastructure, such as designated cycle lanes and cycle parking lots, is being encroached upon. For example, more road space is being allocated to accommodate cars, while cycle lanes are being either narrowed or even eliminated in certain places. Furthermore, Beijing's traditional small alleys (*hutong*), particularly suitable for non-motorised traffic, are being demolished due to make way for urban renewal developments.

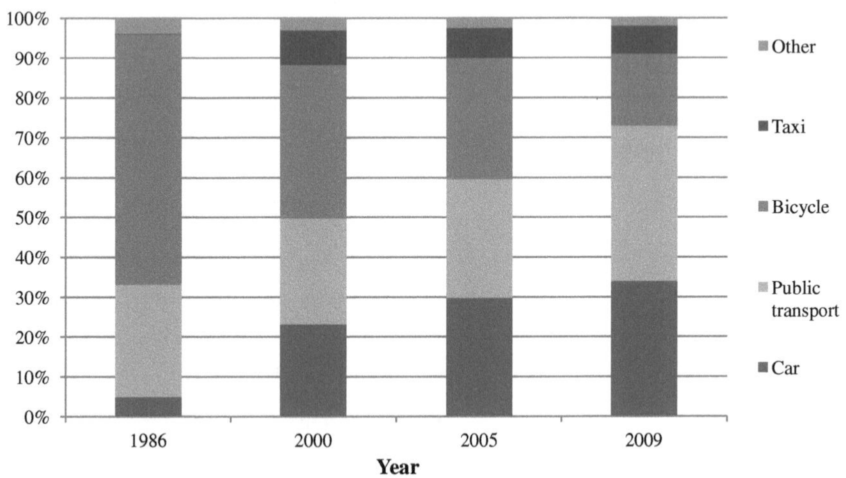

Fig. 5.5 Beijing modal split evolution (*Source* The 3rd Beijing comprehensive travel survey and 2010 beijing transportation development annual report)

One encouraging change in mobility patterns in the past decade is the steady growth of the public transport modal share. This trend can be credited for the most part to the recent dramatic improvements in public transport, especially in the subway system. Although those improvements are partially due to the need for building new public transport infrastructure for the 2008 Summer Olympics games, more importantly, they also reflect a fundamental change in government transport policies. Walk-only trips have been excluded from comprehensive travel surveys; however, it is estimated, based on a sub-survey of the 3rd Comprehensive Travel Survey, that walk-only trips account for about 31 % of all daily trips made in 2005, which indicates that walking is still an indispensable and fundamental way of getting around in Beijing.

Gradually Shifting Transport Policies

Building the Way Out of Congestion

From the mid-1990 to the mid-2000s, the transport policies in Beijing were generally pro-car. On the one hand, these policies were certainly influenced by the national policy during this period. In 1994 the central government published its first Automotive Industry Policy. For the first time, the automotive industry was designated as the 'pillar' of the nation's industrial and economic development. In 2001, the National Tenth Five-Year Plan further made a clear statement encouraging families to have their own cars, and put forward a set of automobile consumption policies to boost private car ownership. On the other hand, the pro-car transport policies also heavily depended on the vision of local authorities for the city. From the 1950 to the early 1990s, Beijing was dominated by non-motorised traffic. Owning a private car was no more than a dream for most people. Back then, the urban form was compact, and the road network was designed mainly for the purpose of accommodating non-motorised traffic. When the city entered a period of rapid motorisation in the mid-1990s, the city's road infrastructure was simply not ready to handle the sudden influx of automobiles. Congestion was the inevitable outcome of these incompatible ingredients. To respond to the deteriorating traffic conditions, the city took a common remedy for congestion relief: supply more roads. Decision-makers were convinced that high car ownership and use was directly related to the high standard of living, and were even tempted to replicate the car-dependent development pattern in North America. The city spent much more money on road infrastructure than on other transport modes. For instance, in 2002 the investment on urban roads and highways accounted for almost 76 % of total investment in transport infrastructure [7]. To some extent, the pro-car policies have served their goals. The city has achieved the building of a world-class road infrastructure system in under 20 years, and both the motorisation rate and car use have surged dramatically. Nevertheless, the harmony between supply and demand was short-lived. The ever-growing car demand outstripped the development of road infrastructure, and congestion, air and noise pollution problems gradually built up. On top of the financial and political constraints faced by most megacities around the world when building more roads in the city centre, additional issues make the problem more acute for Beijing: the need for cultural preservation of the old city makes it difficult and painful to expand new capacity there. The decision-makers gradually realised the fact that it is impractical for Beijing to build its way out of congestion, and that more sustainable strategies have to be sought.

A Renaissance of Public Transport

In 2005, the city published the Beijing Transportation Development Strategic Plan 2004–2020. A strategy of prioritising public transport was first proposed in this strategic plan, as was the prioritisation of transit-orientated development in new satellite towns. In the following year, the city announced a series of preferential policies in terms of land acquisition, funding allocation, road right-of-way assignment and tax subsidy, all for the purpose of promoting public transport development [24]. Furthermore, the public transport service was categorised as one of the basic public welfare items that should be provided by the municipal government. As a result, the transport authority lowered the ticket prices for subway and bus to a flat rate of 2 renminbi (about 30 US cents) and 1 renminbi (about 15 US cents) respectively in 2007, which attracted substantially more public transport users. In 2009, the municipal government went one step further and published Beijing Green Transportation Action Plan 2009–2015 [6]. For the first time, the government clearly announced its vision to build a Transit City.

The city's public transport system has gained significant development in the past decade. Since the XXIX Summer Olympic Games were awarded to Beijing in 2001, the city has accelerated its pace in public transport investment, especially in the mass transit system; the government funding for transport infrastructure began to lean towards public transport projects. In 2009, the city invested more than 70 % of its transport infrastructure budget in the public transport sector. The mass transit system alone received 34.2 billion renminbi (about US$5.4 billion), which accounted for more than 50 % of the total Beijing transport infrastructure budget. In contrast, highway and urban road projects together received only around 22 % of the total budget [9]. The most significant improvement in public transport has been in the city's subway network. To add to the original two subway lines, the city has constructed 13 further subway lines in the past ten years. Currently it has 15 subway lines (with a total length of 372 km) in operation. According to the Beijing 12th Five-Year Mass Transit Plan [4], the city will have a subway system of total length 598 km in operation by 2015, and subway network density will reach 0.5 km/km^2 in urban area, which means that people should be able to walk about 1 km on average to get to the nearest subway station. The new infrastructure and vehicles have greatly improved people's perception of public transport services.

Curbing Vehicle Use and Ownership

Although the strategy of prioritising public transport was established as early as 2005, the city did not take any specific measures to curb private vehicle use until 2008, or ownership until 2011. The tipping point was the 2008 Summer Olympics. To improve the air quality and traffic conditions during the games, the city implemented its first travel demand management (TDM) policy, a road-rationing programme [19]. Based on their number plates, vehicles are allowed to be used only every second day. Enforcement was carried out through an automated traffic surveillance system. The programme was fairly successful in terms of reducing traffic congestion and air pollution. Initially, the government introduced the programme as a temporary measure, but owing to its success the programme was extended indefinitely. Currently, each day cars with either of two particular registration numbers are prohibited from entering the urban area (within the Fifth Ring Road) between 7 a.m. and 8 p.m. In other words, car owners are allowed to use their vehicles four out of five working days.

For a short period of time after the road-rationing scheme was introduced, traffic conditions were improved significantly, but the spare road capacity was soon filled up with newly registered vehicles. The congestion relief effectiveness of the programme started to diminish as time went on, although of course without the rationing programme the traffic congestion would only be even worse. The municipal government was forced to take a series of new TDM strategies at the beginning of 2011 to deal with the by-then almost saturated traffic conditions. These new TDM strategies can be summarised as follows:

- Vehicle ownership restriction

The new car registration quota is limited to 20,000 per month, of which 88 % is reserved for private users. The new quota is distributed to users through an online car registration lottery system—in order to win a new car number plate, users have to pre-register on the car registration lottery website. The lottery takes place once a month, and unsuccessful applicants are required to renew their applications online every three months in order to avoid being treated as having given up. The latest odds of winning were little better than 2 %. As more people are joining the pool every month, the odds of winning could become even smaller. However, if an owner purchases a new car to replace a scrapped one, the new car can continue to use the old car number plate without entering the car quota lottery.

- Out-of-town vehicle ban

Vehicles with out-of-town number plates have to apply for special short-term permits (usually valid for less than seven days) to enter the urban area (within the Fifth Ring Road), and are prohibited from entering the urban area during the rush-hour periods (7–9 a.m. and 5–8 p.m.). This policy bans unnecessary trips through the congested urban area by out-of-town vehicles. In addition, as a policy complementary to vehicle ownership restriction, it prevents people from registering their cars in other places to bypass the restriction.

- Differentiated Parking Pricing

Beijing used to have relatively low parking prices compared with other Chinese megacities, such as Shanghai and Guangzhou. To discourage users from driving into the city centre, a new parking scheme has been instituted, which raises parking fees in the central areas and enforces standard minimum parking rates in other areas. Each urban area is categorised into one of three classes depending on its distance to the city centre, and corresponding parking rates are subsequently determined by the municipal government.

By introducing the aforementioned TDM strategies, the city sent out a clear message about restraining car ownership and use. According to data observed in the first two quarters of 2011 [10], these strategies did improve Beijing's traffic conditions substantially: vehicle ownership growth was reduced by about 72 % compared with the same period of the previous year; the modal share of car decreased for the first time in decades; the modal share of public transport increased to 41 %; peak-hour average speed was improved by 10 %; and congestion time was reduced by 65 min per day. Although all traffic operations data show encouraging signals, those same strategies have also come in for fierce opposition from car owners and pro-car communities. Indeed, although the government is now showing its determination to alleviate traffic congestion, the city's history of ignoring proper demand-side policies in years gone by has placed it today in an unfavourable position when it comes to developing a sustainable transport system.

Towards a More Sustainable Future

Beijing was categorised in the cluster of Non-Motorised Cities in the mobility culture preliminary study, which was based on 1995 data. However, over the past two decades the city has experienced profound changes in mobility culture. Private cars have become the leading power in the motorisation process, and bicycles no longer dominate streets in Beijing. The city has accomplished the building of a surface and underground transport infrastructure system of reasonable quality. In addition, the government transport policy has been modified from encouraging private car purchasing to restricting vehicle ownership and use. No other megacity has ever experienced such rapid and pervasive changes in its transport system within such a relatively short period of time.

The combination of challenges that Beijing has been dealing with are unique and unprecedented: urbanisation, motorisation and economic growth all intertwined so as to create enormous social problems. Furthermore, all these challenges have emerged so quickly that there was insufficient time for the government to develop appropriate transport policies in response. The decision-makers formerly strove to build a city that was modern in the eyes of the Western world. By constructing world-class transport infrastructure, they were eager to prove that China was an emerging power.

During this process, many important and unique characteristics of the Chinese mobility culture were ignored. High-density mixed land use and the bicycle culture, with which Beijing residents were familiar, became alienated. Beijing stood at a crossroads. Traffic congestion and other social issues that resulted from high levels of car ownership and use have attracted great attention from the general public. People started questioning whether it is appropriate to directly emulate Western experience and practice in Beijing, and began trying to find an evolution path to a more sustainable future that takes Chinese characteristics fully into account.

Fortunately, consensus has been reached by the various stakeholders that it is not possible for Beijing to become a car-dependent society like most cities in the United States, and that public transport is the appropriate solution for meeting the ever-growing travel demand. Liu and Guan [23] used simple arithmetic to demonstrate why private car ownership is not the answer for the future of transport in China. Based on their calculations, to reach the level of US car ownership, more than 50 % of the total urban land area would have to be used to accommodate vehicle use and parking, meaning that room would have to found for all other purposes –housing, employment, shopping, recreational activities and so on—in the other (rather less than) 50 %. The municipal government has adopted many policies to promote public transport, and has developed a clear vision for building a Transit City. To achieve this goal, the city has invested massively in the public transport system and has recently introduced a series of TDM policies to discourage the demand for car travel. The combination of these policies is certainly critical to the formation of a Transit City. However, we should also be aware of that none of these measures come cheap, since they all require immense monetary and manpower resources for their promotion, implementation, enforcement and maintenance.

For Chinese cities, it is probably equally important—and more cost-effective—to restore some of the diminishing traditions that are still remembered and cherished by many Chinese. The declining bicycle culture, for instance, is a great example. The bicycle modal share in Beijing dropped to 18.1 % in 2009, and the downward trend is continuing at an astonishing rate of around 2 % of the whole modal share each year. Although Beijing is experiencing a clear trend towards suburbanisation and the city is expanding dramatically, the urban area, and especially the old city, is still relatively compact. In 2005, 45 % of all trips made (excluding those made on foot) were less than 5 km [3], which is a comfortable distance to travel by bicycle. However, the bicycle modal share was 30 % in that year, substantially lower than 45 %. Of all car trips made, around 30 % are less than 5 km. It is obvious that many people have developed a car-dependent behaviour and got used to driving to travel even very short distances that could be undertaken by bicycle. There is no doubt that non-motorised modes (cycling and walking) are indispensable parts of any sustainable transport system [16]. They are not competing against public transport; rather, they serve as modes complementary to it. Public transport cannot provide door-to-door services in the way that private cars can, and has to rely on non-motorised modes to solve the so-called ‘last-mile’ issue. Therefore, how to stop the bicycle mode from continuing to lose users, and how to reduce unnecessary car trips are both questions that pose real challenges. By providing a safe, convenient and comfortable environment for non-motorised modes, the city might be able to restore the bicycle culture and make bicycles become popular once again, which would be a vital step towards creating a more sustainable transport system.

Concluding Remarks

Beijing is a dynamic city. It has a long history, and is presently in the midst of a period of rapid urban growth and motorisation. Since the economic reforms began in 1978, the city has made remarkable progress in every respect. It has changed from being a Non-Motorised City to a nearly Traffic-Saturated City in less than 30 years. The encouraging sign is that the city has developed a clear vision for becoming a Transit City, and has been working on promoting public transport for the last couple of years.

The strengths of Beijing that aid in achieving this transformation are obvious. People in China are more accustomed to a top-down decision-making approach than the population of Western countries. In the context of urban mobility, the evolution of mobility culture is heavily influenced by government transport policies. Currently, the municipal government not only has a vision to promote the prioritisation of public transport, but also the political power and financial capability to actually realise the vision in an efficient manner. The booming subway system in Beijing is an excellent example of this ability, which can also be seen in the recent implementation of a series of rigid TDM strategies restricting vehicle use and ownership. Another strength that is often overlooked is the huge group of new immigrants to the city. It is true that they generate new travel demand, and thus further burden the existing transport system. However, new residents to the city are usually more willing to embrace changes than long-term city dwellers. If they are provided with a good public transport option, they are more likely to adopt it than are the original residents. Herein lies a great opportunity to expand the public transport users group. Therefore, the city should implement public transport-orientated development by providing fast, convenient and inexpensive public transport services to service urban fringes and the new satellite towns.

On the other hand, the city is also facing many obstacles to its transition to a Transit City. The most critical one is the large existing car user group, which has been caused in part by the lack of proper demand management in the past. It would be difficult to persuade the four million or so car owners to give up their vehicles and instead use other modes, because of high sunk costs[1] associated with cars. To change their travel behaviours, or at least reduce the number of car trips, the city should introduce more innovative TDM strategies, such as congestion pricing, in

---

[1] A sunk cost is one that, having been incurred, is impossible then to recover. In this context, a large part of the high fixed cost of purchasing a car is a sunk cost.

addition to making the public transport service relatively more attractive, especially in the central part of the city. Another crucial aspect is the ongoing trend towards suburbanisation. To avoid urban sprawl and an increasing dependence on cars, the government could adjust the quantity of land that it supplies and thereby encourage public transport-orientated development [12]. Furthermore, global economic slowdown could also pose a potential threat and drag down the rate of China's economic growth, which could in turn influence government spending on public transport in the long run, and has the potential to eventually impede the transition.

A successful transformation to a Transit City would not only be important to Beijing residents, but would also create a role model for other Chinese megacities. A more sustainable transport system in China will eventually improve global energy security and reduce greenhouse gas emissions. Further investigations could focus on the impacts of TDM strategies and newly constructed subway infrastructure on mobility behaviours, which would provide us with more concrete evidence to help in predicting the tipping point of transition in Beijing's mobility culture. Although no one can truly predict the future of Beijing, there is no doubt that the city has the potential to deliver when it comes to successful transition to a Transit City.

References

1. Beijing Census Bureau (2011) The 6th population census report
2. Beijing Municipal Bureau of Statistics (2010) Beijing statistical yearbook 2010. China Statistics Press, Beijing
3. Beijing Municipal Commission of Transportation (2007) The 3rd Beijing transportation comprehensive survey
4. Beijing Municipal Commission of Transportation (2010) Beijing 12th five-year mass transit plan
5. Beijing Municipal Commission of Urban Planning (2005) Beijing master plan 2004–2020
6. Beijing Municipal Government (2009) Beijing green transportation action plan 2009–2015
7. Beijing Transportation Research Center (2003) Beijing transport annual report 2003
8. Beijing Transportation Research Center (2006) Beijing transport annual report 2006
9. Beijing Transportation Research Center (2010) Beijing transport annual report 2010
10. Beijing Transportation Research Center (2011) 2011 Transportation operations half-year report
11. Button K, Ngoe N, Hine J (1993) Modelling vehicle ownership and use in low income countries. J Transp Econ Policy 27(1):51–67
12. Cervero R, Day J (2008) Suburbanization and transit-oriented development in China. Transp Policy 15(5):315–323
13. Chen H, Jia B, Lau S (2008) Sustainable urban form for Chinese compact cities: challenges of a rapid urbanized economy. Habitat Int 32(1):28–40
14. Chen J, Gong P, He C, Luo W, Tamura M, Shi P (2002) Assessment of the urban development plan of Beijing by using a CA-based urban growth model. Photogram Eng Remote Sens 68(10):1063–1071
15. China Development Research Foundation (2010) China development report 2010: promoting the human development strategy of China's new cities. China People's Press, Beijing
16. Daniere A (1999) Sustainable urbanization in megacities: role of nonmotorized transportation. Transp Res Rec: J Transp Res Board 1695:26–32

17. Dargay J, Gately D (1999) Income's effect on car and vehicle ownership, worldwide: 1960–2015. Transp Res Part A: Policy Pract 33(2):101–138
18. Feng J, Zhou Y, Wu F (2008) New trends of suburbanization in Beijing since 1990: from government-led to market-oriented. Reg Stud 42(1):83–99
19. Han D, Yang H, Wang X (2010) Efficiency of the plate-number-based traffic rationing in general networks. Transp Res Part E: Logistics Transp Rev 46(6):1095–1110
20. Kenworthy J, Newman P (1999) Sustainability and cities: overcoming automobile dependence. Island Press, Washington
21. Kenworthy J, Townsend C (2002) An International comparative perspective on motorisation in urban China. Probl Prospects IATSS Res 26(2):99–109
22. Liu G (2007) A behavioral model of work-trip mode choice in Shanghai. China Econ Rev 18(4):456–476
23. Liu R, Guan C (2005) Mode biases of urban transportation policies in China and their implications. J Urban Plann Dev 13(2):58–70
24. Liu X (2009) Advancing a scientific development of transportation to construct a transit-oriented Beijing. Urban Transp China 7(6):1–7
25. National Bureau of Statistics of China (1996) China statistical yearbook 1996. China Statistics Press, Beijing
26. National Bureau of Statistics of China (2010) China statistical yearbook 2010. China Statistics Press, Beijing
27. Tian L, Ma W (2009) Government intervention in city development of China: a tool of land supply. Land Use Policy 26(3):599–609
28. Wang D, Chai Y (2009) The Jobs–housing relationship and commuting in Beijing, China: the legacy of Danwei. J Transp Geogr 17(1):30–38
29. Zhang T (2000) Land market forces and government's role in sprawl: the case of China. Cities 17(2):123–135
30. Zhang S, Pearlman K (2004) China's land use reforms: a review of recent journal literature. J Plann Lit 19(1):16–61
31. Zhao P, Lü B, de Roo G (2011) Impact of the jobs–housing balance on urban commuting in Beijing in the transformation era. J Transp Geogr 19(1):59–69
32. Zhou Y, Ma L (2000) Economic restructuring and suburbanization in China. Urban Geogr 21:205–236

Chapter 6
Gauteng: Paratransit—Perpetual Pain or Potent Potential?

Johan W. Joubert

Abstract South Africa is nearly 20 years into its democracy, yet the legacy of apartheid remains evident and the level of inequality steadily rising. Gauteng province, the economic heart of the country that includes Johannesburg and Tshwane, the capital, carries the burden of many of its formerly relocated citizens still living in poverty on the periphery of the large metropolitan areas. Whereas sprawl is commonly associated with low-density, more affluent development on the periphery, Gauteng finds itself with having to provide basic services and mobility to high density, low-income people on its outskirts. The mobility culture in Gauteng is heavily influenced by the socioeconomic disparity. In this chapter we revisit the legislative context that gave rise to racial segregation, and concern ourselves with the impact it had in the evolution of the now-dominant paratransit mode that accounts for more than two thirds of all commuter trips in Gauteng. Although often cited in literature and the media for its violent sectarian conflicts, the minibus taxis, as it is commonly referred to, is much more than a mere mode of transport. Outsiders often perceive the taxi industry to be a chaotic system, but it has evolved into a powerful economic industry with a unique mobility culture, most notably the hand signals—a silent gestural semiotic language in its own right—used by commuters to communicate their desired destinations to passing taxis. Aware of the rising inequality, and not fully understanding or appreciating the exibility it provides, government often perceives paratransit as a necessary nuisance that should be formalised. Improving mobility and accessibility, however, requires both settlement location and transport to be revisited. As such, paratransit may prove to a valuable solution, and not the mere nuisance it is often made out to be.

Introduction

South Africa is nearly 20 years into its democracy, yet the legacy of apartheid remains evident and the level of inequality is steadily rising. As the name suggests, South Africa is the southernmost country on the African continent.

J. W. Joubert (✉)
University of Pretoria, Private Bag X20, Hatfield 0028, South Africa
e-mail: Johan.Joubert@up.ac.za

Institute for Mobility Research (ifmo) (ed.), *Megacity Mobility Culture*, Lecture Notes in Mobility, DOI: 10.1007/978-3-642-34735-1_6,
© Springer-Verlag Berlin Heidelberg 2013

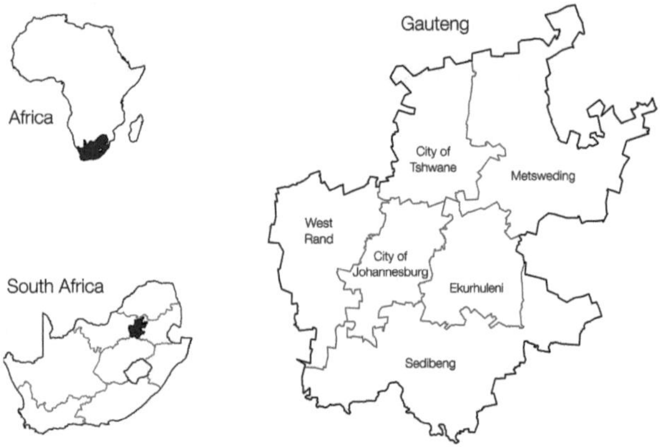

Fig. 6.1 Gauteng in perspective (*Source* generated by author using Quantum GIS Development Team [18])

Figure 6.1 shows the country's location, as well as that of Gauteng, the smallest of the nine provinces. Gauteng is the economic heart of the country, and includes Johannesburg and Tshwane, the latter being the administrative capital of South Africa and formerly known as Pretoria. Many of the province's formerly relocated citizens still live in poverty on the periphery of the large cities. Whereas sprawl is commonly associated with low-density, more-affluent development on the periphery, Gauteng finds itself with having to provide basic services and mobility to a high density of low-income people on its outskirts.

The mobility culture in Gauteng is heavily influenced by socioeconomic disparity and relies on paratransit, which accounts for more than two thirds of all commuter trips in the province. The paratransit vehicles in South Africa are known as 'minibus taxis' and take the form of 15–25-seater minibuses that follow flexible routes without a schedule, and which commuters can board or alight from at any point along the route. These South African minibus taxis are often mentioned in literature and in the media on account of the violent sectarian conflicts that take place over lucrative routes, but minibus taxis in South Africa are much more than merely a mode of transport. Outsiders often perceive the minibus taxi industry to be a chaotic system, but it has in fact evolved into a powerful economic industry with a unique mobility culture, most notable in the hand signals—a silent gestural semiotic language in its own right—used by commuters to communicate their desired destinations to passing taxis.

Aware of rising social inequality, and not fully understanding or appreciating the flexibility that paratransit provides, government often perceives this mode as a nuisance, but a necessary one that should be formalised. Improving mobility and accessibility, however, requires both settlement location and transport to be revisited. As such, paratransit may prove to be a valuable solution—not just a nuisance, as it is often made out to be.

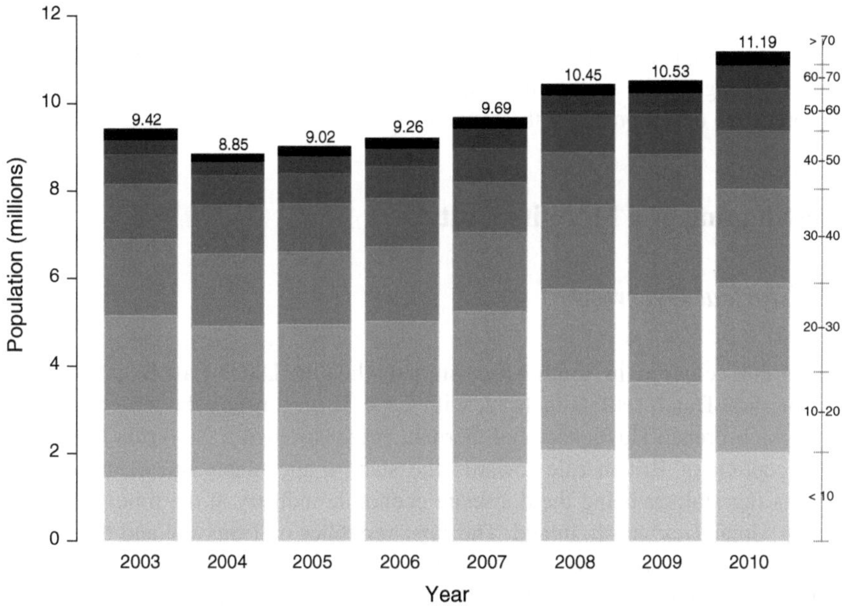

Fig. 6.2 Gauteng population and age distribution (*Source* generated by author using R Development Team [19])

In studying the mobility culture of the region it does not make sense to consider any one of the three major cities—Johannesburg, Tshwane or Ekurhuleni—on its own. The majority of people movement happens *between* these three cities. For example, the N1 highway connecting Johannesburg and Tshwane is the busiest highway in Africa, as many people choose to live in the (mainly eastern) suburbs of Tshwane with its small-town feeling, yet work in Johannesburg, where more job opportunities exist and higher salaries are on offer. The Gauteng population and age distribution is shown in Fig. 6.2, which highlights the recent growth and urbanisation of Gauteng.

The chapter is structured as follows. A brief historical overview of the country, which is known for its former racial segregation in the form of apartheid, is provided in the next section to show how the political environment shaped the mobility culture. We present the status quo with respect to economic equality/inequality, and show how current policy initiatives actually reinforce previous mistakes regarding urban form. In response to the lack of more formal public transport alternatives, paratransit emerged. The section "Paratransit" describes the evolution of paratransit, or minibus taxis as it is more commonly known in South Africa, into the dominant mode of transport. We explain how the industry is organised, how it operates, and the challenges that users face, and end with a discussion of government interventions to formalise the industry. Instead of the general view that paratransit is a necessary nuisance, this chapter aims to show that the mode is actually a very people-centred and social one. The chapter ends by putting the case that, if only it can be appreciated for its ability to adapt and foster social change, paratransit in Gauteng can be seen to have an important and very valuable role to play in the long-term development of the country.

The Shaping of a Mobility Culture

A Historical Overview

The formal economy in South Africa started when the Dutch East India Company sent the first Dutch settlers in 1652, with Cape Town forming the base of the colony, which French Huguenots and German citizens joined a few years later. With the inception of British rule towards the start of the nineteenth century, many farmers (agriculture being the dominant economic industry at the time) embarked on the 'Great Trek' north, inland. The *Boer* republics of Transvaal and the Orange Free State were established, but attracted the attention of the British when diamonds were found in Kimberley in 1870, and some of the richest gold deposits in the Witwatersrand in the Transvaal in 1886. The Witwatersrand is a low, sedimentary range of hills that runs in an east–west direction through the Transvaal.

The British eventually conquered the republics during the Second Boer War, which ended in 1902 with the signing of the Treaty of Vereeniging. Both Transvaal and the Orange Free State were made self-governing colonies of the British Empire. The Union of South Africa was established in 1910, joining the Cape, Natal, Transvaal and the Orange Free State. In 1913 the Natives Land Act was passed, limiting land ownership of blacks to selected black territories. The subsequent Native Urban Areas Act of 1923, and later the Natives (Urban Areas) Consolidation Act of 1945 divided South Africa into 'urban' and 'rural' areas and strictly controlled the movement of (mainly) black males between the two. The control happened through the inspection of pass books, an identity document that all non-whites were required to carry and produce on demand to enforcement officers. Towns and cities became mainly white, since only domestic workers were allowed to live in town. In 1948 the policy of apartheid was officially adopted when the then National Party came into power, and apartheid laws started to be passed, amongst which were the Prohibition of Mixed Marriages Act in 1949, the Immorality Act, Group Areas Act, the Suppression of Communism Act, and the Population Registration Act, the latter officially classifying South Africans as being either 'White', 'Coloured', 'Asian' or 'Native'. The first forced relocation also took place in 1949, when 80,000 Africans were moved to the South-Western Townships, better known by the English syllabic abbreviation, Soweto, on the south-western periphery of Johannesburg in Gauteng. In 1961 South Africa withdrew from the Commonwealth and became a Republic.

Government policy started to move towards restoration in 1986, with pass books being abolished, but not before a national state of emergency was announced.

Apartheid was formally scrapped in 1990, which led to the release of the leader of the African National Congress (ANC), Nelson Mandela, after 27 years in prison. Mandela led the ANC in the negotiations that resulted in the first multiracial democracy in 1994, with Mandela as the first president and the new constitution being signed into law in 1996. The majority of laws have since been rewritten, although economic inequality has increased significantly over the past 15 years. Shortly after the elections the province of Transvaal was divided into four separate provinces: Limpopo to the north; Mpumalanga to the east; North West to the west; and Gauteng, the smallest, to the south.

Inequality

The Gini coefficient is a measure of statistical dispersion and expresses the inequality of a distribution. When applied to income (or wealth) distribution, a value of 0 expresses perfect equality where everyone has an equal income. On the other extreme, a value of 1 expresses maximum inequality, where a single person has all the wealth. To put it into perspective with other countries, namely Germany (0.27 in 2006), United Kingdom (0.41 in 2008/2009), India (0.37 in 2004), the United States (0.45 in 2007), China (0.42 in 2007) and Brazil (0.52 in 2012), South Africa has the highest inequality of these, at 0.63 (in 2009) (figures sourced from Wikipedia, citing secondary sources, particularly CIA World Factbook).

Although inequality has increased since democracy in 1994, apartheid was indeed responsible for the original (re)distribution of people. Figure 6.3 shows the population density distribution [4], Fig. 6.4 the distribution of people below the 'Minimum Level of Living' (MLL) in Gauteng. Forced relocation, which started in 1949, saw large groups of people moved to the periphery of the three main cities: Tshwane, Johannesburg and Ekurhuleni. Although these three cities have since expanded and developed into an amorphous whole, many (mainly poor and black) people remain on the periphery of the cities. The type of developments, however, are not formal high-density high-rise apartments, but rather slum-like: very many, very small informal buildings, in extremely close proximity. Such a distribution of people, and lack of integrated transport infrastructure, has serious implications for mobility.

Urban sprawl is often associated with the middle and upper echelons of society moving to the lower-density periphery of cities. The result is that the people in question, who can afford private cars, commute longer distances to participate in economic (and other) activities. Also, economic opportunities often follow the people with buying power, resulting in multiple economic centres forming. Conversely, in this situation the lower-income citizens reside either (a) closer to economic opportunities, where they rely on public transport that is usually denser in the centre than on the periphery of the city, or (b) close to public transport arteries where they have affordable access to the economic centres.

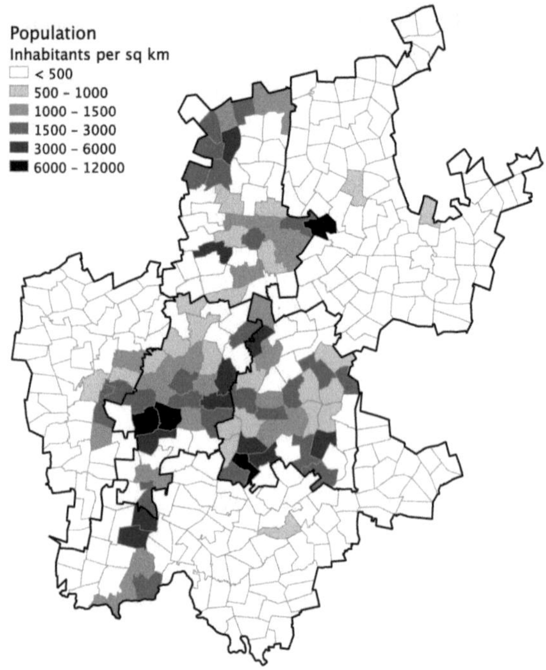

Fig. 6.3 Gauteng population density (*Source* generated from CSIR Built Environment [4] by author using Quantum GIS Development Team [18])

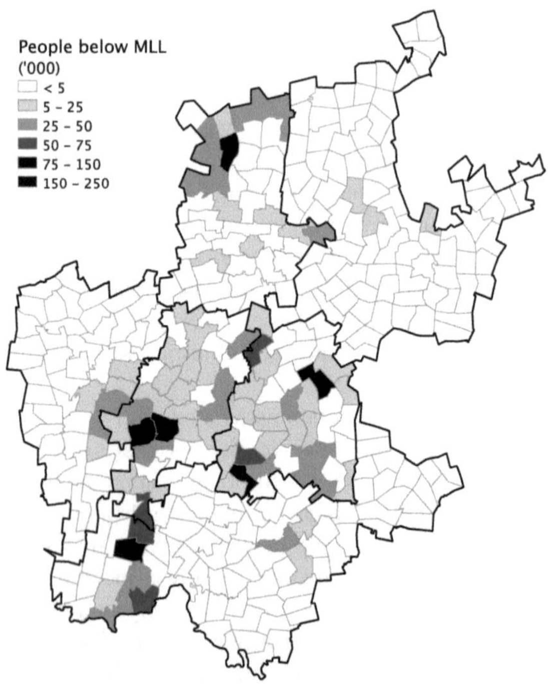

Fig. 6.4 Population in gauteng living below the minimum level of living (*Source* generated by author from CSIR Built Environment [4] using Quantum GIS Development Team [18])

In South Africa, and specifically Gauteng, the opposite is observed: the more-affluent people are living close to centres of economic activity, while the poor are situated on the periphery without access to dense or reliable public transport.

But why has the pattern not been reversed, nearly two decades into democracy? One of the many explanatory factors is the way in which the government approaches economic inequality. Although contributing to economic uplift, the provision of government-subsidised housing has the unintended consequence of reinforcing the apartheid urban form.

Housing Policy Reinforcing Inequality

The Reconstruction and Development Programme (RDP) is an all-encompassing socioeconomic policy framework implemented by the ANC shortly after the arrival of democracy in 1994, in collaboration with various unions, political parties and mass organisations in the wider civil society [20]. The ANC's chief aim in developing and implementing the RDP was to address the immense socioeconomic problems brought about by the consequences of apartheid. Specifically, it set its sights on alleviating poverty and addressing the massive shortfalls in social services that existed across the country. Some of the social achievements of the RDP include improved healthcare, clean water close to or inside homes, electrification, public works, and—most importantly for mobility—housing. On the basis of Section 26 of the Bill of Rights, each South African has the right to adequate housing, and "the state must take reasonable legislative and other measures, within its available resources, to achieve the progressive realisation of this right". To that extent, government has embarked on building vast amounts of 'RDP houses': small low-cost dwellings often criticised for the poor quality of building materials and construction (Fig. 6.5).

The processes of allocating tenders for constructing these houses are fraught with irregularities, something that often hits the pages of the news chaters in South

Fig. 6.5 Example of typical RDP houses in Gauteng, and aerial view of an RDP scheme (*Source* of latter university of pretoria department of geography)

Africa. In the 2009/2010 fiscal year in Gauteng alone, 12,819 serviced building sites have been completed (64,362 nationally), and 33,654 houses (161,854 nationally). In the 2010/2011 fiscal year, 21,866 serviced sites (63,546 nationally) and 25,117 houses (121,879 nationally) were completed. The Community Surveys [24] indicated that the percentage of households now living in formal dwellings increased from 64.4 % in 1996 to 70.5 % in 2007. Nearly 20 % of South Africans live in state-subsidised housing.

The RDP houses are built on small plots mainly on the outskirts of the cities where government can obtain and develop land cheaply. No sooner have the houses been finished than the new owners start subletting the available space on their small plots for the construction of informal shacks, a process now known as 'backyard filling', with the occupants of these informal shacks being called 'backyard dwellers'. Although it could be argued that this has one positive consequence, namely densification, the basic service infrastructure for water, electricity and sewerage was not designed for the kind of densities that result.

Densities may justify formal public transport, but provision to connect the periphery townships across such large distances is not economically viable. As a result, the train and bus transport services became more infrequent, and the quality of the services deteriorated dramatically. Such conditions are fertile ground for the establishment and growth of paratransit.

Paratransit

The concept of paratransit is not a new one. Early work such as that by Orski [17] and Roos and Alschuler [21] describes it as 'personal public transport', a mode that originally serviced the diffuse areas that resulted from suburbanisation. Paratransit, as a concept, is an overarching term for a variety of service offerings that provide different levels of spatial and temporal flexibility, as illustrated in the early taxonomy depicted in Fig. 6.6 [21]. *Fixed route* represents schedule-based bus services found in many developed countries. An example of *vanpool services* is an employer-subsidised programme in which scheduled services are provided through small vans shared by its subscribers. Whereas the vans are

Fig. 6.6 Taxonomy of paratransit (*Source* [21])

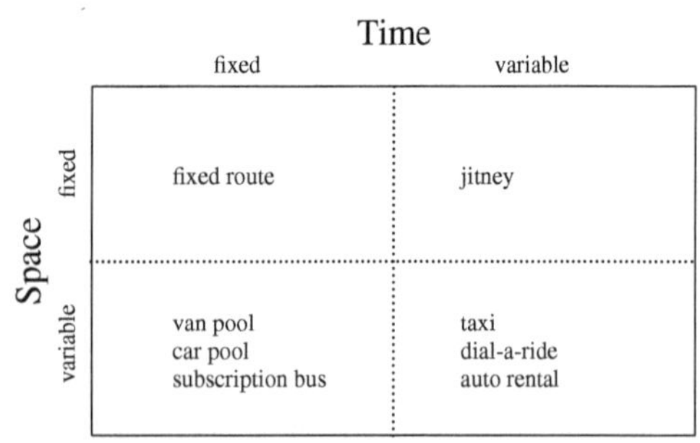

owned and operated by the *van-pool* service provider, a similar and less formal *car-pool* service sees people sharing their own vehicles, and is hence also known as car sharing, ride sharing or lift sharing. In a *dial-a-ride* service, the user makes a reservation with the service provider beforehand, or in the case of a traditional *taxi*, just hailing the service provider on the street. In both cases the user is taken directly from his/her origin to destination without having to share the vehicle. The *jitney* services follow predetermined routes, although they may vary, but do not adhere to fixed schedules.

The majority of the examples of paratransit are formal, and complementary to public transport services. They include provision of services to frail, disabled, elderly or more rural communities. In many developing countries, informal paratransit often exist in the form of jitney-like services in which vehicles carrying between 8 and 30 passengers are used. These services are often in competition with more formal public transport offerings [2, 22]. In such systems, customers would hail a vehicle without prior scheduling.

In the case of South Africa, the form of paratransit system that has developed goes by the name 'minibus taxis', and started as a direct result of diminishing public transport services to black townships. Train services were limited, and many would-be commuters often found themselves stranded as a result of too little capacity. Buses were slow and unreliable. In response, those fortunate enough to own a vehicle started providing lifts to stranded commuters, and compensation soon followed. Originally only sedan-type vehicles were used, but a loophole in the 1977 Road Transportation Act allowed for eight passengers to be carried without being subject to strict safety regulations associated with buses. As a result, minibuses started flooding the market. Spurred on by a sudden surge of deregulation, the minibus taxi industry in the country grew exponentially from 3,782 vehicles in 1985/1986 to 39,604 in 1987/1988 [14]. Currently there are more than 130,000 minibuses nationally providing mobility to millions daily.

From the Department of Transport's National Household Travel Survey [5] we know that for commuting trips in Gauteng, the minibus taxis (with 31 % of trips) were in 2003s only to the private car (39 %) in terms of modal share. When all trip purposes are considered, minibus taxis (32 %) actually exceeded private car use (25 %).

Unfortunately the mode is associated with a lack of safety, and reckless—even ruthless—drivers. As noted by Terblanche [26], taxi owners are caricatured as "lawless thugs at worst, rude business owners with bad customer relations at best". Rival taxi associations compete for the most lucrative routes, and such competition frequently result in violent clashes. As an onlooker to one incident put it, "it is not like war—it is war" [23]. If the minibus taxi industry, then, offers such a violent and unsafe mode of travel, to what can be attributed its widespread use, and—dare we say—'success'? Literature on the South African minibus taxis is sparse, but does offer [7, 14, 22, 28, 30] reviews of various aspects of the industry. As in Woolf and Joubert [30], a more people-centred perspective is given in this chapter to shed light on the industry, which we start by describing from an operator point of view. This has significance, since much of the services are provided on a 'supply push' rather than a 'demand pull' basis. We therefore deem it necessary to first understand how the industry is structured and operates, as this influences how users perceive and interact with the operators.

An Organisational View

For non-white citizens living during the apartheid years, real estate ownership was severely limited. Many families resorted to buying vehicles as an investment which could then earn them a return by being deployed as paratransit. Although the scene was marred by violence, it was still fairly easy to enter the paratransit market. A group of taxi owners in the same geographic area would organise themselves into what is now known as a taxi association. Although the industry started out with owner-drivers, the business model changed so that the majority of drivers are now contracted. It is very common, even the norm, for a taxi to be driven by a member of the family that owns it.

Operations

A taxi owner, being a member of an association, would apply for an operating licence from the provincial government. An operating licence indicates, albeit ambiguously, the route that the vehicle will be allowed to travel, describing it only by the origin and destination rank. In Johannesburg alone 440 ranks were identified [3] of which 41 were formal and the remaining 399 informal; 154 were off-street facilities while the other 286 were on-street. It is up to the different associations to negotiate among themselves the more precise details of the route to ensure an equitable distribution of the business. Formal marshals employed by the associations ensure that only licensed vehicles operate on the specific routes. Illegal operators are referred to as 'pirate taxis'. Taxi passengers pay a flat fare for a given route, set by the associations, regardless of whether they travel the full length of the route, or whether they board or alight along the way. Different fares are applied to different routes, and are based roughly on distance. An analysis of fares against actual route distances revealed that the overall minibus taxi fare structure in the City of Johannesburg was R2.35 (€0.232 or US$0.285 as at July 2012), plus 6.4 cents (€0.0063 or US$0.0078) per kilometre of the route length [3]. Although this may seem very cheap, long commuting distances result in many of the commuters paying in excess of 25 % of their household income on transport, the national average being 20 %. The fare structure is flattish, so shorter routes are probably more expensive on a per-kilometre basis than the longer routes. For the operators (taxi drivers), long routes are therefore less lucrative than short routes. One unintended consequence of this is that many poor, those travelling furthest, have to take multiple, more expensive, short trips as part of a single journey. The reason for this is that the taxi industry offers multiple short—and more profitable—routes, rather than less profitable direct long-distance connections.

Employment Practices

Once an operating licence is obtained, an owner can hire a driver for the vehicle. The owner, taking into account the route and its expected demand, vehicle operating costs, and working hours, calculates the 'check-in' that the driver must earn in a week. This is a colloquial term used to describe the gross profit: fares collected less fuel expenses. In the majority of contracts, the driver hands over the check-in amount at the end of the week, from which he or she is paid a commission of typically 30 %. All money earned over and above the check-in amount is for the driver's own pocket. An alternative business model exists whereby drivers are given a check-in amount to chase, but from which no commission is earned. Rather, once the check-in amount is achieved and handed over to the owner, the driver has the benefit of the vehicle to earn own income for the remainder of the week.

Both business models are very similar and have two unintended consequences. Firstly, drivers minimise cost if income potential is low. As the majority of trips are work-related, taxi drivers provide their services mainly during the morning and afternoon peak, the times of day when many commuters demand mobility. For the remainder of the day the vehicles either (a) leave from origin to destination only once the vehicle is filled completely, resulting in potentially long onboard waiting times; or (b) stop entirely to be serviced or washed, significantly reducing the transport supply during off-peak periods. This in turn reinforces commuter behaviour to travel mainly during peak hours, as they don't have a reliable alternative during the off-peak periods even if their journeying requirements allow for more flexibility.

The second unintended consequence is a lowering of safety standards. Taxi drivers are ruthless in chasing their targets. Giving heed to traffic regulations is the exception. The City of Johannesburg [3] reported *average* morning and afternoon peak occupancy figures of 96 and 100 % respectively. So vehicles are often overcrowded, and drivers don't obey traffic signals or speed limits, often overtaking in dangerous areas such as emergency lanes and on the shoulders of the road, all in an effort to get to their destinations faster—and more often than not putting the occupants at risk.

Working Your Way Up

In a country where unemployment is running at 25 % [25], finding work is extremely difficult. In the larger cities it is common for someone trying to enter the taxi industry to start off as a freelance vehicle washer at or around a taxi rank, while saving enough money for driver training, and eventually a driver's licence. As an individual gets to know the routes originating from and arriving at the specific rank, he or she may choose to become an (unofficial) marshal. Taxi ranks are more often than not informal, with little or no signage indicating the routes serviced by the ranks. Vehicles, too, are not marked. It is the prerogative of entrepreneurial marshals to attract approaching pedestrians and direct them to the right taxi. Once a marshal has filled a taxi, the driver remunerates him (in cash) for his services.

As time goes on and an aspiring driver gets to know the various taxi drivers, either as a washer or a marshal, he stands a chance of being approached by one of the drivers as a stand-in, or locum, on occasions where the driver will be absent from service. Being an occasional driver builds one's reputation, and opens doors for the future, leading eventually perhaps to a more permanent position as a driver. Long-distance taxi associations, those who provide services between cities, have a slightly more formal approach to dealing with vacancies. Candidate drivers put their names on a waiting list with the association. In addition to considering available family members, scouting owners would then select drivers from the waiting list.

Hailing Your Ride

Each route stipulated by an operating licence usually indicates the start and ending rank. But it is common knowledge in South Africa that only about 40 % of all trips originate, or end, at a rank. The remainder originate and/or terminate along the route. Potential customers stand next to the road showing a hand signal associated with their destinations. Taxi drivers then recognise hand signs associated with their route and, capacity allowing, stop to pick up the individual(s). Other, unassociated, hand signals are—or at least should be—ignored.

Some hand signals are common throughout Gauteng, and even the entire country. For example, an index finger pointing downward (Fig. 6.7a) indicates any *local* destination relative to the location of the signaller. When a circular motion is added (Fig. 6.7b), the signal means *very local*, and implies a stop within a few blocks from the origin. Another universal sign is an upward pointing finger suggesting *town* (Fig. 6.7c).

Less universal hand signs include the 'T' (Fig. 6.7d) usually associated with a specific major *T-junction* on the route. The downward-pointing fingers of (Fig. 6.7e)

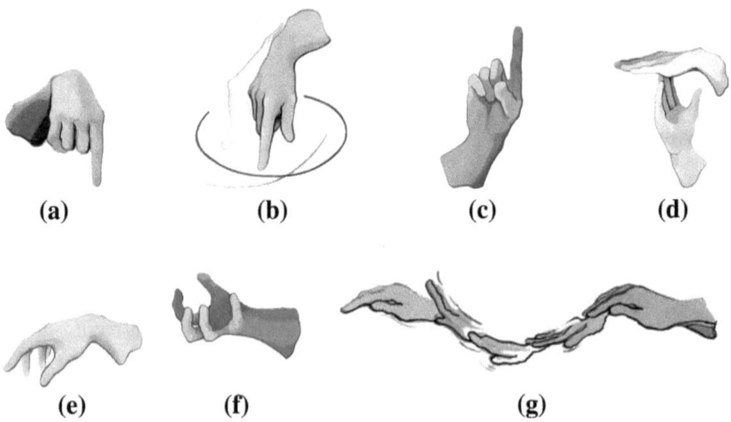

Fig. 6.7 Selection of taxi hand signs (*Source* [29])

represent a collection of *shops*, and this hand sign is associated with major shopping centres. Sometimes hand signs are wholly unique and have a specific historical meaning. The cupping hand (Fig. 6.7f) is associated with *Orange Farm*, a township south of Johannesburg. The curvy motion of (Fig. 6.7g) is associated with more than one destination, but most notably with Diepsloot, a township on the boundary between Johannesburg and Pretoria (Tshwane) whose name is directly translated from Afrikaans as '*deep ditch*', referring to its bumpy and hilly terrain.

The gestural language acts as a semiotic code between drivers and commuters, but it is mostly undocumented. With the exception of the 26 hand signs captured by Woolf [29] for the greater Johannesburg area, very little is documented explicitly. More than 50 signs have been identified in the Johannesburg area alone. Knowledge transfer happens mainly through word of mouth, and this in itself is a challenge in a country which has 11 official languages. The barrier to entry, from a user point of view, is very high. Many users, once they've been able to decipher the hand signals required to get from their origin to destination, do not venture to adapt their routes or experiment with shorter possibilities, since it is very difficult to learn to navigate the network. As a result of their lack of in-depth knowledge, which takes time and trouble to acquire, many people overpay for their mobility. Also, using the wrong hand signal may indicate to (informed) bystanders that you are lost, making you vulnerable in a country known for its high crime rate.

Implications for Commuters and Other Modes

It could be argued that since the minibus taxi industry is self-regulated, route optimisation with the objective of maximising connectivity for commuters is self-destructive for the industry. Routes are kept as short and multi-legged as possible to maximise income. The consequence for commuters with long journeys is that they often have to take several trips, with the total cost being higher than what a single trip would have cost. No formal route planning actually exists, and market forces have not succeeded in shaping the paratransit network. To their credit, new taxi services spring up to provide mobility in a new development long before the local government even thinks about putting out a tender for bus services there. It actually seems that government has already acknowledged, even relies on, the agility and responsiveness of the taxi industry. To give an example of the industry's flexibility, the minibus taxi industry's national body, SANTACO, has even announced that it will launch a low-cost airline in its efforts to diversify and service rural areas.

The minibus taxis in South Africa, as in many other developing countries [2], are in competition with formal public transport. Being more flexible and proactive, paratransit often outsmarts its more formal and bureaucratic counterparts. In response, many of the bus operators have also begun to deviate from their planned routes in an attempt to scavenge patrons before continuing on their trip. Ultimately the commuters are carrying the cost. Routes become longer, and subsequently more expensive, burdening the users with both time and cost expenses.

Formalisation

Acknowledging the high cost of transport for poor citizens, government, being the custodian for mobility service provision, has attempted a number of interventions to formalise the minibus taxi industry in South Africa, some more successful than others. As in other developing countries, competing paratransit services are often viewed as a necessary nuisance, and not as a strategic, even valuable, part of the mobility portfolio.

The Taxi Recapitalisation Programme aimed to replace old and unsafe vehicles by providing taxi owners with a once-off reimbursement for their old vehicles. In return, owners had to convert their very ambiguous area-based operating licences to less ambiguous route-based operating licences. From the discussion in Woolf and Joubert [30], we conclude that the main downfall of the programme was that government approached the taxi industry as a single body, trying to enforce a top-down initiative without acknowledging the inherent bottom-up structure of the industry, whereby local associations vote representatives to regional and provincial councils, who in turn vote national representatives into power.

With the success of Bus Rapid Transit (BRT), mainly in South American countries, the South African government quickly got on the bandwagon. The first African BRT system was launched in 2009 in the form of the first phase of the Rea Vaya (Sotho for "we are going") project in Johannesburg. In essence, the BRT system is owned and operated by PioTrans, a company wholly owned by the 'taxi industry'. McCaul and Ntuli [15] elaborate on the negotiations that led to the formation of PioTrans by the former taxi owners. Woolf and Joubert [30] highlight some of the seemingly controversial and inconsistent aspects of the deal, especially since government used to insist on dealing *only* with SANTACO, but in the case of the Johannesburg BRT local government chose to continue with the rival National Taxi Alliance (NTA) to avoid negotiation deadlock.

Still, the fact that the Johannesburg BRT was successful in actively involving the minibus taxi industry suggests that the fundamentally different bottom-up approach—i.e. involving the individual taxi owners and bypassing the national and regional taxi bodies—is proving to be more appropriate. Venter [27] argues that this success can also be attributed to the taxi industry's own urge to diversify and reinvent itself, and not so much to the efforts of government.

Another force driving government to attempt formalisation is loss of income. Although sizeable, the minibus taxi industry forms part of the country's 'second economy'. The dualistic economy [6] was first popularised by the then president Thabo Mbeki. The 'first economy' is a modern one: formal with a well-documented structure in terms of receipts, records, a credit system and legally enforceable rights and remedies. The second economy is, in short, everything the first is *not*: informal, self-regulated, and fuelled by small undocumented and taxless cash transactions. It is a fact that the minibus taxi industry is non-subsidised yet makes money—this, however, may be attributable to its tax evasion. The taxi industry was traditionally an anti-government body during the apartheid years, and their sentiment is captured in this excerpt from a statement about BRT by SANTACO:

> For more than a century to date, the taxi industry successfully moved millions of people without subsidy competing against the state subsidized buses and trains, yet we survived and grew in strength. We endured the pains of the imposed deregulation era that led to violence and loss of lives because of the self-centred interests of the apartheid regime, 14 years into democracy we are now confronted with a roundtable systematic plot to nicely get rid of us… NO MORE.

One cannot but agree that the industry is the most original example of post-apartheid black empowerment. Taxis are a major group of road users—some may say *mis*users. Significant damage is incurred by the road surface, especially by the shoulders as a result of minibuses pulling off next to the road for passenger pick-up or drop-off. Notwithstanding their contribution to taxes through fuel levies and (limited) toll fees, the major means of funding the public purse, income tax, is generally avoided. On the proposed introduction of a tax amnesty to get the taxi industry to contribute to income tax, Terblanche [26] writes:

> Taxi owners also marvel at the strength of the industry they have built up over the past two or three decades without any help from government… Black entrepreneurs scraped together what savings they could and bought claptrap vehicles to repair because banks would not finance new ones. Through trial, error, chaos and conflict they mapped out viable routes more thoroughly than any BMW full of government-employed transport engineers could ever have hoped to… 'And now,' the feeling goes in the taxi industry, 'the government wants to come and tax us.' Of course, there are irrefutable arguments as to why they should join the tax net and, intellectually, most taxi owners agree. But emotionally, it is a bitter pill.

An Opportunity Waiting

Those who do not use paratransit in South Africa often fail to understand the seemingly chaotic system, and thus view it with antagonism—it is easy to dislike that which one does not understand. The minibus taxi industry relies on knowledge acquired through experience, making it very difficult for people to get to know and understand it. Many simply do not have the choice of an alternative mode, whether for reasons of availability, reliability or affordability, and hence must learn to navigate the minibus taxi network through word-of-mouth advice and local knowledge.

Even amidst the perceived chaos, one cannot dismiss the extent and sheer capacity of the industry that has evolved, and what Terblanche [26] refers to as "the amazing phenomenon that is the minibus taxi industry". But what is its niche? And if minibus taxis do indeed have a role to play, what can be done to enable them to achieve this role in a constructive manner?

Deploying Flexible Transport

With high levels of unemployment and poverty remaining, many of the RDP houses remain unoccupied, and in many others the occupants run into arrears in attempting to pay for basic services such as water, electricity and sanitation, as a result of affordability issues. Moreover, to achieve their housing delivery targets, politicians often ignore the long-term urban planning challenges associated with their decisions, and the large-scale RDP housing schemes thus end up in the same location as, or in close proximity to, the former shanty towns which they supposedly replace, i.e. on the periphery of the cities where there are very limited mobility options. The self-regulating taxi industry is the first to evolve and adapt, and to begin to service these areas.

Access to decent housing is important, but one must be very careful that a short-term solution does not become a long-term nuisance. It is imperative that South Africa in general, and Gauteng specifically, starts to plan more comprehensively, and take both land use and, at the same time, transport into account. Decision support tools are emerging that are capable of evaluating the total cost of being mobile. One may choose to live further from one's economic and leisure activities to benefit from cheaper housing, but then the additional travel cost needs to be accounted for as well [1, 13, 28].

In this chapter we argue that it is indeed the peripheral high-density areas that provide a niche for the minibus taxis. It is the opinion of the author that providing formal public transport like rail and BRT, which require expensive fixed infrastructure, across such vast areas like Gauteng, would be fiscal suicide. High-capacity modes have many benefits, but need to be constructed with a long-term view as the capital outlay is extensive. If such infrastructure reinforces a bad urban form, as is likely to be the case in South Africa with its apartheid past, it becomes very difficult to break free from and/or correct the wrongs of the past. Developing nations cannot, and should not have to, afford to make such costly decisions. During the transitional developmental phase, with the housing infrastructure already in place in many areas, minibus taxis provide a viable, yet flexible means of providing mobility and accessibility for peripheral dwellers. One study has shown that it need not remain beyond the realms of possibility to see minibus taxis feed into a bus network, which in turn feeds the rail backbone of the public transport network [8].

But to achieve such ambitious network designs, fundamental changes of attitude are required—firstly in road users, and secondly (although this is often overlooked) in urban and transport planners.

Changing a Mobility Culture

A recent television commercial in South Africa confirmed—even promoted—the perception that car ownership is a symbol of progress and wealth. A young black gentleman, clearly a recent graduate now in the job market and dressed in a business suit, was shown commuting on the bus. He clearly knew his fellow commuters as casual acquaintances, in the manner in which people who frequent the same bus do vaguely know one another. When he alights, there is a somewhat suppressed emotional scene in which an elderly woman hands him a braided key holder as parting gift. He also hands what seems to be his mobile number scribbled on a scrap of paper to a young lady. The next scene shows him walking into a car dealership, and receiving his keys to a new, entry-level compact hatchback. With a contented smile he drives off the showroom floor, and shortly after overtakes the same bus he, just a few minutes ago, was commuting on. The message is quite clear from the advertising campaign's slogan: "Move on. Move up." Car ownership lets you leave *that* old world behind, liberating your mobility.

Clearly this is a concern, as it creates the impression that unless you own a car you haven't escaped poverty and joined the 'New South Africa'. A similar message is also given by government, as it views transport subsidy as a *poverty alleviation instrument* rather than a means to provide mobility and all-day-long accessibility *to all*. This echoes the general sentiment that public transport is for the poor. One can partially understand, then, why many taxi drivers behave the way they do in order to pursue their targets, serving mainly poor commuters.

The animosity towards minibus taxis from the public in general, and private car users specifically, may be warranted. Still, it remains an overgeneralising cliché to make taxi drivers out to be violent and lawless thugs.

Because of crime levels, which are partly due to lack of education, high levels of unemployment and inefficient law enforcement, many citizens are sceptical and negative about the country and its future. There are a number of initiatives to change the minds of the general public. South Africa may be one of many countries with a Bill of Rights stipulating what citizens can, and should, expect from their government, but it is probably the first to adopt a 'Bill' of Responsibilities setting forth a list of responsibilities that underline a set of values which flow directly from the Bill of Rights. The Bill of Responsibilities is an educational resource produced jointly by Lead SA (a partnership between Primedia Broadcasting and the Independent Group of Newspapers that aims to encourage South Africans to "stand up and lead" in a constructive manner) and the Department of Basic Education, which aims to include the Bill in primary education.

More specific to the taxi industry, Holland [10] calls on citizens in a recent *Daily News* editorial to "dump" the "anti-taxi attitude". Imperial Logistics, one of the largest global logistics and supply chain divisions in southern Africa, also recently launched its *I–Pledge* road safety campaign, acknowledging the need to change personal attitude and behaviour in a country with a dismal road safety record. Two of the campaign posters are shown in Fig. 6.8. The first depicts a stereotypical white business woman who commutes as a single-occupant car driver. She probably won't use paratransit any time soon, but the poster depicts the aggressive and antagonistic relationship between private car and taxi drivers. Private car drivers seldom, if ever, yield to public transport vehicles, and often deliberately try to cut them off—while taxi drivers, in response, will cut into traffic and intersections in unsafe ways, often causing unnecessary congestion. The second shows a young trendy black youth listening to the loud kwaito music played in many urban minibus taxis and urges taxi users to put safety before more superficial considerations when choosing a taxi to travel in.

Fig. 6.8 Posters published as part of the *I–Pledge* campaign (*Source* [11])

The other set of attitudes requiring attention is that of the authorities. Two perceptions, in particular, need challenging. The first is that public transport is for the poor only, and that transport subsidies are merely poverty alleviation instruments. The second is that taxis are a necessary nuisance which must be formalised.

Both these perceptions can be attributed in part to the current state of urban and transport planning. Over and above the planning procedures itself, the nature of the technology used in South Africa for planning purposes does not allow for a varied, non-uniform use of land. Also, on the transport side, the dynamics inherent in the taxi industry are not accounted for at all. The result is that the decision support tools used to help government decide on how to spend money to improve accessibility and mobility simply ignore, or at best misrepresent, the means and modes of the majority of the citizens for which infrastructure is planned. In most cases these are the poor and marginalised, those who most need the benefits of improved infrastructure.

As in the case of changing citizens' attitudes, progress has already been made, and a number of initiatives are noteworthy. The Gauteng Provincial Department of Roads and Transport launched state-of-the-art planning in 2011 and started to implement it to address both land use and, simultaneously, transport.

In works such as Fourie and Joubert [9], Joubert and Fourie [12], and Neumann and Nagel [16], the inclusion of paratransit in planning is very promising. It allows for business cases to be developed that show to all stakeholders—government, public transport operators, private car drivers and the minibus taxi industry—that pareto-optimal solutions are feasible and achievable. That is, for one party to provide better mobility does not imply that another party must give up some of its share. On the contrary, if more navigable networks can be designed, and mode changes become more seamless with the introduction of electronic and integrated fare collection, public transport as a whole will become a viable option for more commuters.

Conclusion

The racially segregated history of South Africa has made a deep and obvious impression on the mobility culture in Gauteng. Although the country became a democracy nearly 20 years ago, the urban landscape still reflects the past. Tension is rising among those on the periphery who are left behind in increasing economic inequality as they wait for basic services and the promise of a better future.

In this chapter we have tried to provide a comprehensive view of the minibus taxi industry as a major stakeholder in, and architect of, the mobility culture. Marred as it is by inefficiencies, bad publicity, and a lack of understanding and therefore of appreciation, the paratransit mode is indeed a ‘pain’. But a perpetual one? No, not if we establish an awareness of its unique and valuable contribution to South Africa’s transport. We acknowledge that changing the mobility culture is slow, but various initiatives have been launched, with success, and the technology is ripe for building on the momentum. The issue of whether paratransit is a pain or a mode that has potential points to a perceptual challenge that must be addressed in both government and among the public at large.

The minibus taxi industry is proactive, creative and dynamic—all key attributes that will allow it to make a unique and very valuable contribution in a society that could, or at least should, see a dramatically changing spatial landscape in the near future. The minibus taxi has a key role to play in the further development of what Nelson Mandela dubbed the Rainbow Nation.

References

1. Blais P (2010) Perverse cities: hidden subsidies, wonky policy, and urban sprawl. UBC Press, Vancouver
2. Cervero R, Golub A (2007) Informal transport: a global perspective. Trans Policy 14(6):445–457
3. City of Johannesburg (2004). Integrated transport plan: 2003/2008
4. CSIR Built Environment (2011). Geospatial analysis platform and NSDP spatial profiles. http://www.gap.csir.co.za/

5. Department of Transport (2003). National household travel survey: Republic of South Africa
6. Development Policy Research Unit (2008). Poverty and the 'second economy' in South Africa: an attempt to clarify applicable concepts and quantify the extent of relevant challenges (DPRU policy brief series PB 08–20). Cape Town
7. Dugard J (1996) Drive-on? An analysis of the deregulation of the South African taxi industry and the emergence of the subsequent "taxi-wars". Master Thesis. University of Cambridge, Cambridge
8. Fletterman M (2008) Designing multimodal public transport networks using metaheuristics: industrial and systems engineering. Master Thesis. University of Pretoria, Pretoria, South Africa
9. Fourie PJ, Joubert JW (2009). The first agent steps in agent-based transport planning. In: Proceedings of the 28th Southern African Transport Conference (SATC 2009), Pretoria, pp 43–52
10. Holland H (2011). Let's dump anti-taxi attitude. Daily News, 10 October 2011
11. Imperial (2012) I-Pledge campaign. http://www.ipledge.co.za
12. Joubert JW, Fourie PJ (2010) Growing better transport solutions: let the commuters do the thinking! Innovate 4:70–71
13. Litman T (2010) Transportation affordability: evaluation and improvement strategies. Victoria Transport Policy Institute, Canada
14. McCaul C (1990) No easy ride: the rise and future of the black taxi industry: South African Institute of Race Relations, Johannesburg
15. McCaul C, Ntuli S (2011). Negotiating the deal to enable the first Rea Vaya bus operating company. In: Proceedings of the 30th Southern African Transport Conference (SATC 2011), Pretoria
16. Neumann A, Nagel K (2011) A paratransit-inspired evolutionary process for public transit network design: working chapter 11–15, transport systems planning and transport telematics, Technical University of Berlin. http://www.vsp.tu-berlin.de/
17. Orski C (1975) Paratransit: the coming of age of a transportation concept. Transportation 4(4):329–334
18. Quantum GIS Development Team (2011). Quantum GIS geographic information system. Open source geospatial foundation project. http://qgis.osgeo.org
19. R Core Team (2011). R: A language and environment for statistical computing. R Foundation for Statistical Computing, Vienna, Austria. http://www.R-project.org
20. Republic of South Africa (1994). White chapter on reconstruction and development. General Notice 1954 of 1994
21. Roos D, Alschuler D (1975) Paratransit—existing issues and future directions. Transportation 4(4):335–350
22. Schalekamp H, Behrens R (2010) Engaging paratransit on public transport reform initiatives in South Africa: a critique of policy and an investigation of appropriate engagement approaches. Res Transp Econ 29(1):371–378
23. Smillie S, Eliseev A, Mbongwa L, Sapa (2007) This is not like war—it is war. The Star Newspaper, 29 May 2007
24. Statistics South Africa (2007) The RDP commitment: what South Africans say
25. Statistics South Africa (2011). Quarterly labour force survey, quarter 1, 2012
26. Terblanche B (2006). Roaring taxi industry beat its own path. Business Day, 19 April 2006
27. Venter C (2011) The lurch towards formalization: lessons from the implementation of BRT in Johannesburg, South Africa: Thredbo 12. In: International conference series on competition and ownership in land passenger transport
28. Venter C (2011) Transport expenditure and affordability: the cost of being mobile. Devel South Afr 28(1):121–140
29. Woolf S (2007) Taxi hand signs: documenting South African taxi and signs for the blind and sighted, Johannesburg
30. Woolf S, Joubert JW (2011) When paratransit evolves to become the answer, taxi hand signs point the way! working chapter 019 Centre of Transport Development

Chapter 7
São Paulo: Distinct Worlds Within a Single Metropolis

Marcela da Silva Costa

Abstract This chapter discusses how social inequality has framed São Paulo's mobility culture and has led to a stressed mobility system. To demonstrate this, figures of inequality as it relates to mobility are given, as well as indicators that illustrate the principal mobility problems. Social, economical, cultural and political issues that lie behind the current situation are also discussed, and stories from typical São Paulo citizens are presented in order to reveal more clearly the impact of inequality and transport policies on daily life. In conclusion, this chapter presents the measures currently being taken in São Paulo in order to address the most visible mobility problems. Moreover, it calls for further action to promote policy that is more balanced, so as to reduce the gap between the wealthy and the poor, and pave the way to a more sustainable city.

Introduction

São Paulo is a city with 11 million inhabitants, and forms the core of a metropolitan area in which more than 19 million people live [1]. It is the largest city in South America, and one of the ten largest in the world. It is also the richest Brazilian city, with the highest total GDP. This wealth was supported in the past by its industrial development, and nowadays stems from commerce and business. No other city in Brazil has the same sheer immensity, nor attracts so many people and such a volume of goods, as does São Paulo. More than 23 million passenger trips (by all transport modes) take place daily within the city [12]; annually, more than 33 million passengers pass through its airports [13]; and its car fleet already surpasses 4.7 million [14]. São Paulo lies at the intersection of some of the country's main highways, being connected by them to the major cities in the southern, central western and north–eastern regions. The city also acts as a gateway to the principal port of Brazil, the Port of Santos.

M. da Silva Costa (✉)
Vetec Engenharia LTDA, Rua Olimpíadas 100, 2° andar, Sao Paulo, SP 04551-000, Brazil
e-mail: marcelas_costa@yahoo.com.br

Institute for Mobility Research (ifmo) (ed.), *Megacity Mobility Culture*, Lecture Notes in Mobility, DOI: 10.1007/978-3-642-34735-1_7,
© Springer-Verlag Berlin Heidelberg 2013

São Paulo has gone through two significant periods of growth, which were pivotal in its historical development. The first major cycle of population growth took place between the end of the nineteenth century and the beginning of the twentieth century, when the city attracted a significant number of European immigrants. They came mostly from Italy to work in the coffee plantations, in order to replace the slave labour force that had been exploited in Brazil for over 300 years. Between 1890 and 1900 the population rose from 65,000 to 240,000 [1].

Revenues from coffee production provided the basis for subsequent industrial development, afterwards driven forward by the two World Wars. During this period industrial production was expanded to produce goods for domestic consumption. From the 1950s to the 1980s, at the height of its industrial development, São Paulo underwent a new cycle of population growth, attracting migrants from the north-eastern region of Brazil who were looking for better opportunities in the more advanced south-eastern region. This internal migration was responsible for the population explosion in São Paulo that took place at that time, causing it to rise from 2.2 million to almost 8.5 million inhabitants [1]. To this day, job opportunities, cultural and leisure pursuits, shopping centres, and the huge variety of domestic and international flights on offer continue to attract people from all regions of the country for tourism and business, and to relocate to the city.

This population growth, especially inasmuch as it has been concentrated in the periphery, has had dramatic impacts on the city's development and its mobility system. However, other issues lie behind the current urban mobility situation in São Paulo, which is deeply marked by severe congestion, an overcrowded transport system and a high number of road traffic-related fatalities. São Paulo's mobility culture is greatly influenced by the city's historical and cultural background, by the socioeconomic issues inherent in a developing country, and by political decisions framed over decades. For these reasons, São Paulo also illustrates the social inequality that still characterises Brazilian cities, notwithstanding the economic achievements of the past 10 years. The Gini coefficient, as a measure of inequality of income or wealth, gives us a way of evaluating the gap between the rich and poor in Brazil. According to the United Nations, in 2009 the country ranked 12th most unequal in the world, with a Gini coefficient of 0.54. This inequality has been particularly heightened in São Paulo by political decisions that have created two entirely different worlds, two different cities within what is ostensibly a single urban entity—one for the wealthy people and another for the poor.

This chapter will explore the main features of inequality in São Paulo related to mobility. It will also outline the main economic, social and political issues that have resulted in a city which has a mobility system designed predominantly to meet the needs of higher-income groups. The issues considered include the modal distribution of transport infrastructure investment benefiting the rich and the poor, as well as urban and housing policies in terms of how they affect mobility patterns. The chapter also discusses the steps that have been taken in São Paulo in order to address the most visible mobility problems. Finally, it calls for new ways of promoting a more socially balanced approach to transport policy, in order to reduce the mobility gap between the rich and the poor, whether residents or workers. (The term 'mobility gap' essentially indicates the difference in the ability of the rich and the poor—who are usually not car-owners and can also even lack access to public transport—to travel, resulting in poverty-related mobility impairment: an unmet need for travel, and typically a lower 'class' of travel, amongst poorer citizens).

A City of Contrasts

As a result of its history and strategic location, and of the nature of its economic activity, São Paulo is a melting pot of accents, cultures and lifestyles that make it unique in Brazil. However, this intensely active, modern, elegant and culturally vibrant city—this mirror of recent Brazilian development, with its immense resources—is fully accessible only to people with significant income.

São Paulo is a place of contrasts. The city is among the world's ten most expensive to live in [10], has the second largest helicopter fleet (surpassed only by New York), and is home to luxury armoured cars that are now increasingly seen on its streets. The opulent 'wealthy city', with its modern skyscrapers lighting up the Berrini Avenue and the banks of the River Pinheiros, stands in contrast to the sprawling 'poor city' that spreads out to the margins. The differences between the wealthy city and the poor city go further: there is also an uneven distribution of green spaces, urban services, and cultural and leisure facilities between the two. Therefore, two distinct cities effectively coexist within São Paulo, their boundaries defined by the vastly differing income and political power of the social groups dwelling in each.

The impacts of inequality on urban mobility in São Paulo are clearly reflected in the socio-spatial segregation, and the accessibility gap (the mobility gap, but looked at particularly in terms of ability/inability to reach important destinations, affected in part by where they are located), between the wealthy and the poor. On the one hand, a sizeable minority enjoy a city with freeways, tunnels and bridges, a modern subway system, and accessibility to their workplaces, which tend to be highly concentrated in the city centre and in the vicinity of their housing areas in the southwest quadrant of the city. On the other hand, the large majority of the populace—the marginalised groups—are pushed out to the periphery, where they face scarcity of jobs, a lack of basic infrastructure, and low-quality transport services (Figs. 7.1, 7.2, 7.3). This means that the people who can afford cars and higher house prices are already close to the job-rich areas. The 'expanded centre', which is where the main core of commerce and services is to be found, as well as the rich neighbourhoods in the south-western quadrant, together account for only 15 % of São Paulo's population. However, more than 50 % of the city's jobs are located there. While the average family income in these areas is around R$4,200 per month, in the extreme east side of the city it is R$1,600 [12].

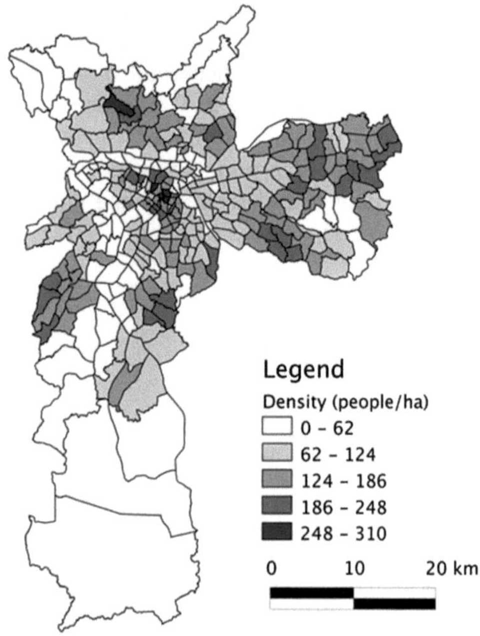

Fig. 7.1 Urban density (*Source* OD survey, METRO/SP)

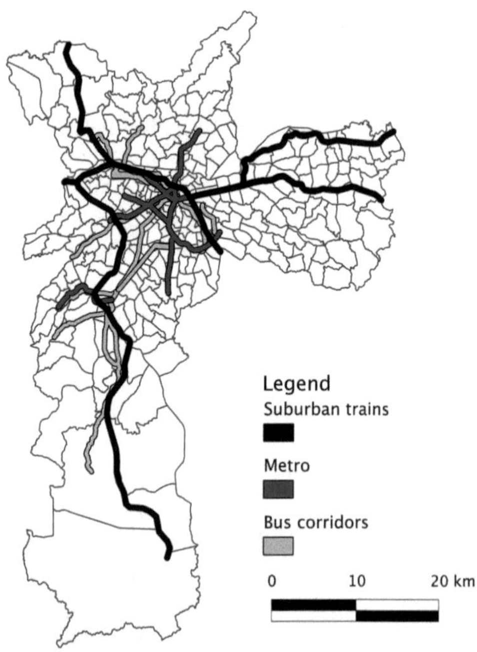

Fig. 7.2 High-capacity transit (*Source* CPTM, METRO/SP, SPTrans)

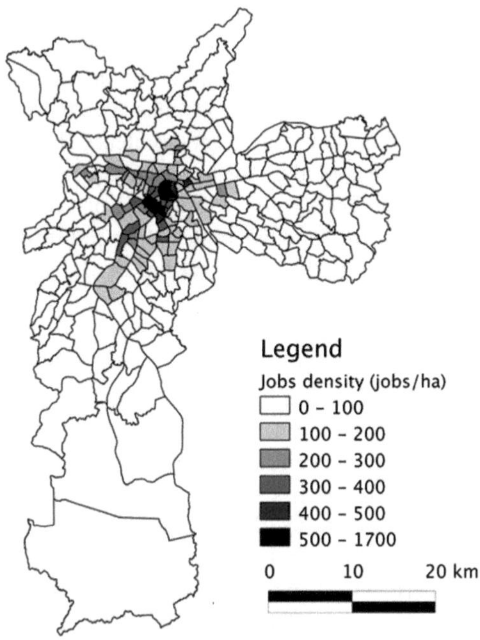

Fig. 7.3 Job density (*Source* OD survey, METRO/SP)

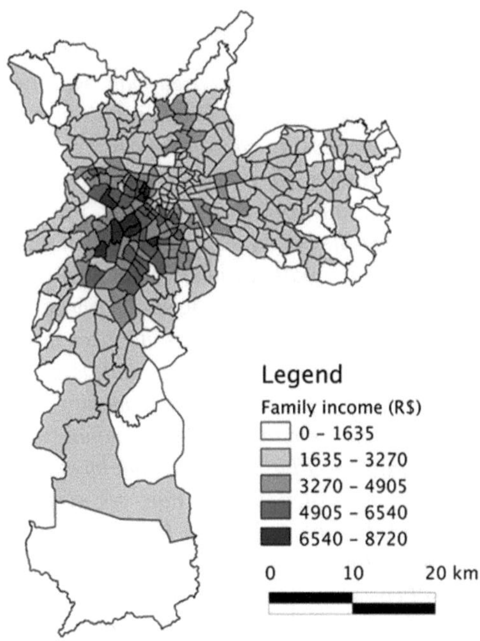

Fig. 7.4 Average family income (*Source* OD survey, METRO/SP)

The economic inequality in São Paulo, with its causes and effects, has been examined by a number of urban planners, geographers and sociologists, from the point of view of its impact on urban development and mobility [24, 25, 26, 17, 22, 23]. Now that the subject has been thoroughly discussed, the underlying issues behind this inequality become clear. The first is an upper-class predilection for certain areas of the city and, consequently, for the properties to be bought there. Secondly, there is the willingness of successive governments to pander to their wishes in order to guarantee their political support. Thus, with the government support, the property market has in effect decided where development is going to take place in São Paulo.

We have thus seen over the years the abandonment of the city centre—or the old town—which lost a significant number of residents, especially between 1980 and 2000. This has been followed by the redistribution of the high-income population to within the new borders determined by the property market as the location in which to concentrate its investments: the south-western quadrant of the city [9]. This area has been chosen by the higher-income groups as their preferred location in which to live, work and enjoy leisure activities (Fig. 7.4). At the same time we have seen the unruly growth and simultaneous deterioration in urban quality of the periphery, which, so far as public investment is concerned, has been passed over. Since the 1980s, the population of the extreme southern, eastern and northern regions has increased more than 70 % due to internal migration, in contrast to the declining numbers living in central areas of the city [9]. Several factors have encouraged this development: housing policies, urban legislation, unbalanced public investment in transport and infrastructure, and private investment focused only on satisfying the consumerist lifestyle of the high-income groups.

Living in an Unequal City

João is a 32-year-old married man, with two kids named José and Maria. Since they didn't get vacancies for their kids in a public day-care centre, his wife, Nadir, stays at home taking care of the children. Thus it is only his low-paid job that supports the family. He has been working as a valet parking attendant for 5 years, after giving up work as a 'motoboy' when the kids were born. The valet parking service operates in an elegant office building in the western side of the city. Living in the extreme south, around 25 km away from his job, daily life is not easy owing to the lack of good, fast transport services. He leaves home at 5.20 a.m. and walks for 5 min to a bus stop, where he takes a minibus. In the early morning the minibuses service is high-frequency and he gets one quickly. However, the streets are poorly maintained and narrow, making the journey longer and uncomfortable. After 30 min, he reaches a bus terminal and takes a second bus to a train station, which runs on a bus corridor. The bus services which meet the city of São Paulo are fully integrated, thus he does not need to pay the fare again. When he reaches the train station at around 6.10 a.m. to take the Line 9, which runs parallel to the Marginal Pinheiros Freeway, it is already crowded. The trains are low-frequency and often he cannot get on the first one. The platforms are crowded with people. After waiting for 10–15 min, he finally boards a train. It takes 20 min to reach the

Vila Olímpia station, where he alights. He still has 10 min more walking to finally reach his job. It is almost 7 a.m. and he has a long day ahead. Some customers arrive early because of the 'Peak Hour Operation' restrictions. Joao needs to be punctual to park their cars.

Fernando, a 36-year-old married man, works in the same office building that the valet parking operates in. His first kid, Beatriz, was born last year. With the support of nannies and with good day-care facilities close to home, his wife, Helena, is able to keep her job in a marketing office on Berrini Avenue. Every day, at the same time, they leave their apartment in a high-class neighbourhood in the western side of the city, each in their own car. She leaves the baby in the day-care centre on the way to work and he goes directly to his job. They work in the same area, around 8 km from home, but they prefer to use different cars. Fernando never knows what time he will be able to leave work, so Helena needs to be able to take the baby home from the day-care centre at the end of the day. Therefore, public transport is not an option for them. Besides, they think it unsafe and uncomfortable. Every day Fernando faces the menace of congestion and experiences anxiety about safety on the roads. He is an engineer and works in an engineering design office, which occupies an entire floor of the building. He has recently been promoted to department manager. He usually leaves home at 7.45. Before this, however, he checks the traffic information on his smartphone and chooses his route to work. Living in an area with good road infrastructure, he has various route options to help him avoid congestion. He can take either the Jânio Quadros Tunnel or the Sebastião Camargo Tunnel, in neither of which buses are allowed to travel, and which both lead to Juscelino Kubitschek Avenue. The journey to work takes around 30 min if traffic conditions are normal, otherwise it can take as much as 50 min. But either way, Fernando is comfortable in his new, well-equipped car. When he reaches the office building he leaves his car with the valet parking attendants. A free parking spot is part of his salary package now that he has been promoted. Thus he no longer needs to pay to park his car, which is one more incentive to drive it every day.

Urban Policies Reinforcing Inequality

It is high land prices that are responsible for pushing the low-income segment of the population to the periphery of São Paulo. However, the city's peripheral development has been consolidated by the way that the state government has planned housing provision. In choosing the city outskirts to implement large housing developments for those on low incomes, it has further exacerbated the socio-spatial segregation.

In the early twentieth century, urban planning began to be used as an instrument to establish a spatial order in São Paulo along the lines of class segregation. The city was socially divided between high-lying and low-lying areas, with the wealthy settling in the higher central districts—the locations of formal urban interventions—and the poor on the floodplains and along the railways. Until the 1980s, the pattern of socio-spatial segregation continued to be one of 'centre and periphery', with the middle and upper classes gravitating towards infrastructure-rich neighbourhoods, and workers confined to the peripheral areas characterised by illegally built accommodation and irregular public services. From the 1980s onwards, the centre–periphery model continued to typify the city, but the processes that produced this pattern of urbanisation changed. The emergence of apartment blocks in the peripheries and, at the same time, the proliferation and increasing density of *cortiços* (overcrowded rented slum tenements of a precarious nature, with inadequate shared amenities) in the central areas resulted in the situation whereby different social groups lived very close to each other, but separated by walls and security equipment [7].

The popular housing development named Cidade Tiradentes, built in the 1980s at the eastern end of the city at a distance of almost 25 km from the city centre, neatly exemplifies the approach to housing policy in São Paulo. Today, more than 230,000 people live in its core and the area immediately surrounding it, occupied by slums and illegal settlements constructed on private land. When it comes to public investment in infrastructure, this area has been overlooked, and lacks basic services and high-capacity public transport. In the absence of any economic/policy incentives and of suitable infrastructure, private businesses—which might have brought with them local job opportunities—have avoided this area, further aggravating the structural social exclusion.

At the same time as housing policies have led to the removal of low-income families to the outlying areas, urban legislation has come into being which separates off and protects the wealthier areas of the city [6, 15, 16]. This separation has been effected by means of two distinct approaches: demarcating certain areas for single-family housing, and establishing high standards for construction. These two measures have been applied for the most part in the south-western quadrant of the city. In practice, they have served to preserve these areas exclusively for residential use and local business activity, and to maintain high property prices in them, as well as preventing their occupation by 'undesirable' social groups.

The Role of Transport Planning in Creating Two Different Cities

Among the processes that have contributed to the creation of these two different cities within São Paulo, and have consequently increased the mobility gap, is the imbalance between the two of public investment in transport and infrastructure: this has further reinforced the boundaries between the wealthy city and the poor city. While the wealthy areas have received abundant investment, the periphery has undergone very few improvements. Transport plans, instead of investing in rail-based services connecting the periphery to the city centre, have focused on improving the road network and extending the metro network precisely in the highest-income areas. As a consequence, suburban trains do not comprehensively penetrate the densely populated areas located on the periphery, which suffer from low-frequency and irregular services. Currently, only 13 % of São Paulo's population live within 1 km of a metro station. The same is true of the dedicated public transport lanes, which are also highly concentrated in central areas of the city.

Not only has public investment been concentrated on the areas preferred by high-income groups, but it has also focused on private transport infrastructure. The car has established its position as the status symbol of the upper classes, as well as being the means by which they are able to keep their distance from the poorest classes [5]. For more than four decades, the central focus of transport planning in São Paulo has been to improve car traffic flow. From 1967 to 1977, 27 % of the total city budget was used for this end. Investment in road infrastructure remained high in the 1980s; however, the most substantial investments were made between 1990 and 1997, mostly in building tunnels and expressways [26]. Among these, two projects are notable for being implemented exclusively for the benefit of cars: the Ayrton Senna Tunnel, concluded in 1996 and located in the affluent neighbourhood of Moema; and the cable-stayed Octavio Frias de Oliveira Bridge, completed in 2008 and located on the banks of the Pinheiros. Neither in the tunnel nor on the bridge are pedestrians, cyclists, buses or heavy goods vehicles allowed to travel.

Taking advantage of the ease of access and high-quality transport infrastructure there, private investors have shifted their attentions towards the south-western quadrant. Their investment activities, however, have met the needs specifically of high-income consumers. Luxury offices and housing, as well as shopping centres dedicated to international brands, beautify the region. In some areas—such as along the Faria Lima Avenue—the frequent improvements, together with property speculation, have led to very high rents, currently exceeding those of cities such as Frankfurt and New York. According to the Global Office Real Estate Review published by Colliers International in 2010, São Paulo ranked as the eighth most expensive in the world for rental prices for 'Class A' offices (state-of-the-art high-quality prestigious buildings of a 'significant' size and in a prime location) [4].

Following the investment in road infrastructure, car use in São Paulo has become increasingly widespread. However, this is not only a consequence of the increase in vehicle affordability and of long-term infrastructure policies—it also results from other public policies (or a lack thereof) and market forces, which have created ideal conditions for the rapid expansion of car use. These 'stimulus packages' have included benefits at both federal and local level. At the federal level, car ownership has been facilitated by lower interest rates and greater availability of credit. In addition to the effect of new car manufacturers entering the Brazilian market, tax-reduction policies applying to new vehicles have also resulted in a slight lowering of prices. At the municipal level, the stimulus packages have focused on improving traffic flow for cars, lowering parking costs, and ensuring the wide availability of roadside parking.

Despite the efforts of the local administration, these constant alterations to the road system to improve car traffic flow have not yielded the expected benefits, since their effects have often been offset by an increase in the number of cars on the road. However, these projects have had a significant impact on road safety and the quality of the urban environment. When São Paulo chose to invest significantly in infrastructure for private transport—especially in tunnels, bridges and freeways for the exclusive use of cars—it created a very unfriendly environment for pedestrians and cyclists.

High-Income Groups' Attempts to Control Transport Decisions

The improvements to the areas occupied by the wealthy minority have only been possible because of the huge gap in political power that exists between high- and low-income citizens. Furthermore, the former group plays a strong role in political decision-making in São Paulo, especially when it comes to provision of road infrastructure. This reflects a situation common throughout the cities of Brazil: the wider the gap between social classes' economic and political power, the stronger the resulting upper-class segregation, and the stronger the role that their segregated area will in turn play in shaping the whole of the urban spatial structure [27].

Not content with exerting a noticeable influence on the road infrastructure provision, the high-income groups have also tried to use their power to control the provision of public transport. The main manifestation of this can be seen in the expansion of the metro network. The most recent episode involved the design of Metro Line 6, which runs through a well-to-do neighbourhood close to the city centre. The wealthy residents requested the relocation of a planned station away from the centre of the Higienópolis District. They feared an influx of 'objectionable types' into the area, stemming from the accessibility provided by the proposed station. The way that the Metro Company of São Paulo initially complied by changing the station's location unleashed a wave of debate on virtual social networks, and protests in the city. These manifestations culminated in an animated demonstration on the streets of the district, involving placards deriding the attitude and lifestyle of the groups opposed to the construction of the new station [2].

The implementation of cycle lanes has also faced some resistance. A number of these have been implemented in parts of the city, including some of the upper-class localities. In the Moema district, shopkeepers became concerned about their business. They feared that their customers would disappear as a result of cycle lanes in front their shops leading to narrowing of the streets and difficulties in finding parking spaces. Their complaints, too, became the subject of widespread discussion, and received much media attention [21].

The 'metro episode', as the first event has been labelled by the media, together with the strong reaction against the implementation of the cycle lanes, shows how São Paulo's elite has tended to act to protect the boundaries of the wealthy city. Fortunately, in both cases the will of the majority won the day by means of popular demonstration. The metro station is now planned for a site a few metres from the one originally proposed. The cycle lanes have been implemented in the Moema district, despite the complaints of shopkeepers. These episodes, at least, show that the city's general populace is becoming increasingly engaged in issues relating to mobility, and also that discussion of these matters has expanded to now involve diverse sections of society. Moreover, they have raised deeper questions, which can help us to rethink how mobility policies have been framed in the city and how civic society has been affected by these decisions.

A Stressed City

The economic inequality and political decisions discussed above, in pushing people of low income to the periphery and creating such imbalance between the location of jobs and housing, have created an unnecessary demand for travel. The resulting explosion in traveller numbers, whether by private or public modes, has put extra pressure on this already unbalanced transport system. São Paulo's mobility system is thus profoundly stressed. It faces daily the threat of collapse arising from severe traffic congestion or even total gridlock; intensely overcrowded public transport; very long travel times; and shockingly high levels of traffic-related fatalities, especially of pedestrians and motorcyclists.

The increase in the affordability of vehicle ownership over the last 10 years has not in itself been enough to reduce the social and economic disparity that still blights Brazilian cities. Moreover, it has had a devastating effect on urban mobility, especially by making car ownership easier. Since 2000, the numbers of cars and motorcycles in São Paulo have increased by 25 and 36 % respectively. During 2010 an average of 12,000 cars and 4,000 motorcycles were registered per month in the city [14]. The number of motorcyclists, in particular, has been expanding rapidly in the hunt for ways to avoid congestion. Currently, motorcycles form the backbone of urban logistics for small-load transportation, owing to their ability to 'pierce' congestion. However, the image of the motorcyclist is associated with indifference to the rules of the road, and with accidents, death and criminality, because of the increasing numbers of armed robberies involving motorcycle riders.

Stressed Drivers

Congestion is fast becoming the defining feature of São Paulo, whether in the context of the rest of Brazil or the whole world. Traffic and transport are frequent subjects in the media and favourite topics of day-to-day discussion. Since

1997 there have been a raft of measures implemented to reduce peak hour traffic in São Paulo. The 'Peak Hour Operation' consists of the restriction of traffic within the limits of the area defined as the expanded centre, effecting a ban on certain vehicles during the peak hours (7–10 a.m. and 5–8 p.m.) depending on the last digit of the number plate and the day of the week. It is estimated that around 20 % of the car fleet do not move through the expanded centre during this time. However, this system has not been enough to relieve peak hour congestion.

Congestion figures in São Paulo are reported by length of gridlocked roads. Currently, only 868 km of roads are monitored, but they represent the most important arteries of the city. Even short extensions of the total length of gridlocked roads means longer travel times. In 2009, the congestion extension broke the record twice in one day, reaching 294 km. Over the last decade the average length during the peak hours has been 118 km. From 2000 to 2008 the average traffic speed during the peak hours was 19 km/h [17]. Over more or less the same period, the average travel time by car rose from 27 to 31 min [12].

Congestion has affected both the economy and the logistical efficiency of the city. The annual loss due to traffic congestion in São Paulo is estimated at US$21 billion; this includes time lost by workers, increased fuel consumption by fleets, a value placed on health problems related to air pollution, and the cost of delays in the delivery of goods [3]. Congestion has also made a deep impact on people's everyday life. The social losses of congestion are poorly documented, but it is estimated that São Paulo's citizens spend on average 27 days per year stuck in congestion [17]. This time could have been spent on education, leisure, cultural activities, or simply with the family.

Stressed Public Transport Users

Those on lower incomes need to travel much longer and further to overcome the spatial segregation imposed by their housing location and by the lack of good transport services. São Paulo has no more than 350 km of high-capacity public transport, distributed between suburban trains, metro and dedicated public transport lanes. This adds only 2 % to the road network, which is over 17,000 km long.

About 95 % of São Paulo's population lives within 500 m of a bus stop [20]. The bus service, despite its wide coverage, is unreliable, offers limited connections with the rail-based services, and is often dogged by traffic problems. Between 2000 and 2008, the average time spent per person travelling by public transport rose from 59 to 67 min [12].

The design of high-capacity public transport only began in the 1960s, when the city already had almost 4 million inhabitants and traffic congestion was already fairly commonplace. In 1975 the first metro line began operation in São Paulo, 17 km in length and with 20 stations, linking the northern and southern regions of the city. Since then, only 54 km of new metro lines have been built in the city. Despite the low coverage and the present overcrowding, the metro network in São Paulo is considered to be an international point of reference by dint of the quality and cleanliness of its stations. It also enjoys a wide acceptance by the population of the city.

As a result of the slow pace at which high-capacity public transport has been expanding, measures of public transport infrastructure per inhabitant in the metropolitan area have decreased considerably over the years. In the metropolitan area, which takes in the city of São Paulo and 38 municipalities, the total length of public transport infrastructure per inhabitant fell from 38 km/million people in 1967 to 24.7 km/million people in 1997 [24], and currently stands at 24.2 km/million people. The consequence is congestion levels that spiral ever upwards and a public transport system that suffers intolerable overcrowding.

Public transport occupancy and crowding levels have also attracted attention, even internationally. In 2009, suburban trains and metro lines together carried almost 1.3 billion passengers. In 2010, as a result of high demand and limited supply, the metro lines in São Paulo reached intolerable occupancy levels. According to the CoMET—a programme of international railway benchmarking—São Paulo's metro system is the most crowded in the world [8]. This excessive demand, together with poor network connectivity, has overloaded the main stations and made the daily lives of passengers in São Paulo very difficult. The bus system, too, has faced overcrowding, in addition to long travel times due to the lack of dedicated bus lanes. For all these reasons, people still see the car as the best option for their trips, despite congestion.

The effects of these mobility-related problems, however, are felt unevenly by the poor and the rich. Despite the growth in mobility over the last decade, the poorest travel less frequently than the wealthiest. While the latter make on average 2.7 trips a day, the former make only 1.5 trips a day. However, since they use public transport, the travel time of these poorer citizens is on average 20 % longer than that of the rich. The burden of transport costs as a proportion of income also varies between these two groups: while the poorest people spend on average 30 % of their income on transport, the wealthiest people spend only 18 % [12].

A Stressed Mobility System, and Stressed People

The consequences of living in such an unequal and stressful environment, taken together with a well-established social intolerance [5], are plain to see in the attitudes manifested on the road traffic and public transport infrastructure and in urban spaces. Cars, motorcycles (which usually weave in and out of the lanes of traffic) and pedestrians compete for space on the crowded streets of the city. In this lop-sided conflict, pedestrians and motorcyclists have proved the most vulnerable. In 2010, the victims of 80 % of all traffic fatalities and injuries that occurred on São Paulo's streets were pedestrians and motorcyclists (evenly split between the two) [18]. The present situation in São Paulo is fairly grim, then, and is even more stark for those of low means, who do not enjoy a privileged position in a mobility system that is designed for the rich and powerful.

Travelling in a Stressful City

It is 4 p.m. and João is allowed to leave work after an 8 h day. However, he is used to staying on for an extra 2 h every day so he can clean customer's cars. This gives him some extra money with which to support his family. On the other side of the city, his wife tries to distract José and Maria during the day with TV programmes and their few toys. When she needs to go out for whatever reason, she relies on her neighbour to look after the kids for while. She usually goes to a small neighbourhood market, where she can buy make some purchases for which she doesn't have to pay until the following month, when João gets his salary. To get there, she needs to cross a very busy road. The time allowed for crossing the street is short, so she needs to be fast. Because of this, she prefers keep the kids at home. For any other trip purpose, Nadir needs to take the bus to get around, since there are not many amenities close to home. When she reaches home, she greets her neighbour and her kids in front of the house. She can spend some time talking to them, along with some other neighbours who pass by that way.

In the western side of the city, when João leaves his work, it is already 6 p.m. The streets are full of people and cars. He walks to the train station. The pavements are irregular and poorly maintained, even in this wealthy neighbourhood. Such is their poor state of repair that many times he has to walk in the road. When he reaches the train station for Line 9, the entire length of the platforms are crowded with people. It is common for people to suffer minor injuries while waiting for the train. At last he catches his train, but the 20 min trip seems to take an hour, such are the uncomfortable conditions. He gets out of the train and goes to the bus terminal. In the evenings, the bus corridor, which links the terminal to the southern region, is always busy. By the time he gets off the bus he has already travelled for almost one and a half hours, but João is still far from home. For the last part of his trip, he takes a minibus. When he finally reaches home it is already 8 p.m. Tired, and aware that his next journey will start very early tomorrow morning, João spends only few minutes with the kids. His wife takes the opportunity to watch her favourite soap opera. During the breaks, she sees many advertisements for cars. She wonders when Joao will finally be able to afford a car, and hopefully arrive home earlier and less tired.

It is 6 p.m.and Fernando thinks about going home. He checks the traffic conditions on the Internet. There are 90 km of gridlocked streets, 10 km of them on the western side of the city. Fernando decides to stay in the office for longer and to wait for better traffic conditions. This has become common for the last 2 years. He would rather spend more time working in the office than stuck in congestion. When it is 7 p.m., Fernando checks the traffic again and decides to leave, but not before also looking up the best route. Even though the traffic is still heavy, he has options. He decides to take the Marginal Pinheiros Freeway then the Cidade Jardim Bridge. He leaves the office and drives through the neighbourhood towards to the freeway. The traffic is slow and the drivers are impatient. Many people are walking on the road. After 10 min, he reaches the freeway. He needs to be careful because many motorcyclists are riding fast between the lanes. Traffic accidents are common on this section of road. When the traffic slows down, he looks out of the car window and sees the trains running on Line 9. The train is completely crowded and he wonders how people can cope with it. After 10 min on the freeway he reaches the bridge and passes over the Pinheiros. Only 10 min more on the local roads and he is home. Not so bad for a weekday, thanks to the traffic information and the weather conditions. It's not raining this evening—if it were, goodness knows how long he would take to get home. Before reaching home he stops in a bakery and gets some pastries. There are many options for buying delicious food, as well as an abundance of essential day-to-day services close by. There are also plenty of restaurant delivery services, from which the family can order food without the effort of going out to get it. It is about 7.40 p.m. when Fernando meets his family, so he can spend some time with them before going to bed. Tomorrow morning he needs to wake up earlier because it is Thursday and his car is not allowed to move between 7 and 10 a.m. because of the Peak Hour Operation.

The Beginnings of Change

As demonstrated above, the mobility problems of São Paulo have worsened rapidly, a situation that has been reinforced by past omissions and mistakes in urban and transport planning. However, a broad response seems to be taking place. This has been seen in a very large-scale public transport programme, with steps taken in parallel by the state and the local governments. By means of this process, they have attempted to overcome some of the institutional barriers and thus to develop more coordinated strategies.

Shifts in government's perception of the issue have been crucial in supporting these measures. Firstly, it has been understood that strategies centred on road network improvements and traffic engineering measures are not working to solve the congestion problem in São Paulo. Secondly, it has been recognised that endlessly supporting private transport is no longer sustainable, since this has led to a stressed mobility system. Consequently, the crucial role of multimodality and public transport in addressing the city's mobility problems has been recognised. However, before it can become a real alternative to private transport, the public transport network needs to be both expanded and enhanced so as to meet certain quality standards.

With these perceptions as foundational, a comprehensive programme to improve the transport system in São Paulo is being developed. Its backbone is the extension of the metro network, and the improvement of the suburban rail system in order to expand its transport capacity. After 8 years without any expansion of the metro, the Metro Company of São Paulo inaugurated in 2010 the extension of Line 2, following in 2011 with Line 4, linking the city centre with the western region. Extension of existing lines and also construction of further new lines are at design stage. The goal is to move from the current 74 km of metro lines to 163 km by 2020, and to improve the quality of the railway lines [11].

Dedicated public transport lanes are also undergoing design and implementation. In some bus corridors, low-emission fuels have been tested. Other medium-capacity systems to connect the periphery to the city centre are also being designed. Some technologies that have been chosen, however, are seen as somewhat controversial—for example monorail, as it will not be able to provide the necessary capacity to meet the high demand in these areas, and because of its urban impact. Nevertheless, these measures do at least have as their aim the promotion of a more balanced distribution of infrastructure and transport services within the city, in contrast to the policies adopted over many decades in São Paulo.

The perceptions of non-motorised modes of transport have also been changing, motivated in part by the alarming rate of traffic accidents, but also by new and emerging concepts of transport in general. In São Paulo, as in Brazil as a whole, the image of cycling as an activity associated with poverty is now being reconsidered, not least because of the achievements of cycling organisations and their involvement in promoting 'green' ideals and a healthy lifestyle. Moreover, the government has tried to sell the idea of cycling as a new means of transport, in tune with the concepts of sustainability that are gaining ground worldwide.

This has resulted in specific initiatives aimed at non-motorised modes. The plans include providing the city with an extensive cycling network. In less than 4 years, the city's network has been transformed from consisting of a few kilometres of cycle paths, confined to parks and green areas, to 47 km of cycle paths distributed throughout the city. Cycle lanes in shared streets add another 22 km to the system, but these are open only on weekends and need special operation by the authorities to function (their boundaries have to be marked out and their use controlled by enforcement officers). The number of cycle facilities integrated with bus terminals and train stations has also been increased. However, in spite of this progress in terms of infrastructure provision, the coexistence of bicycles and cars on the road network in São Paulo still faces great challenges, due to the features of the road system and also driver and motorcyclist behaviour.

To improve walkability is also on the city's transport agenda. However, while provision of good infrastructure is the focus for bicycle users, for pedestrians the methods centre more on public awareness campaigns—enforcement and road safety—with the aim of improving driver/pedestrian interaction and creating a safer environment for non-motorised road users. Established in 2010, a 'Pedestrian Protection Program' has created 'Zones of Maximum Pedestrian Protection' (ZMPPs), distributed between the city centre and Paulista Avenue, the financial heart of the city. In these areas, both specific signposting and enforcement officers directed the drivers to give way to pedestrians on street crossings. After some weeks of being shown the warnings, the drivers began to be fined for any traffic violations involving pedestrians. A comprehensive awareness campaign has been conducted by media as well. Small physical interventions have also been made in these areas, including illuminating pedestrian crossings, removal of obstacles on pavements, and action to ensure that all the crossings are accessible for users with special needs [19]. Unfortunately, these interventions can only be seen in the central areas of the city—the periphery continues to be ignored when it comes to such measures.

Separately from the response drafted by the public administration, the population seems to be taking its own initiative in looking for new forms of participation in the decision-making process; virtual social networks and the media have played an important role in keeping people informed about and engaged in mobility-related matters. The metro episode is in part a reflection of this transformation. However, São Paulo's population is still a long way from being able to access a truly participatory system whereby they could effectively influence the policy-making process, either on mobility issues or urban affairs. Policymaking remains strongly influenced by the property market and class structures, representing the interests of the higher-income groups.

It is still too early to see what all these processes achieve. However, an initial accomplishment was the reduction of daily trips by private modes from 53 % in 2002 to 45 % in 2007 [12]. Traffic accidents involving pedestrians and cyclists have also decreased in the past 2 years, but they still represent a significant proportion of victims. Mortality rates of motorcyclists, however, have not ceased to rise. Therefore, progress is being made at a slow pace, is under constant threat as a result of changes in the political situation, and relies on institutional coordination between local and metropolitan authorities and on a permanent funding framework to ensure long-term investment in transport projects.

A Long Way to Go

In São Paulo, there has been a gap in transport provision, whether in terms of meeting the needs of a megacity in transformation, or those of the various social groups that live in it. The most recent initiatives by the public administration have attempted to overcome this gap by intense investment in public transport. In addition, there has been a set of misguided policies that have entrenched economic inequality and, consequently, increased the mobility gap between rich and poor. To remedy this, both state and local governments have tried to promote a more balanced public investment across the city's regions and among all transport modes. Thus, the action taken in city development has aimed to address the most visible mobility problems, those that have sprung from the wrong decisions of the past. These measures have been proposed on the basis of a traditional transport planning approach: investment in infrastructure and transport services.

The historical and cultural background of the city have also played a key role in the making the mobility framework in São Paulo what it is today. However, these issues have not received adequate attention in the mobility strategies developed in

São Paulo—nor indeed in any other Brazilian city even to this day. Since mobility culture involves objective dimensions (transport supply, socioeconomics and policymaking) and subjective dimensions (perceptions, attitudes and lifestyles), action to really change it has to go beyond the typical improvements to public transport that are currently underway. Therefore strategies that can translate into an effective change in habits and perceptions towards mobility must embrace ideas that have not hitherto been considered in traditional transport and urban planning in Brazilian cities.

The key issue for São Paulo in creating a more sustainable city is bringing about a greater balance between land use and transport, as well as a balanced mobility system where the various transport modes each play their different roles effectively and are well integrated. This involves people having genuine transport options to choose from, without any kind of social, geographical or economic constraints. It also involves the embracing of a multimodal system in which the complementarity of transport modes is recognised.

The interfaces between private and public transport, and those between motorised and non-motorised modes, should be improved, in order to encourage more efficient multimodal travel. Further, integrated transport facilities and fare policies should be jointly considered so as to promote combined transport services. The benefits could be seen in reduced travel times and costs for all transport users, but there would also be environmental and economic benefits related to the lower fuel consumption that would result.

Communication strategies also need to be considered. These are crucial in spreading new concepts which encourage a more rational car use and promote multimodality. Among these concepts are mobility services, car-pool programmes, and car and bike sharing, which are still not widespread in Brazilian cities. For this to happen, it is essential to establish partnerships with the private sector in order to foster these mobility concepts. Additionally, virtual social networks and the media should be used to disseminate these ideas.

Innovative strategies to tackle the congestion problem should be carefully analysed. Congestion charging has been one of the alternatives discussed, though the idea has lost influence in recent years. The main objection raised to it is the current inability of the public transport system to absorb any increase in demand that might result from implementation of congestion pricing policies. However, some public and private stakeholders are in favour of a comprehensive discussion on the issue. They argue that it could result in foreseeable advantages such as reducing travel time for bus and car users, and could provide revenues with which to improve the public transport network [28].

So far, however, there has not been any discussion in São Paulo about controlling growth of the car fleet. To limit—or discourage—car ownership in a country where the population has experienced socioeconomic achievements in the past few years, and where car ownership implies social status, would certainly prove very unpopular. More than that, it would be counter to the federal policies under development, which have encouraged the motor industry. This would thus be a political risk that no local government would be willing to take on.

Another big challenge for São Paulo is that of uncoupling public transport, and non-motorised modes of transport, from their association with poverty and underdevelopment. For this to happen, the image of the public transport system needs to be changed. The first step in achieving this has been accomplished with the extension of the railway-based system. However, the bus service, which serves the largest part of the city, needs to be improved. To make it a genuine alternative to private transport, several changes will be essential: expanding the availability of information in bus stops; improving the vehicles' comfort features; testing and giving qualifications to bus drivers; and implementing operational improvements to make the service both faster and more reliable. At the same time, cycling needs to be recognised as a means of transport suited to all social classes and lifestyles, and pedestrians must be respected on the roads in order to improve safety standards.

The geographical mismatch between housing and jobs also needs to be addressed by effective policies to promote, on the one hand, housing programmes in central areas and, on the other hand, a more balanced distribution of jobs and amenities throughout the city. Besides improving housing conditions, it is necessary to review the pattern of public housing location in the periphery and to increase public housing supply in the expanded centre, where the jobs are located. This has to be followed by mechanisms for land-price control and social rental programmes to ensure affordable housing for low-income people in areas that are well serviced by public transport and infrastructure.

'Recycling' abandoned buildings for residential purposes and creating special zones of social interest are strategies already being considered for inclusion in the city's plans. However, it is also necessary to develop strategies for the periphery, founded on incentives which will attract economic activity and so expand the number, and quality, of jobs there. Housing measures already have their legal basis in the City Statute, a federal law passed in 2001. This law provides consistent legal backing to the municipalities, to enable them to tackle urban, environmental and social problems, using innovative guidelines and tools for urban planning and management. Even though the City Statute has been on the books for more than 10 years, to translate the tools that it provides into concrete action is still a goal yet to be achieved by either São Paulo or other Brazilian cities.

There is thus a long way yet to go to in the struggle to change mobility culture in São Paulo. A new way of thinking about the city and its mobility system will be required. If São Paulo does not wish to go the way of the typical megacity in a developing country, it will need to embrace new tools and policies in relation to land use, car use, and a more democratic use of public space. These steps, however, should not be confined to the sphere of the public administration. São Paulo is looking for a new urban mobility *identity*, which requires the people of the city also to rethink some of their labels and prejudices. If the populace of São Paulo will not start questioning its traditional habits and perceptions, the city will be prevented from moving on to become a more balanced, fairer and more sustainable community.

São Paulo: A Vision of the Future

Twenty years have passed since São Paulo reached intolerable mobility problems, which led to a massive backlash throughout the city. Since then, deep changes in urban and mobility features are evident, which have led to a very different scenario.When José was born 23 years ago, his family, living on the perimeter of the city, did not enjoy a good life. However, he grew up to see the railway line being extended right into the neighbourhood, and various infrastructure and housing improvements, which helped to change his family's life and that of many other families in the locality. Taking advantage of a line of credit from the bank, he and his family moved from their small house into a more comfortable flat closer to the expanded centre. In this area, well serviced by public amenities, both he and his sister Maria could attend a better school. Now, he is on the point of graduating with a degree in Business Studies, and can already help support his family out of his salary as intern in a bank on Paulista Avenue. His daily trips, both that from home to university and the one from university to workplace, take a maximum of 30 min each. This is only possible thanks to a comprehensive public transport improvement programme undertaken over the past 20 years. High-capacity public transport currently covers a large part of the city. They are more than 330 km of rail-based lines spreading out to the metropolitan area, with high-frequency comfortable trains. New stations have been built in the highest-density areas, and many others have been redesigned to meet accessibility requirements. Dedicated public transport lanes have also been expanded through the city, making travel by bus quicker. Metropolitan public transport is now fully integrated, meaning that fares are affordable and connections faster. Train and metro stations in the periphery, and in strategic points throughout the city, are equipped with parking and bike storage facilities, making multimodal travel easier. All of these improvements have greatly enhanced the image of public transport, and it is now widely perceived as comfortable, safe and reliable.

Beatriz, in turn, grew up on the western side of the city, attending good schools and with access to all sorts of public and private facilities. During her childhood, her parents were preoccupied by the need to protect her from the city's dangerous streets, and so took her to all her activities by car. This started to change when she began to attend a preparatory course at college. Seeing the transformation which the city was undergoing, her parents became more relaxed about safety, and started to let her travel on her own by public transport. Beatriz and José study in the same university in São Paulo and have similar daily routines: go to the university, and then on to an internship in the city's financial centre. Sometimes Beatriz hitches a ride with her parents to the train station, then takes a train to the university. On other occasions, she rides her bicycle and keeps it in a bike storage facility in the train station. From her parents' perspective, riding a bicycle on the busy and dangerous streets of São Paulo was impossible a few years ago. Now cars, motorcycles and bicycles coexist on safer streets. Many years of infrastructure improvements for non-motorised modes of travel, together with enforcement of the law and awareness campaigns, have created a safe environment for all cyclists.

New concepts of mobility have been promulgated as well. Universities, public organisations and large companies have developed car-pool programmes, whereby two or more people drive together in a privately owned vehicle. Some of Beatriz's classmates usually car-share to get to the university. Other mobility services offer different mobility alternatives for those who don't own a car. Beatriz got her driver's licence last year. Now she is able to enjoy all these services, which are by now familiar to the people of São Paulo. She can even obtain the use of a car from her parents. However, she is perfectly satisfied with the transport ser vices in São Paulo and not discontent with the option of riding her bike on the city streets with safety.

There is no reason why this sketch should not be a realistic representation of the daily lives of two people of differing backgrounds in São Paulo one day, if steps are taken to reduce the existing social and economic chasms—and consequently bridge the mobility gap—between poor and rich, between centre and periphery. This is the future that all São Paulo's citizens are looking for, and the legacy that both government and society should leave to the generations to come.

References

1. Brazilian Census Bureau (2010) Population census. http://www.ibge.gov.br/english/
2. Cimino J (2010) Moradores de Higienópolis, em SP, se Mobilizam Contra Estação de Metro. Folha de São Paulo. http://www1.folha.uol.com.br/cotidiano/782354-moradores-de-higienopolis-em-sp-se-mobilizam-contra-estacao-de-metro.shtml. 13 Aug
3. Cintra M (2008) Os Custos Dos Congestionamentos Na Capital Paulista. Revista Conjuntura Econômica 1:1
4. Colliers International (2010) Global office real estate review: first half of 2010. http://www.colliers.com/Country/UnitedStates/content/globalofficerealestatereviewmidyear2010.pdf
5. DaMatta R (2010) Fe em Deus e Pé na Tabua—Ou Como E Por Pue o Trânsito Enlouquece no Brasil. Rocco, Rio de Janeiro
6. Feldman S (1996) Planejamento e Zoneamento: São Paulo 1947–1972. Thesis, São Paulo
7. Fix M, Arantes P, Tanaka GM (2003) Understanding slums: case studies for the global report 2003. São Paulo. http://www.ucl.ac.uk/dpu-projects/Global_Report/cities/saopaulo.htm
8. German E (2011) São Paulo has World's Most Crowded Metro. Global Post. http://www.globalpost.com/dispatches/globalpost-blogs/bric-yard/sao-paulo-has-worlds-most-crowded-metro. 25 Apr
9. Hughes PJA (2004) Socio-spatial segregation and violence in the city of São Paulo: references for policymaking. http://www.scielo.br/scielo.php?pid=S0102-88392004000400011&script=sci_arttext
10. Mercer Consulting (2011) Cost of living survey. http://www.mercer.com/costoflivingpr
11. Metro Company of São Paulo (2006). Essential metro network by 2020
12. Metro Company of São Paulo (2007) OD Survey: Pesquisa Origem e Destino. http://www.metro.sp.gov.br/metro/numeros-pesquisa/pesquisa-origem-destino-2007.aspx
13. National Civil Aviation Agency (2009) Annual statistics. http://www2.anac.gov.br/estatistica/anuarios.asp
14. National Traffic Department (2010) Statistics. http://www.denatran.gov.br/frota.htm
15. Nery Junior JM (2005) O Zoneamento Como Instrumento de Segregação em São Paulo. http://www.cadernosmetropole.net/download/cm_artigos/cm13_68.pdf
16. Rolnik R (1997) A Cidade e a Lei: Legislação, Política Urbana e Territórios na Cidade de São Paulo

17. Rolnik R, and Klintowitz D (2011) Mobilidade na Cidade de São Paulo. Estudos Avançados
18. São Paulo Traffic Company (2010) Traffic accidents: annual report
19. São Paulo Traffic Company (2011) CET 35 Anos
20 São Paulo Transport (2011) General Information about Bus Service
21. Spinelli E (2011) Nova Ciclofaixa Cria Polêmica na Região de Moema, em SP. Folha de São Paulo. http://www1.folha.uol.com.br/cotidiano/996142-nova-ciclofaixa-cria-polemica-na-regiao-de-moema-em-sp.shtml. 25 Oct 2011
22. Taschner SP, Bógus LMM (2001) São Paulo, uma Metrópole Desigual: EURE (Santiago)
23. Torres HG (2006) Residential segregation and public policies: Sao Paulo in the 1990s. Revista Brasileira de Ciências Sociais, São Paulo
24. Vasconcellos EA (2001) Urban transport, environment and equity—the case for developing countries. Earthscan, London
25. Vasconcellos EA (2005) Transport metabolism, social diversity and equity: the case of São Paulo, Brazil. J Transp Geogr 13:329–339
26. Vasconcellos EA (2005) Urban change, mobility and transport in são paulo: three decades, three cities. Transp Policy 12:91–104
27. Villaça F (2000) Segregation in the Brazilian metropolis: international seminar on segregation in the city. Cambridge, MA
28. World Business Council for Sustainable Development (2009) Mobility for development: São Paulo, Brazil. http://www.wbcsd.ch/DocRoot/St8vWLi78ZMluEDgeUCi/M4D_S%E3oPaulo_080509.pdf

Chapter 8
Atlanta: Scarcity and Abundance

Laurel Paget-Seekins

Abstract The Atlanta, Georgia metropolitan region exists solely as the result of the intersection of transport infrastructure—railways, interstate freeways, and the world's busiest airport. However, neither vehicle drivers nor public transport passengers are happy with the state of their mobility. Racism, poor planning, and rapid economic growth have worked together to create a low-density urban fabric and a transport network that limits accessibility. Atlanta's development cannot be separated from its history of racial and class segregation. Mobility in Atlanta suffers from an abundance of vehicles on the roads and a scarcity of public transport services. Atlanta built roads to serve its low-density suburbs, and refused to invest in public transport. Drivers are inconvenienced, but the public transport passengers, mostly poor and Black, have limited access to the majority of the region.

Introduction

Atlanta, Georgia is a tale of two regions—one in which people complain about traffic, and one in which people complain about public transport services. No matter what their method of mobility is, Atlantans are unhappy about it. The traffic report is often the top story on the website of Atlanta's daily newspaper. Reports warn of when to expect abnormal levels of congestion, and when an interstate is blocked owing to an accident, it is breaking news. Mutual frustration at the traffic situation provides Atlantans a basis for unity, and a convenient topic of conversation.

Public transport passengers are, likewise, brought together by their frustration at the quality of the service. If strangers start a conversation, it will probably be to complain about how the bus is always late, or to wonder out loud why the train is not moving. The buses and trains are full of advertisements for used-car dealers hoping to take advantage of that frustration.

L. Paget-Seekins (✉)
Departamento de Ingeniería de Transporte y Logística, Pontificia Universidad Católica de Chile, Vicuña Mackenna 4860, Macul, Santiago, Chile
e-mail: laurel@gatech.edu

Institute for Mobility Research (ifmo) (ed.), *Megacity Mobility Culture*, Lecture Notes in Mobility, DOI: 10.1007/978-3-642-34735-1_8,
© Springer-Verlag Berlin Heidelberg 2013

While both groups are unhappy, their actual experience of mobility is very different. Atlanta is designed for car drivers; yet they are inconvenienced because of their abundance. The public transport passenger suffers from scarcity; the Atlanta region has deliberately not invested in public transport. These issues are not unconnected, and are both linked to the history of Atlanta's urban development.

While Atlanta's mobility woes can be attributed to poor planning, some of that planning was, in fact, intentional. Atlanta's development of transport infrastructure and urban form cannot be separated from its history of racial and class segregation, and has resulted in a megacity with limited mobility, especially for people without a vehicle.

This chapter is divided into five sections. The first section presents the history of Atlanta's development and of its transport infrastructure. This is followed by an analysis of Atlanta's spatial and demographic patterns. The third section examines the available transport infrastructure and mobility behaviour. The fourth section examines the policies and perceptions which shape that behaviour. Finally, concluding thoughts are presented.

A Brief History of Atlanta

Atlanta is a major metropolitan region for no reason other than the location of transport infrastructure. Atlanta has no navigable waterway and no stock of natural resources, nor indeed any other feature that explains its location. The city sprang up at the nexus of several railway lines, and was officially named Atlanta (from a suggestion by the Chief Engineer of the Georgia Railroad) in 1847. It was a railway city until it became a car and airport city. It remains an important freight railway hub, but also hosts the world's busiest airport and is intersected by three major US interstate highways.

Atlanta became a main railway hub and supply centre during the US Civil War (1861–1865). Despite being burned by Northern troops, the city continued to grow as a regional centre after the war. By the turn of the century it had 15 railway lines, with over 100 trains a day passing through it. In 1889, Atlanta opened its first streetcar (tram) line and electrified it two years later; by 1901 it had 214 km of streetcar track. As the city grew, the streetcar network continued to grow with it, reaching 354 km of track in 1924. However, at the urging of the city government, the streetcars were replaced by cars, buses, and trackless trolleybuses. Ironically, the city leaders thought that this would *reduce* congestion and modernise the city [3, 12]. The last streetcars ran in 1949, but now the city has started building a new streetcar line in the downtown.

The 1950s brought significant federal legislation, building the interstate highways and subsidising housing mortgages, paving the way for urban sprawl. It also brought Atlanta its first year of hosting the world's busiest airport, as a consequence of the leadership of Mayor Hartsfield [7]. Urban growth was facilitated in the first instance by the streetcar network, but then amplified by the growth of car

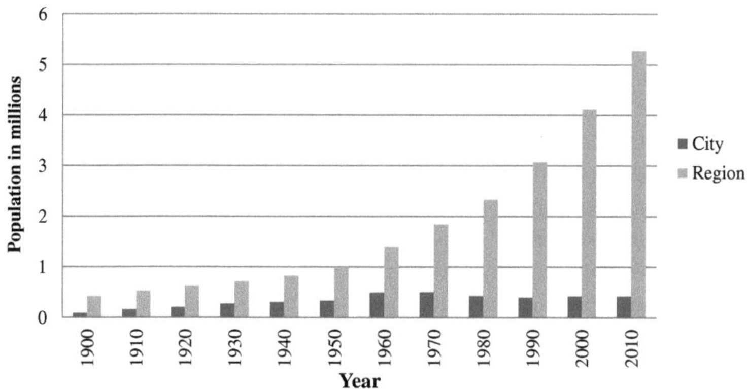

Fig. 8.1 Atlanta population over time (*Source* US Census Bureau [18])

ownership. As seen in Fig. 8.1, growth in the regional population took off in the 1950s.

The growth of the suburbs was also driven by 'White flight' from the City of Atlanta. The US civil rights movement in the 1950s and 1960s, which was headquartered in Atlanta, forced desegregation of public facilities, such as schools, swimming pools and restaurants. In response, many White residents left the cities for newly constructed suburbs. By 1970 the City of Atlanta was majority Black, as over 60,000 White residents had fled the city during the previous decade.

Despite the rising political power of Blacks in the City of Atlanta, the construction of the interstates and associated 'urban renewal' projects left a permanent mark on their communities. Transport projects were in some cases actually designed to divide the Black and White communities. The building of the 'Downtown Connector' (the main interstate freeway through the city) cut off the thriving Black business district from the White downtown [8]. The legacy of segregation is still evident in Atlanta in the street names—the names of major streets were changed as they crossed the boundaries between Black and White neighbourhoods, in order to create an impression of separation and allow people's race to be identified by their address [9, 15].

Racism also impacted the development of the public transport system. A rapid rail system was proposed in 1960 (just 11 years after the last streetcar was removed). The state legislature created the Metropolitan Atlanta Rapid Transit Authority (MARTA) to build and operate public transport in the five core counties in the region. However, the funding measure was approved only in the two counties that contain the City of Atlanta. Much of the anti-public-transport sentiment centred around fears that public transport would allow poor Blacks access to the suburbs and bring in crime and 'urban' problems. The Whites who fled the city for car-only suburbs did not want the people they had just left behind to be able to access their new communities [8]. MARTA took over the privately operated bus service and opened its first rail service in 1979, but it is still confined to the two counties around the central city.

Despite the impact of racism on its transport infrastructure and land-use patterns, Atlanta pressed on with its economic growth, adopting the image of 'The city too busy to hate'. The slogan was developed by civic and business leaders during the civil rights movement in an attempt to distinguish the City of Atlanta from the racial turmoil taking place in other southern cities in the USA at the time. A compromise between Black political leaders and the downtown business community kept racial tension in the background and economic growth in the foreground. The region's business climate, good weather, and network of transport infrastructure attracted new businesses and new residents. Without any physical barriers or land constraints, Atlanta's land area kept growing until it became one of the lowest-density metropolitan areas in the USA.

In order to keep up with its growth and the increasing problem of traffic congestion, Atlanta kept building highways and runways, but not public transport. The Atlanta airport (Hartsfield-Jackson International) opened the world's largest terminal in 1980 [7]. Highways were built and expanded, in some places doubling from three lanes to more than six in each direction. But MARTA received no new source of regional or state funding; public transport was seen as a social service for poor Blacks, not a major transport investment.

Atlanta's growth spurt, and the final touches to its image as a city, were completed when it won the bid to host the 1996 Summer Olympic Games. Ironically, this bid would probably have not been successful without the MARTA rail system, and MARTA played a major role in transport during the games.

The first major crack in Atlanta's attempt to build its way out of congestion appeared in 1998, when the US government declared the Atlanta metropolitan region to be out of compliance with the Clean Air Act. The region was ineligible for any federal transport funds until it demonstrated that it could reduce its creation of ozone and other pollutants to a level that met the national standards. This crisis led to the first attempt at creating a regional public transport system—a network of commuter bus services. A local bus service was started in each of the remaining three core counties that originally voted out of MARTA.

A growing coalition of civic, business, and political leaders are currently calling for increased investment in public transport for environmental reasons, the infeasibility of building more roadway capacity and a desire to keep Atlanta economically competitive. Despite the growing political support for public transport, actual services in the Atlanta region are decreasing. In 2010 a lack of funding led to the cancellation of all bus services in one of the surrounding counties, and forced MARTA to cut 10 % of bus service and 14 % of rail services.

People and Places

A Small City in a Sprawling Region

The actual City of Atlanta is very small, in terms of both area and population, when compared to the entire metropolitan region. The geographic boundaries of the Atlanta

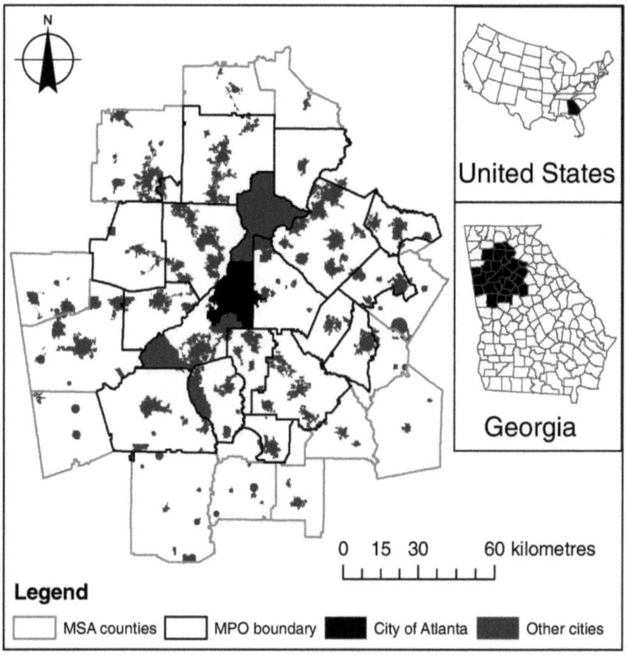

Fig. 8.2 Map of Atlanta region (*Source* created by author using data from Atlanta Regional Commission, US Census Bureau [18])

region vary depending on how it is defined. The United States government uses several different methods to define metropolitan regions. The Metropolitan Statistical Area (MSA) is determined by urbanised areas with economic integration. There are 28 counties and 144 cities in the Atlanta MSA. The MSA covers 21,690 km^2, of which only 347 km^2 (or 1.6 %) is in the City of Atlanta [18].

However, transport planning is undertaken by Metropolitan Planning Organizations (MPO) whose boundaries are determined by urbanised area and air quality regulations. The boundaries of the Atlanta MSA and MPO can be seen in Fig. 8.2.

The region possesses no physical boundaries to slow its growth. There are no navigable rivers, no mountains of any size, and no coastline. However, geography is only part of the reason why Atlanta was able to sprawl so far. The region lacked policy and planning controls on growth. Land-use planning is the responsibility of local governments: all 144 cities and 28 counties in the region control their own planning. The federal and state governments kept on allocating funding to build highways, so there were few incentives to limit growth.

Recently Atlanta's MPO, the Atlanta Regional Commission, has started to provide incentives for denser development, and local elected officials have begun to advocate increased density, recognising the economic and environmental benefits. However, there is a very negative perception of density amongst Atlanta residents. The preference is for single-family housing on large plots. Even in neighbourhoods in the City of Atlanta, proposals to increase density are opposed by residents worried about the impact that this would have on property values, traffic, and the quality of life. Even increasing density along proposed new public transport routes is opposed by neighbourhood groups.

A Region Divided

The Atlanta MSA region had a population in 2010 of 5,268,860, but only 8 % lived inside the City of Atlanta [18]. As seen in Fig. 8.1, the regional population continues to grow, while the city population, having declined in the 1970s and 1980s, at present remains steady. The region's density in 2010 was 2.43 persons per ha (hectare, 0.01 km^2). The density of the region is at its highest in the centre, and decreases toward the edge; over 65 % of residents live in the five core counties [18].

As seen in Fig. 8.3, Atlanta is a segregated region: the north side of the region and outer-ring suburbs are majority White, while the inner-ring suburbs, especially

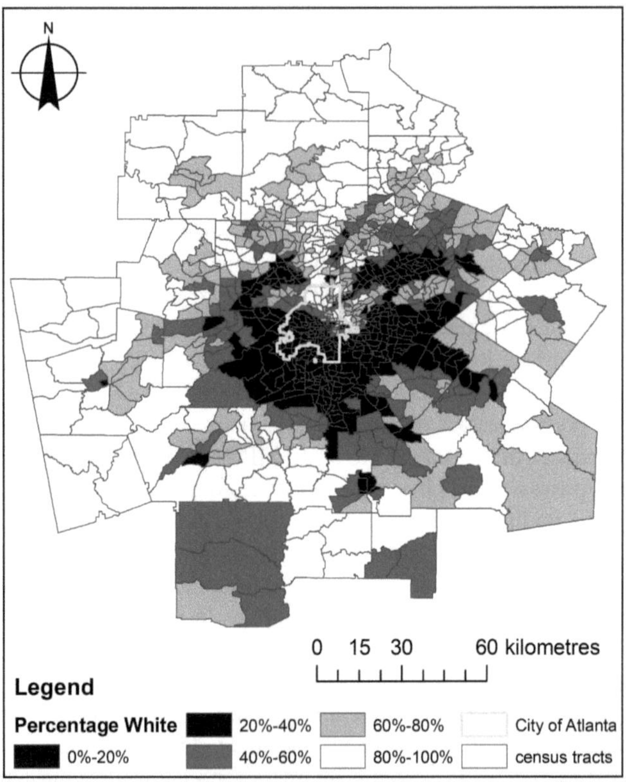

Fig. 8.3 Atlanta 2010 residential racial concentrations (*Source* created by author using data from Atlanta Regional Commission, US Census Bureau [18])

on the south side, are majority Black. There are pockets of immigrants from Asia and Latin America in the north-eastern suburbs, and resettled African refugees in the eastern suburbs. According to the 2010 census, the region is 54 % White, 30 % Black, and 10 % Hispanic. The City of Atlanta is 53 % Black (an 8 % point decline since 2000), 36 % White, and 5 % Hispanic [18].

In a reversal of the previous pattern of White flight, the City of Atlanta has been gaining White residents since the 1990s. City neighbourhoods are being gentrified, and the city has demolished all of its public housing projects. Over the last decade the city gained 22,000 White residents and lost 30,000 Black ones [18]. At the same time, the number of people of colour and people in poverty living in the suburbs has increased. By 2010, 85 % of Atlanta's poor were living in the suburbs [4].

The City of Atlanta is divided by income: high-income households, mostly Whites, live on the north and east sides; low-income households, primarily Black, live on the south and west sides. It has the highest income inequality of any city in the USA, with a Gini coefficient of 0.57 [19]. In 2009, over 22 % of City of Atlanta residents lived below the official poverty line. In contrast to the city, the region as a whole has *less* income inequality than the USA overall, and only 13.4 % of its population living below the poverty line. For a reference point, the poverty rate in the USA in 2010 was 15.1 % [5].

Infrastructure of Abundance and Scarcity

Streets and Cars

The City of Atlanta is at the crossroads of three major US interstate highways. Two north–south highways run at an angle and merge for the length of downtown Atlanta. Called the 'Downtown Connector', the combined highway is 16 lanes wide at points. In the centre of Atlanta, the combined highway intersects an east–west interstate highway that runs across the southern USA. The city is also surrounded by a ring road. Less than 2 % of Atlanta's 42,312 km of road centre line (multiple lanes are not counted) are freeways. Local streets comprise the vast majority. But 36 % of the vehicle kilometres travelled (VKT) are on the freeways [1]. It is worth noting that this includes traffic generated outside of the region and travelling through Atlanta to destinations in other parts of the country.

Because of availability of data (or lack of it), the transport infrastructure supply for the Atlanta region is given for the slightly smaller MPO region, rather than for the MSA. According to the 2010 Texas Transportation Institute Urban Mobility Report, 74 % of peak VKT is driven under congested conditions, a situation which lasts for five hours a day. This results in an annual waste of 35 gallons (132 L) of fuel and 44 hours of delay per peak car traveller. The travel time index is 1.22, which means that it takes 22 % longer to travel during peak hours than in non-peak hours [17].

When people complain about traffic in the Atlanta region, they are for the most part complaining about the freeways and the major roads. Owing to the nature of Atlanta's street network, there are often few alternative routes. The suburbs were built largely on a local, collector, arterial model; this system funnels all the traffic onto the main streets. Because of the prevalence of cul-de-sacs, in some places you have to drive a number of kilometres to get somewhere less than one kilometre away in linear distance. There is a small grid network in downtown Atlanta, but the city is divided by interstates and railway tracks. There are limited numbers of streets that cross all of the barriers.

This network design creates even more difficulties for pedestrian and bicycle trips. Not only are trips longer, but the limited routes available are dangerous because of high traffic volumes. There are few bicycle facilities, and sidewalks are in short supply along many major arterials, and in residential neighbourhoods too.

Compounding this lack of routes is the fact that people also drive long distances. In 2009 the VKT per person per day was 46; this is above the average of 40 VKT per person per day for the largest 100 metropolitan areas in the USA [1]. This can be attributed to the low density of the region, the lack of mixed land uses, and the spatial mismatch between jobs and housing.

The City of Atlanta contains the primary central business district of the region; however, there are major employment centres outside the city, primarily in the northern suburbs along the freeways, and to the south of the city at the airport. The job density for most of the region is extremely low. Despite having the highest-density job districts, only 17 % of the 2.2 million jobs in the region are in the City of Atlanta [5].

The majority of people in the region cross county boundaries to work. Fewer than half of the employed residents of the City of Atlanta work in the city itself, and only 18 % of people employed in the city live there [5].

The Atlanta region in 2009 had 3.6 million registered personal vehicles. This equates to an average of 2.2 vehicles per household in a region with an average household size of 2.7 people [1]. The abundance of vehicles contributes to more than just congestion in Atlanta—as noted above, the Atlanta metropolitan region is not in attainment with the USA's air quality standards for ozone and particulate matter.

Over the past decade, Atlanta has averaged 24 days per year exceeding the standard for ozone, with a high of 40 days in 2011 [20]. High levels of ozone are unhealthy, and can trigger respiratory illnesses such as asthma. Asthma is the number one reason for admission to the Children's Healthcare of Atlanta Pediatric Hospital, and 10 % of children in Georgia suffer from asthma. In addition, transport sources are responsible for 36 % of the greenhouse gas emissions in Georgia [14].

Public Transport

Atlanta does not have the public transport infrastructure to serve the majority of journeys people need to make—many trip origins/destinations are in locations not served by public transport. There is only 77 km of railway line in the inner two counties, serving 38 stations. Local bus services are available in only four of the 20 counties in the MSA, and in one city outside those four counties. There are peak-hour-only commuter bus services from 12 counties to downtown and midtown Atlanta.

Only 12 % of the Atlanta MPO region is within half a mile (0.8 km) of a public transport service. That 12 % of the land area contains 58 % of the jobs and 39 % of the population. Travelling by public transport, the average Atlanta resident can reach only 22 % of the region's jobs within 90 min. This is 7 % points below the average for the 100 largest regions in the USA; in a combined job and housing public transport accessibility index, Atlanta ranks 91st [2].

In addition to having limited coverage, the Atlanta public transport system runs on very low frequencies. The trains run every 15 min during peak hours and every 20 min during off-peak hours and on weekends. MARTA buses run on average every 30 min during peak hours, and every 45 min during off-peak hours. The buses in suburban counties are even less frequent. Some bus routes have no service at all on Sundays.

Given the lack of access to public transport, and the inconvenience of what access does exist, it is not surprising that very few trips are made by public transport in Atlanta. The average number of weekday public transport boardings in 2009 was 550,000 [1]. In a 2011 travel survey in the region, the personal vehicle was used for 83 % of trips, and public transport for 3 % [13].

The majority of people travelling by public transport in Atlanta are doing so because they have very few other options. On average, public transport passengers in the Atlanta region come from households earning 37 % of the median household income. Half of MARTA passengers report "no car available" as their main reason for using public transport. MARTA passengers are 76 % Black, 13 % White, 8 % Hispanic, and 4 % other races [11].

Both the lack of access to a vehicle and low income limit the mobility of a household. Lower-income households make fewer trips than higher-income households. In the latest regional travel survey, households bringing in less than US$10,000 annually make 9.3 trips per day, compared to 12.7 trips per day for households whose income is over US$150,000 per year [13].

As a consequence of the limited number of destinations accessible by the system, together with the low frequency of the service, it is difficult to live in Atlanta without a vehicle. This means that majority of low-income households own cars, despite the great expense. Low- to medium-income households spend on average 32 % of their income on transport, higher than the average 29 % that they spend on housing [10].

There is a public transport option for high-income commuters who want to avoid the traffic. A Georgia state agency operates an express bus service from the suburbs to downtown and midtown Atlanta. Almost half (49 %) of the express bus passengers have a household income of over US$75,000, compared to just 7 % of MARTA passengers. This public transport service only serves the peak-hour work trips, and 97 % of the passengers own at least one car [16].

Politics and Perceptions

There is a feeling in Atlanta that you cannot survive without a car—that no one in their right mind would choose to live there without one. In part this is due to the poor reputation of public transport in general and of MARTA in particular. The common wisdom amongst public transport promoters in Atlanta is that MARTA has an 'image' problem. The general public thinks that MARTA is unsafe, inconvenient and poorly managed. This perception is not always linked to reality, but persists because of media reports, the demographics of MARTA passengers and employees, and insufficient funding for public transport.

Historically, the public transport system in Atlanta has been seen as a social service for the poor, rather than as transport infrastructure. Over three quarters of MARTA passengers are Black. Perhaps in part because of this, MARTA is considered to be unsafe. The media perpetuates this narrative by highlighting any crime committed on the public transport system.

One of the original arguments against extending public transport to the suburbs in the 1970s was that it would bring crime with it. Even in 2011, the Atlanta newspaper the *Atlanta Journal-Constitution* finds suburban leaders who are willing to say, "Criminals catch that kind of transportation into our county… and I'm not going to support anything that works toward increasing our crime either [6]". Code words such as 'criminals' and 'urban problems' are used to mask blatant racism. While it still occurs, it is no longer socially acceptable in Atlanta for people to express explicitly racist opinions or discriminate against individuals based on their race.

However, racism is still ingrained in the institutions and funding structures of transport. MARTA is the largest public transport system in the USA that receives no funding from the state in which it is located. The Georgia legislature, which is dominated by rural and suburban White conservative politicians, funds highways, but refuses to fund public transport. At the same time it maintains control of the finances of MARTA, dictating how much of its own tax revenues the public transport agency can spend on operations and how much on capital expenses. This limits the level of service that MARTA can provide to its mostly low-income and Black passengers.

In the early 1960s, three regions in the USA started to plan new heavy rail subway systems. The new systems in Atlanta, Washington DC and San Francisco were considered as the renaissance of heavy rail in the USA. Owing to limited local funding, Atlanta only built 77 km of track. Meanwhile San Francisco and Washington DC continue to expand their systems. The Bay Area Rapid Transit (BART) system in San Francisco is 167 km in length, and the Washington DC Metro 171 km. Given the same opportunity for federal funding to develop a regional rail transit system, it was perceptions and funding policy that led to a different outcome for Atlanta.

In addition to funding structures, the policymakers in Atlanta are orientated toward solving the problems of drivers, and not those of public transport passengers. The dominant transport narrative in Atlanta is that the problem with the region's transport system is congestion. When the problem is defined as congestion, the solutions considered are increased capacity, intelligent transport systems, and peak-hour public transport that serves commuters who own cars. A competing narrative is emerging that defines the transport problem as the lack of choice. However, this narrative does little to improve the mobility of those dependent on public transport, because it still focuses on providing public transport that serves the needs of car owners.

Conclusion

Atlanta is a transport city. It exists solely as a result of the intersections of transport infrastructure—first railway lines, and now interstates and the world's busiest airport. However, ironically, despite the transport infrastructure, no one in the region is happy with their mobility. Both drivers and public transport passengers are frustrated by the lack of reliable travel times.

Atlanta is identified as a Hybrid City in the preliminary study (see "Chap. 2"), but it has the mentality of an Auto City. Drivers complain about long trips and slow traffic, but the option of not driving is much worse. Public transport services are limited and infrequent, entire neighbourhoods have either no sidewalks or a severe shortage of them, and bicycling is difficult because of the long distances involved and the limited number of alternative routes. Moreover, the low-density land-use patterns and indirect street network inhibit mobility regardless of the mode considered.

Atlanta did not somehow end up as the epitome of urban sprawl and car use just by chance. 'The city too busy to hate' was indeed busy, as the region grew at a fast pace without any physical impediments. However, this growth took place in the context of land-use/transport planning that permitted low-density development and spent large sums of money on building new roads and expanding existing ones. And while the battles of the civil rights movement were not fought in Atlanta, the impact of racism cannot be denied. White flight from the City of Atlanta fed the growth of the suburbs. The term 'urban' became code language for Black, poor, and unsafe. The funding and extent of public transport services was limited in order to contain the spread of 'urban problems.'

This lack of investment in public transport created a barebones system that serves primarily people who—for reasons related to disability, income or age—have no other options. This leaves a segment of the population, who are primarily poor people of colour, with limited mobility and access to opportunity. It also makes owning or travelling in a private vehicle practically a necessity for the vast majority of Atlanta residents. The scarcity of public transport contributes to the abundance of vehicles, causing congestion and air pollution.

It has come to the point where business and political leaders recognise that in order to remain economically competitive, the region needs to invest in public transport. However, Atlanta's period of rapid population growth was orientated to highways, and has left a low-density landscape that is less conducive to public transport. In addition, Atlanta now has to catch-up in an environment characterised by fewer federal funding opportunities and higher construction costs.

If Atlanta wants to restore balance to its mobility, and end the twin problems of scarcity and abundance, it will have to change its perceptions of public transport and higher-density development—but this will require the underlying racism and inequity in the region's mobility to be addressed.

References

1. Atlanta Regional Commission (2010) 2010 transportation factbook. Atlanta Regional Commission, Atlanta
2. Brookings Institute (2011) Missed opportunity: transit and jobs in metropolitan America. Brookings Institute, Washington
3. Carson OE (1981) The trolley titans. Interurban Press, Glendale
4. Chapman D (2010) Poverty moves fast to suburbs. Atlanta Journal-Constitution. Cox Media Group, Atlanta
5. City of Atlanta (2011) Community assessment economic development. City of Atlanta Comprehensive Development Plan, Atlanta
6. Hart A (2011) Mass transit: a must-have or a no-win?. Atlanta Journal-Constitution, Atlanta
7. Hartsfield-Jackson Atlanta International Airport (2011). Airport history. 2011. http://www.atlanta-airport.com/Airport/ATL/Airport_History.aspx
8. Keating L (2001) Atlanta: race, class, and urban expansion. Temple University Press, Philadelphia
9. Kruse K (2005) White flight: Atlanta and the making of modern conservatism. Princeton University Press, Princeton
10. Lipman B (2006) A heavy load: the combined housing and transportation burden of working families. Center for Housing Policy, Washington
11. MARTA (2008) General riders, half-fare program, mobility program demographics and system usage profiles. Metropolitan Atlanta Rapid Transit Authority, Atlanta
12. Martin J (1975) From mule to MARTA, vol I. Atlanta Historical Society, Atlanta
13. NuStats PTV (2011) Regional travel survey: final report. Atlanta Regional Commission, Atlanta
14. Olivares E (2010) Taking the temperature: transportation impacts on greenhouse gas emissions in the Atlanta region. Atlanta Regional Commission, Atlanta
15. Potteiger M, Purinton J (1998) Landscape narratives: design practices for telling stories. Wiley, New York City
16. Public Performance and Management Group (2008) Georgia regional transportation authority rider survey. Georgia State University, Andrew Young School of Policy Studies, Atlanta
17. Texas Transportation Institute (2010) 2010 annual urban mobility report. Texas Transportation Institute, College Station
18. US Census Bureau (2010) 2010 Census and 2005–2009 American Community Survey Data. US Census Bureau, Washington
19. Weinberg D (2011) US neighborhood income inequality in the 2005–2009 period. American Community Survey Reports. US Census Bureau, Washington
20. Zimmer-Dauphinee S (2011) Ambient monitoring program. Georgia Department of Natural Resources. http://www.georgiaair.org/amp/

Chapter 9
Los Angeles: A Transit Metropolis in the Making?

Sylvia Y. He

Abstract The Los Angeles region epitomises polycentric urban form with highly fragmented and devolved governance. For a long time, the region has been cited as an example of urban sprawl. However, its pattern of employment is not really dispersed—as expected in sprawl—but, rather, organised in many decentralised 'employment centres'. These centres have grown and evolved over a long period. In the emerging urban form, the density gradient has multiple local maxima coincident with the employment centres. The multimodal population density function and the underlying travel demand have gradually changed the way in which people travel, and how we make transportation plans so as to accommodate such traffic demand. We are witnessing a comeback in mass transit, both in initiatives and investment. Movement of people in this car-centric region, although still heavily reliant on its extensive highway network, may well be increasingly undertaken by public transport in the near future. Is the traditional notion that Los Angeles is a low-density, car-orientated city being challenged? A history review, a contemporary analysis, and an examination of the outlook for the future might lead us to an answer.

Introduction

The Los Angeles region, or the Los Angeles Consolidated Metropolitan Statistical Area (CMSA), is the largest region in Southern California. It is made up of five counties: Los Angeles, Orange, Riverside, San Bernardino, and Ventura. This five-county region has a total population of approximately 18 million as of 2010 [45]. At the hub of the region is the City of Los Angeles. It is ranked as the second-largest city in the US in terms of population, having about 3.8 million inhabitants. Los Angeles has the fifth busiest airport in the world, the busiest port in the country, and probably the most congested freeways in the world.

S. Y. He (✉)
The Chinese University of Hong Kong, Shatin, N.T., Hong Kong
e-mail: sylviahe@cuhk.edu.hk

Institute for Mobility Research (ifmo) (ed.), *Megacity Mobility Culture*,
Lecture Notes in Mobility, DOI: 10.1007/978-3-642-34735-1_9,

© Springer-Verlag Berlin Heidelberg 2013

The mobility culture of Los Angeles formed in the context of America's transport industry revolution, and was influenced by California's unique culture and quick embracing of innovations. In 1876, the Southern Pacific Line, the first railway in California, reached the city. Since that time, regional train lines have continued to supply most of the long-distance transport and have dominated interstate transport. In the 1910s, America's first coast-to-coast road for car travel—the Lincoln Highway—connected San Francisco with New York City. The transport systems back then included horse-drawn wagons, cable cars (on cable railways in the streets), streetcars (trams), trolleybuses, petrol- or diesel-powered buses for short-distance travel, and cars and trains for long-distance travel [31]. The development of the US railway system was catalytic to the state's main population growth. The first major influx of migrants occurred after oil was discovered in California. By 1923 the oil and petroleum industry was booming, and made California the largest oil producer in the country. Because of the pleasant climate and booming economy, many Americans began to view California as the 'Golden State'. This jewel on the West Coast became a symbol of free living, and continued to be viewed this way as Southern California entered the age of the automobile.

In this chapter, I first review the history of transport development in Los Angeles. I then analyse the region's inherent resistance to using public transport from several aspects related to the four dimensions of mobility culture [20]. Following that, I introduce 'place-based' planning, and synthesise Los Angeles' mobility culture with this method, making use of a downtown streetcar revitalisation plan to exemplify the reconnection and revitalisation of downtown through investment in a vintage, yet innovative, transport mode. This project showcases the values of history and culture in the resurrection of public transport. For public transport development in the region, I present a conceptual regional network that gives special consideration to polycentric urban form and fragmented governance. I conclude that identifying local agglomeration economics and the underlying traffic demand between and within employment centres, and incorporating them into transport planning, could be a viable strategy for making Los Angeles a transit metropolis.

The History of Los Angeles' Transportation Development

In the early twentieth century, rail dominated the transport system in America. According to former US Senate antitrust attorney Bradford Snell [41], in 1920 nearly 90 % of all city and intercity trips were undertaken by rail, in which the majority of intra-city trips were undertaken by streetcar. By contrast, only one tenth of Americans owned a car. The first streetcar system in Los Angeles started as a route just 2.5 miles long. The streetcar later developed rapidly. In 1898, it changed its name to the Los Angeles Railway Corporation (LARY), one of the largest rail transit networks in the world. During its peak in the 1920s, the network stretched to almost 600 miles and owned over 1,200 vehicles. Whilst the 'yellow cars' run by LARY served the urban area, the 'big red cars' run by the Pacific Electric Railway linked suburban areas such as Long Beach and Orange County [29]. The Los Angeles streetcar system was the largest streetcar system ever constructed in history, and played a pivotal role in shaping the urban form in downtown Los Angeles [10].

The fate of the streetcars changed when the population started to spread out from city centre to the suburbs, a process facilitated by single-family land-use zoning, the advance of the car industry, and the proliferation of highway construction. The lack of concentrated downtown population settlements and business establishments meant that streetcar patronage fell ever closer to unsustainable levels, and system maintenance was insufficiently funded. After a period of prosperity, the streetcar companies suffered substantial financial losses. To cut losses, they began to drop their services in the lightly patronised long-distance routes, and started replacing them with buses. By the 1950s, most of the major rail systems had been bought out and dismantled. On the verge of bankruptcy, the Los Angeles Railway and the Pacific Electric Railway were both acquired in 1958 by a state-established agency—the Metropolitan Transit Authority (MTA). Shortly after this acquisition, the remaining tram lines were closed. The Pacific Electric last operated trams in 1961, and the Los Angeles Railway was abandoned in 1963—a 90 year operation had drawn to a close [10, 29].

Some people believe that the abandonment of the Los Angeles streetcar was accelerated by an alleged nationwide monopolisation scheme—the 'General Motors streetcar conspiracy' [2, 29, 41]. When streetcars were still the primary transport mode, cars were to the urban systems the new transport alternative. One of the main challenges facing this new means of transport was the need to gain more street space. In 1936, General Motors, in partnership with several oil and tyre companies, established the National City Lines, a holding company that bought tram lines across America. They converted many streetcar lines to bus routes, and replaced the streetcars with buses built by General Motors [29]. In the 1940s, National City Lines purchased the Pacific Electric and then the Los Angeles railways before they were handed over to the MTA. Meanwhile, a series of similar acquisitions swept the country's streetcar systems [41]. This conspiracy was even featured in the documentary *Taken for a Ride*, and a Hollywood film, *Who Framed Roger Rabbit?* The story of the General Motors streetcar conspiracy disclosed some monopolistic manoeuvrers during the era of the downturn of streetcars. However, it would be an overstatement to say that the conspiracy destroyed the streetcar system in Los Angeles [2, 29, 42]. The streetcar systems and operations had intrinsic inefficiencies; the companies began to have fiscal struggles more than a quarter of a century before General Motors established National City Lines [42].

Before their disappearance, streetcars fought another battle against private vehicles over the use of public space—that of parking restrictions. Some researchers have pointed out that the large number of parking spaces is, together with underpriced parking, an important but often overlooked determinant of mode choice. One of the leading scholars in the sphere of parking is Donald Shoup, professor of urban planning at UCLA (University of California, Los Angeles). In his book *The High Cost of Free Parking* [40], he presented his concerns about free parking and suggested a number of institutional changes. He proposed that there should be changes at an institutional level regarding parking costs, to effectively reduce car use. The debate as to whether managing parking supply and pricing can reduce driving and thereby ease traffic levels is, however, not new. Back in the 1920s, the rail systems suffered from unprecedented congestion, along with inefficient operations and revenue losses [7]. The Railroad Commission and the Board of Public Utilities jointly conducted an investigation of the rail company's problems. The report pointed out that one main reason for the congestion was the rapid increase in the number of cars in the city. The cars parked on the street were beginning to hold up both other cars and streetcars. On the basis of this report, the Railroad Commission proposed a ban on downtown parking, claiming that it was necessary to clear the congestion. The City Council responded with a proposed ordinance that would eliminate parking in the business district from 9.00 a.m. to 6.20 p.m. from Monday to Saturday [7]. The ordinance went into effect on 10 April 1920. Because many stakeholders were involved, this parking ban soon became controversial, especially for car dealers and retail merchants in the downtown area, because of its impact on their business. There were soon protests against it. Opponents of this new legislation argued for the right to use the roads that their taxes had paid for. Only 19 days after the initial ban was put in place, the mayor signed an amendment that allowed parking up to a 45 min limit on downtown streets from 10.00 a.m. to 4.00 p.m., but banned parking from then till 6.15 p.m.

During the deliberations on the no-parking legislation, alternatives to congestion relief were also brought up. Several of these ideas, such as one-way streets and restrictions on commercial loading, materialised later in Los Angeles' development. In particular, one suggestion significantly changed the downtown streetscape—the construction of car parks, intended to help retain downtown business. Even before the enactment of the parking ban, the downtown business community had feared that this new legislation would weaken the control exerted by the central business district (CBD) over the local economy, if shoppers were prohibited from using cars in the downtown area [7]. Building car parks and designating off-street parking spaces was suggested as one way to relieve traffic congestion while at the same time retaining business that relies on car use. This rationale has been embedded in the region's transport-land-use regulations. Nowadays, many business buildings in commercial zones are required to provide a minimum level of parking spaces. In Los Angeles County, for example, the minimum requirement is one parking space for each 250 sq. ft. of floor space for general commercial use, or one parking space for each 400 sq. ft. of floor space for non-medical office uses. Even for residential land use, a minimum of two covered parking spaces per single-family residence is required [27]. These parking requirements, and the resulting overabundance of parking spaces, have to some degree contributed to both Los Angeles' car use and its sprawl [47].

After the arrival of cars, another historic milestone significantly impacted the decentralisation and car dependence of American cities—the National Interstate and Defense Highways Act of 1956 ('the 1956 Interstate Highway Act'). According to a study that ranked the top ten influences on the American metropolis of the past 50 years, the response "The 1956 Interstate Highway Act and the dominance of the automobile" was ranked top of the list by members of the Society for American City and Regional Planning History [12]. This legislation authorised US$25 billion for the construction of the interstate highway system. The federal government covered 90 % of the cost of the construction, while local governments paid for 10 % [12]. During the interstate era, local governments had great incentives to concentrate on freeway construction [12]. With the completion of the interstate highway system, highway expansion in many metropolitan areas plateaued by the 1980s [15]. By the end of the interstate era, Los Angeles encompassed many interstate and intercity highways and became a dispersed 'autopia' (a term coined by Banham [6], the British architectural critic and writer, to describe one of four 'ecologies' of Los Angeles as an 'automobile utopia'). The massive 41,000-mile highway system was originally intended to deliver central cities from congestion and to provide high-speed intercity travel. Nonetheless, the highway system has induced much more traffic than expected and has become increasingly congested [15]. According to the 2009 Urban Mobility Report [43], Los Angeles is ranked the most congested metropolitan area.

If *money* can explain why freeways were built in cities [42], then the same fiscal reasoning can account for the end of freeway construction and the resurrection of public transport. US public transport (transit) systems in the ten years after the Second World War were mainly privately owned. They underwent substantial decline, with an annual drop in passenger numbers of over 7.5 % [42]. Short of private investment, many transit systems became near-bankrupt. Impelled by metropolitan political pressure, the federal government gradually converted the privately owned transit systems to public transit systems. By the late 1970s, the federal government had become the principal funding resource for public transport. At the beginning of the 1980s, the federal government started to withdraw federal operating subsidies, while maintaining the original capital subsidy level [42].

Nowadays, public transport in America is still heavily subsidised. To partially offset the operating cost, transit agencies are permitted to increase fares. However, such proposals are sometimes rejected on the basis of income inequity. In 1994, the Los Angeles County Metropolitan Transportation Authority was sued over a proposed fare increase by a group of public transport passenger advocacy organisations because the transit agency was pouring a large amount of capital into new rail lines and commuter transport (which are both more likely to be patronised by high- and middle-income and non-ethnic minority groups) rather than local bus services (which are more likely to be used by low-income and minority groups) [42]. The lawsuit was settled in 1996. The transit agency was restricted from imposing fare increases for the next ten years. In addition, the agency was mandated to improve the bus systems over the same period [42]. Even with upgraded vehicles and low fares, the public transport patronage in the region is rather low.

Inherent Resistance to Public Transport: A Mobility Cultures Perspective

To understand the inherent resistance to using public transport in the Los Angeles region, I will now analyse several aspects related to the four interacting dimensions of the mobility culture framework [20]. This framework was originally proposed by Götz and Deffner. Through simplification of the transport system, the mobility culture framework focuses on the cultural and contextual constraints that have shaped the mobility patterns of an area. There are four important and interrelated components:

1. spatial structure and transport supply;
2. perceptions and lifestyle orientations;
3. mobility behaviour;
4. policymaking and governance.

Spatial Structure and Transport Infrastructures

With the arrival of the Southern Pacific Railroad, the growth of the Los Angeles region took off [7]. Ever since then, the region's urban growth has never stopped. Coincident with this urban growth was rapid motorisation and suburbanisation. For example, between 1980 and 1990, the percentage of workers commuting by car increased by 22 % in the central county—Los Angeles County—and by 50 % in suburban counties—Orange, Riverside, San Bernardino, and Ventura [8]. As suburbanisation continued, the central county started to see employment levels fall. From 1990 to 2000, Los Angeles County was the only county in the region in which employment decreased. By contrast, Riverside County experienced the fastest growth in both jobs and population, while San Bernardino (SB) and Riverside counties experienced growth in jobs that outstripped population growth, an indication of their transformation from dormitory communities to urbanised areas [18].

The Los Angeles region has one of the densest highway networks in the country (Fig. 9.1). Across this region, the historical Los Angeles CBD has the highest network accessibility. The central core of the network encompasses a large area of approximately 250 square miles. From the urban economics perspective, this area offers firms the highest network accessibility in the region. As these highway networks stretch from the region's centre to suburban counties, the network accessibility gradually decreases [19].

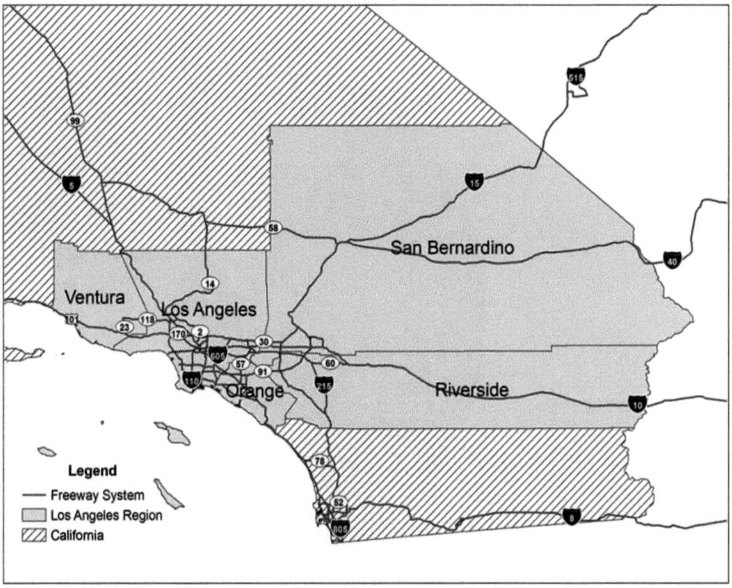

Fig. 9.1 Highways in Southern California (*Source* US Federal Highway Administration and US Census Bureau; map created by the author)

Compared to highway network, the region's rail network is much sparser. It was not until 1990 that Los Angeles opened its first light rail line—the Blue Line. Since its opening, other rail lines have subsequently opened. The Red Line—the most patronised—opened in 1993, and later had three extensions added; the Green Line opened in 1995; the Gold Line opened in 2003 and was extended in 2009. Although the growth of rail system-wide passenger numbers has been slow, a couple of rail lines have experienced a substantial growth in patronage. For example, the Gold Line had roughly 1.1 million total boardings as of March 2012, a 24 % increase over a two-year period.

Compared to the region's small rail network, the Los Angeles area has good bus service coverage and much larger bus passenger numbers. The average weekday bus passenger count is more than three times higher than the total figure for rail [5].

Considered as a long-term strategy to improve the region's mobility, public transport has been prioritised in the government's infrastructure plan. For example, in the 2009 Long Range Transportation Plan (LRTP), the Los Angeles County Metropolitan Transportation Authority has a budget of US$297.6 billion for the fiscal years 2005–2040 (Fig. 9.2). Buses, in particular, have been allocated US$101.9 billion in the plan, more than 34 % of the total budget. Development and maintenance of highways, streets, roads, and multimodal will cost US$94.4 billion (31.7 %). The third-highest expenditure will be on rail and public transport corridors, which makes up another 20.8 % of the proposed expenditure. Expenses on bus and rail account for approximately 55 % of the total budget.

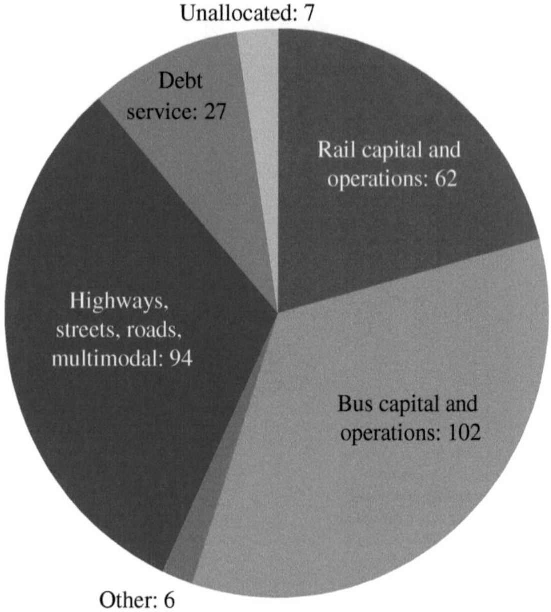

Fig. 9.2 Use of funds for fiscal years 2005–2040 (billion US$) (*Source* Los Angeles County Metropolitan Transportation Authority)

Voters in this region strongly and enthusiastically support the building of more rail lines and improvements to bus services. A recent example is the approval of Measure R appropriation, which is a ballot measure to invest US$40 billion in traffic relief and transport upgrades throughout the Los Angeles County over the next 30 years. Measure R has various goals: (1) rail expansion; (2) street improvement; (3) traffic reduction; (4) public transport enhancement; (5) enrichment of quality of life. In the proposed expenditure plan, approximately 65 % of the revenue will be spent on rail expansion projects, Metrolink and rail facilities, and bus operating improvements [31]. In order to pay for transport projects and improvements, county sales taxes will be raised by 0.5 % for the next 30 years. An ordinance, known as the Traffic Relief and Rail Expansion Ordinance, was created to administer the transaction and use of the tax. The ballot was approved in November 2008 by a two-thirds majority. Measure R is by far the most supported legislative appropriation of funds in recent years. Since 1968, there have been seven countywide transportation bonds and taxes; Measure R received the highest percentage of the vote (67.9 %) and the most votes (2,039,214) [28].

Spatial Inequality and Perception of Public Transport

Even though public transport projects are popular among voters, many middle-class families in the Los Angeles region are not reliant on this mode. They choose to live in the suburbs and commute to the city centre by car. This phenomenon, known as job–housing spatial mismatch [21, 22], is the root of the long-distance commute [9, 17]. The transport cost of long-distance travel by car has, of course, been significantly lowered by the aforementioned extensive and well-connected highway network. In addition, commuters—and this is of especial relevance to those who come from households with children—benefit from safer neighbourhoods and better school districts when they choose a suburban residential location.

Many neighbourhoods in the City of Los Angeles are considered dangerous on account of robberies, gang crime and drug trafficking, to give but a few of the reasons. And this perception is borne out by the facts: the city is safer than only 26 % of the cities in America, and the chance of becoming a victim of crime is 1 in 174 for a violent crime and 1 in 41 for a property crime [32]. Over the years, some neighbourhoods in the inner city of Los Angeles have become run down. For example, in the Central City East lies 'Skid Row', where the largest group of homeless persons in the US reside, with thousands living in cardboard boxes and tents along the pavements. With continuing efforts to improve public safety, however, the crime rate has been declining. In 2009, it reached a 50-year low [1, 26].

Another unfavourable aspect of the city is its poorly performing school districts. In spite of several education reforms which have attempted to reduce the inequalities in school finance, substantial and continuing performance gaps still remain. In particular, suburban districts are better off because they receive significantly more revenues from property taxes and other types of financial aid from their local governments than do their urban counterparts.

The government has made continual efforts to revitalise the downtown economy. The latest successful example of this effort is 'L.A. Live', a multipurpose development that hosts various kinds of pop culture and sports events. This development includes the Nokia Theatre, the Grammy Museum, an ESPN broadcasting studio, the Ritz-Carlton and JW Marriott Hotels, and Ritz-Carlton-branded condominiums. Other high-end condominium developments have recently been completed in nearby areas. However, these types of real-estate projects have not yet changed the socioeconomic profiles of the majority of downtown dwellers.

A large proportion of the households in the central city are low-income. Some of them remain in the downtown area either because they cannot afford suburban housing or because they lack adequate transport. People's residential location depends, to a large extent, on their car ownership. Not owning a car usually implies that they are not able to locate their home very far away from downtown, where access to public transport services is easier than it is anywhere else. According to the 2006–2010 American Community Survey 5-year estimates, the city of Los Angeles has a larger proportion of low-income households (defined as those with incomes under US$15,000) than the national, state, or county have (Table 9.1). The median income of a household in the City of Los Angeles in the past 12 months (in 2010 inflation-adjusted dollars) is US$49,138.

Over time, the public transport systems in Los Angeles have come to be associated with an unsafe and tarnished image. Compared to the bus systems, the rail systems have a relatively more favourable image with respect to cleanliness,

Table 9.1 Household income in the previous 12 months

Income level	US	California	Los Angeles County	Los Angeles City
Under US$15,000	12.75 %	10.41 %	11.99 %	14.65 %
Under US$25,000	23.56 %	19.87 %	22.52 %	26.55 %
Under US$50,000	48.22 %	41.68 %	45.42 %	50.67 %
Median income	US$51,914	US$60,883	US$55,476	US$49,138

Source 2006–2010 American Community Survey 5-year estimates

speed and comfort. While some perceptions of public transport can be changed in the short term (by, for instance, upgrading bus fleets and improving bus and rail operations), others, such as those concerned with in-vehicle safety and the time loss/inconvenience involved in changing train/bus, might take longer. It should be noted that the formation of perceptions can be a long-term process. It is postulated that children who are car dependent may continue this behaviour into adulthood, thereby limiting their alternative travel modes for much of their lives [37, 44].

Travel Patterns

The lack of public transport infrastructure in place, people's perception of the public transport systems, easy access to freeways, and low driving costs have all contributed to a high level of car use. According to the American Community Survey of 2009 (1-year estimates) [4], workers over 16 rely heavily on private vehicles (Table 9.2). Over 85 % of the commuters in the region go to work by car. In addition, different travel patterns are observed between households from the central county (i.e. Los Angeles County) and those from the suburban counties. For

Table 9.2 Means of transport and average commute time to work (workers 16 years and over)

County	Workers	Car, truck or van				Public transport (excluding taxicab)		Average commute time
		Drove alone		Carpooled				
LA	4,283,494	3,082,024	71.95 %	464,076	10.83 %	308,797	7.21 %	28.6
Orange	1,390,898	1,088,304	78.24 %	145,563	10.47 %	40,219	2.89 %	25.9
Riverside	832,355	651,841	78.31 %	103,139	12.39 %	12,359	1.48 %	31.1
SB	781,988	576,704	73.75 %	135,828	17.37 %	14,608	1.87 %	29.0
Ventura	380,895	286,029	75.09 %	51,856	13.61 %	5,426	1.42 %	24.0
Total	7,669,630	5,684,902	74.12 %	900,462	11.74 %	381,409	4.97 %	28.2

Note Summary for workers who walked, or used other commuting means, or worked at home is not reported here. Average commute time is in minutes
Source American Community Survey [4] 1-year estimates

example, the central county has relatively low car use (83 %), compared to suburban counties (90 %). This urban/suburban contrast is seen in use of public transport as well. In Los Angeles County, more than 7 % of the commuters use public transport; in the other four counties, public transport modal split ranges between 1.4 and 2.9 %. For the five-county region taken as a whole, a mere 5 % of the workers commute by public transport. The situation is slightly better in the central area of the region: for the Los Angeles–Long Beach–Santa Ana Metropolitan Statistical Area, the proportion of commuters who drive to work (including 'drove alone' and 'carpooled') accounts for 78 %, while the proportion of workers who take public transport has reached 11 %. As for commute time, workers living in the central county do not necessarily have the shortest commute. The average commute time for Los Angeles County is 28.6 min, which is longer than the commute time for either Orange or Ventura counties. This is not surprising, because some suburban workers might choose to work in firms located in the suburban employment centres instead of in the downtown CBD, thus saving commute time (and costs).

Dependence on the car, as evident in the aforementioned commuter trips, also exists in general trips. According to the 2009 US National Household Travel Survey [46], 84 % of the trips in the Los Angeles region are undertaken by car. Trips on foot account for the second-largest percentage (12 %). Public transport accounts for an even lower percentage (2 %) for general trip purposes, compared to its modal share for commuter trips (5 %). Bicycle is a rarely used mode choice, accounting for only 1 % of all trips (Fig. 9.3).

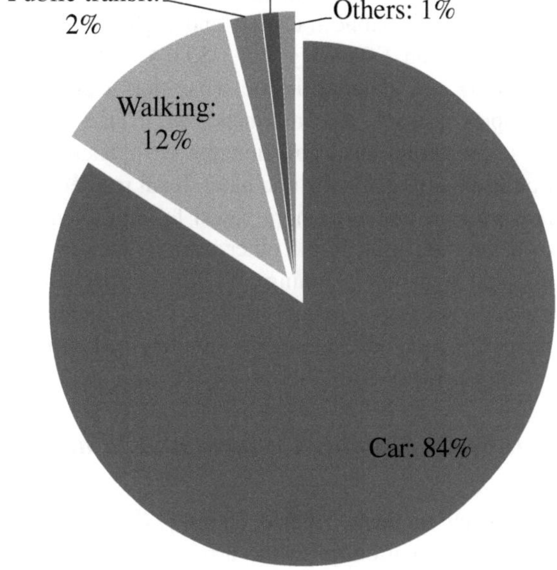

Fig. 9.3 Modal split in the Los Angeles region (all trip purposes) (*Source* US Department of Transportation [46])

Fragmented Governance

In this socioeconomically diverse region, transport planning is by no means an easy task. The Los Angeles County alone has 88 incorporated cities. To enable coordination of transport planning across these political boundaries, planning organisations were established. The Los Angeles County Metropolitan Transportation Authority, commonly known as the Metro, was established as both public transport operator and transport planning agency. It develops its own LRTP [30]. Its long-term priorities have been included in the Metropolitan Planning Organization's (MPO) Regional Transportation Plan for the region, which will ensure that these priorities are eligible for receiving funding from the federal government.

The region's MPO is the Southern California Association of Governments (SCAG). It was created to coordinate the decision-making of governments in Southern California and to guide regional planning. SCAG is the largest MPO in the nation and covers six counties (Imperial, Los Angeles, Orange, Riverside, San Bernardino and Ventura) and 190 cities. SCAG is charged with putting regional plans together and is responsible for centralised technical works regarding regional land use, community development, and the regional transportation plan. The policy at SCAG is guided by an 83-member Regional Council from 67 districts [37]. Since SCAG represents a large number of local authorities, reaching consensus can be time-consuming, and even result in delays in decision-making.

Federal legislation requires regional transport planning to be made through MPOs. Most of the federal funds go through the California Department of Transportation or SCAG. Nevertheless, transport planning in Southern California has been decentralised to the local county level [15]. In particular, public transport funds go directly to local transit agencies. One critical reason for the decentralisation in transport planning is the increased percentage of the local funds set aside for both public transport and highways since the 1991 Intermodal Surface Transportation Efficiency Act (ISTEA). This act was a milestone for transport financing and planning in the sense that it provided flexible funding sources for transport projects. A consequence of ISTEA is the increased funding from, and thus control by, local governments [15]. There are pros and cons to this devolution of planning. On the one hand, local residents pay more taxes to finance transport projects and have more power to align their monetary contribution with their own interests. But on the other hand, it creates a higher degree of complexity in regional transport planning. "[T]he transportation system is a regional system; flows do not stop at county borders. Local preferences are being accommodated at the price of the development of an integrated transportation system" [15].

Synthesis: Mobility Cultures and 'Place-Based' Planning

Compared to many Hybrid Cities [24], which are characterised by high public transport patronage in the urban core and high car use in the urban outskirts and suburbs, the density of rail lines is relatively low in the Los Angeles region.

Since the 1940s, Los Angeles has been developed around highways instead of pre-existing rail lines. The development of modern rail systems in the region has come rather late, with the first rail line opened in 1990. To many residents in the region, this alternative mode is fairly recent. However, just because the region is still car dependent does not mean that this will not change in the future. With significant improvement in public transport accessibility and service levels, it is possible that movement of people in this car-centric region will be undertaken increasingly by public transport.

When planning public transport, there is no one-size-fits-all policy. The region's inherent resistance to using public transport suggests that multidimensional elements will need to feature in its development: political, geographical, social, historical, and institutional areas must all be addressed. Locally embedded constraints should be carefully taken into consideration. Place-based planning might be the best approach for addressing the local issues associated with the public transport systems, as it places importance on understanding the history and origins of the problems. The cultural roots and heritage of the area could act as constraints to applying a universal solution to each city's urban problems. To manage global megacities, urban planners and policymakers need to 'think globally and act locally'. Place-based solution is the outcome of place-based governance, as opposed to the traditional top-down central governance, with legislative bodies that deliberate with limited knowledge of a problem and come up with one-size-fits-all legislation [23]. It functions in the form of collaborative dialogue and intergovernmental cooperation, aiming for deliberative democracy, efficiency and equity. Place-based governance takes into account local government and various stakeholders, and encourages empowered participation [13], social innovation [14] and use of social capital [35]. Because Los Angeles has a decentralised geography, a highly socioeconomically diverse population, complex urban and societal problems, and devolved and fragmented governance, a variety of innovative transport infrastructures will need to be put in place in order to serve the individual travel demands of various activities/purposes, and to serve local government's objectives.

An Example of Place-Based Planning: Revitalising Streetcars

Half a century has elapsed since Los Angeles' downtown streetcars last operated. The streetcars are now once again being brought to the planning table. The initiative of resurrecting downtown streetcar service dates back to 1997. The idea, which was inspired by the historic trolleybuses and cable cars which are a tourist attraction in San Francisco, was to link various elements of downtown by streetcar. "The concept was envisioned as a way to help reinvigorate, connect, and promote downtown by unifying otherwise disparate business, cultural, and entertainment communities" [10]. Bringing it back to the city could be a catalyst for transforming the image of public transport. As of November 2011, the streetcar project was in the alternative analysis stage, which is part of the first stage in the planning process. Although the alternative plans have not yet been finalised, the planning and service area of the streetcar services will cover several important landmarks/ neighbourhoods such as Bunker Hill/Union Station, historic downtown, and South Park [31]. The core service area and important connections (see arrows extending from perimeter) of the streetcar services are illustrated in Fig. 9.4.

The streetcar services have two main purposes—to enhance mobility and the circulation of public transport in downtown, and to support growth and revitalisation. The feasibility of bringing back the streetcar services as part of a broader redevelopment strategy for the downtown area was studied by the Community Redevelopment Agency (CRA) of the City of Los Angeles. In the feasibility report, the streetcar project is enthusiastically supported: "[t]he resurrection of the Los Angeles streetcar is being seen as the next step in this rail transit revival, not only because it can provide an alternative mode of transportation, but because of its community focus. It has the ability to be an economic redevelopment catalyst and help in bringing back the downtown life that once existed when the previous Los Angeles Railway system was still in operation. As a result, there is strong community and regional support for the re-introduction of streetcar service in downtown Los Angeles" [10].

Downtown CBD is the 'glue' of the region and the entry point of a new era of public transport development. In this area, image building, landscape design, infrastructure planning, and transport connectivity improvement are all critical to the success of the new wave of public transport resurrection. The streetcar programme exemplifies

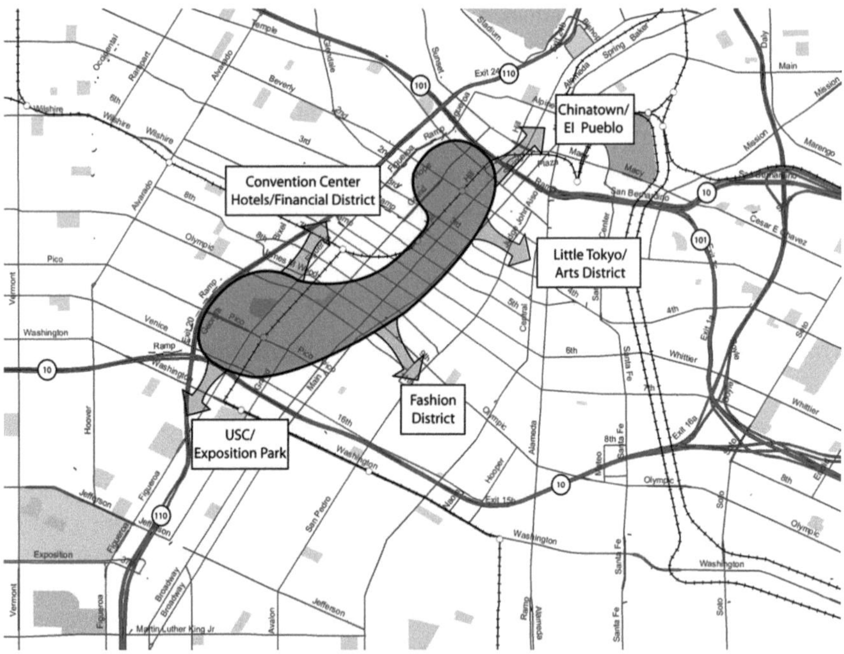

Fig. 9.4 Core service area of the downtown streetcar (*Source* CRA [10])

the importance of connecting various historic and contemporary business, culture and entertainment elements in downtown Los Angeles, while addressing the long-term local economic development goals of the inner city. In particular, this project strategically analysed the fate of Los Angeles' historic streetcars, and placed considerable sentimental and historic value on this long-missed means of public transport.

A Way Forward: Public Transport Planning in a Polycentric Region

The analysis of its mobility culture has helped us to understand the challenges facing public transport development in Los Angeles. Whilst it is important to continuously improve the service and image of the bus system, the impression that it makes on many current car users could take time to change; as for the innovative streetcar project, although it is clean and quiet, its service area is limited to downtown—for a larger service network and greater regional impact, a rail network would be the best option, for at least two reasons.

Firstly, we have witnessed a strong enthusiasm for developing new rail lines and extending existing networks over the years. Even under President Obama's administration, high-speed rail remains the centrepiece of the government's infrastructure vision. Rail mass transit is favoured by planning agencies because of its potential to secure the most outside funding [15]. Rail development is also favoured by local business owners and residents, because they can benefit from the appreciation of the land values and more liveable neighbourhoods that result from development orientated to public transport. A rail line has the potential to attract development along the corridor and boost the population density in surrounding areas, which in turn will induce efforts to improve the built environment and public transport accessibility, and eventually increase the mobility of nearby residents.

Secondly, population density is a crucial land-use indicator when evaluating the feasibility of public transport projects [25, 33]. Residents and business establishments have gradually gravitated towards the many established economic agglomerations, making the region one of the most densely populated metropolitan areas in the United States. The increasing density of the Los Angeles region signals an upcoming change in the underlying travel demand, and a potential increase in the use of public transport, should a network be in place to connect up the region.

Polycentric Urban Form

The Los Angeles region epitomises polycentric urban form, which could have some implications for the design of such a network. Increasing suburbanisation has given the region a long-lasting image: that of a highly dispersed metropolis.

This image is to a large extent associated with the region's spread-out urban form. However, the pattern of employment in this region is not low-density or scattered development as would be expected in sprawl. Rather, it is organised in multiple decentralised 'employment centres' [16, 36].

Among other definitions of employment centres, Giuliano and Small [16] define a centre as a cluster of contiguous zones having a certain minimum employment density (set at ten jobs per acre), and containing between them a minimum number of employees (set at 10,000). The employment subcentres of the urbanised area of the Los Angeles region in 2000, identified in earlier studies by Giuliano and her colleagues [18], are illustrated in Fig. 9.5, which contains a total of 48. The locations of these employment concentrations have remained remarkably stable since 1980, but the size and density of employment has changed substantially. While some subcentres have grown faster than others, the fastest growth is shown to be in the suburbs and exurbia (the 'stockbroker belt') of the region. There is a 'rank size' effect, i.e. there are a few very large but many smaller centres. The largest centre in terms of area, Los Angeles Downtown, spreads over nearly 18,000 acres, while the smallest, Newport Beach, spreads across little more than 600 acres [18]. In the light of the identified local agglomeration economies, a viable strategy for making Los Angeles a transit metropolis could be to recognise the underlying traffic demand between/within these employment centres and to incorporate them into regional and local transport planning and design.

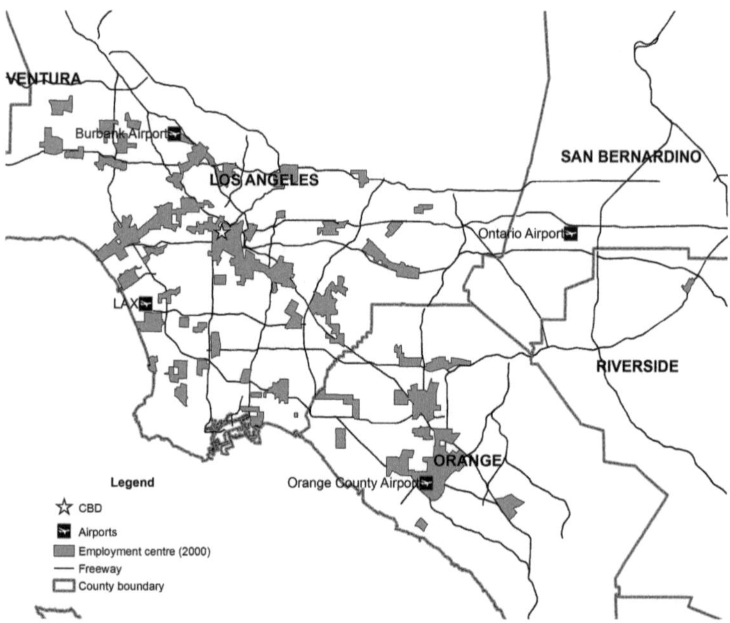

Fig. 9.5 Employment centres in the Los Angeles region (2000) (*Source* Giuliano et al. [18])

A Hub-and-Spoke Regional Public Transport Network

Because agglomeration of employment generates strong travel demand, a network that connects these clusters would facilitate use of public transport. The location of intersections would be based on employment density because business establishments and jobs are much more compactly clustered than residences as a result of bid rent [3] leading to high land prices and also owing to economies of scale. In contrast, residential locations are distributed more sporadically than jobs across the region. Therefore one possible structure would be a star network, consisting of hub and spokes that connect selected employment centres (the hubs) with rail lines. At the lower level of this network, bus services could be used to connect commuters' homes to the rail network. However, considering general perceptions of the bus system and the population's reliance on cars, people might prefer to drive from home to the nearest rail station instead of taking the bus. To accommodate their travel behaviour, park-and-ride facilities would play a crucial role by feeding car users into the rail system.

A 'hub-and-spoke' transport network (also termed a 'star network') has emerged in recent decades as a centripetal form of network structures [38]. In contrast to the point-to-point structure, the hub-and-spoke structure is a transitional stage of network development, where the limited traffic volume between the hub and individual nodes means that they are connected with limited routes. The advantage of hub-and-spoke structure network design is economies of scale at the hub, which acts as a distribution centre for the incoming and outgoing traffic (Fig. 9.6). Furthermore, the lower transport costs and higher infrastructure quality at the hub will lead to economies of scope (average cost economies stemming from a more diverse use of existing assets) in shared use of transhipment facilities [38].

The commonest adoptions and adaptations of the hub-and-spoke network structure have been in air traffic and freight traffic at the international, national,

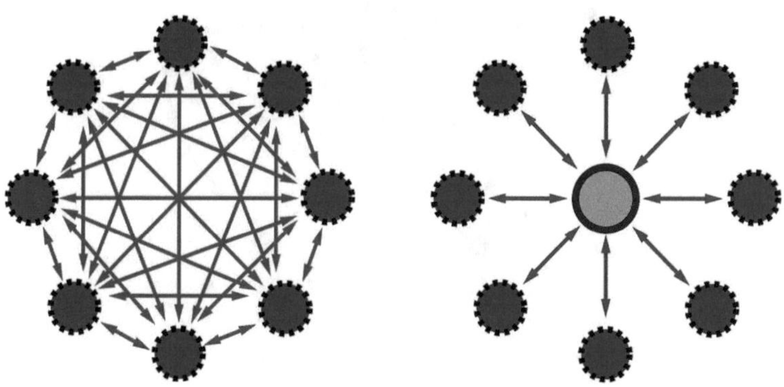

Fig. 9.6 Point-to-point (*left*) and hub-and-spoke (*right*) network structures (Author's representation)

and regional levels [38]. It is also applicable to transporting people, at the regional and metropolitan level. For example, Newman and Kenworthy [34] illustrated how an automobile-orientated city can be redesigned through a hub-and-spoke infrastructure network. In their plan, railways are used for intercity connections, whereas bus and light rail transit are used for connections between a town centre and local centres. The intermodal reconstruction plan, which is based on Sydney, could work for other metropolitan regions with a similar pattern of population density and business activity intensity (e.g. CBD, town centres, and local centres) (Fig. 9.7).

This type of settlement pattern is obvious in the Los Angeles region, which has a CBD and a number of employment subcentres of different sizes. Within the 48 employment centres in the region [18], the CBD in downtown Los Angeles would be kept as the network hub, whereas the other employment centres could share the same level of importance lower down the network hierarchy. However, since there are 47 other centres, the network would be more efficient if some of these centres were prioritised over others. The priority criteria would be based on employment size, growth rate, intermodal linkage, or a combination of the above. For example, Fig. 9.8 illustrates such a conceptual hub-and-spoke network that

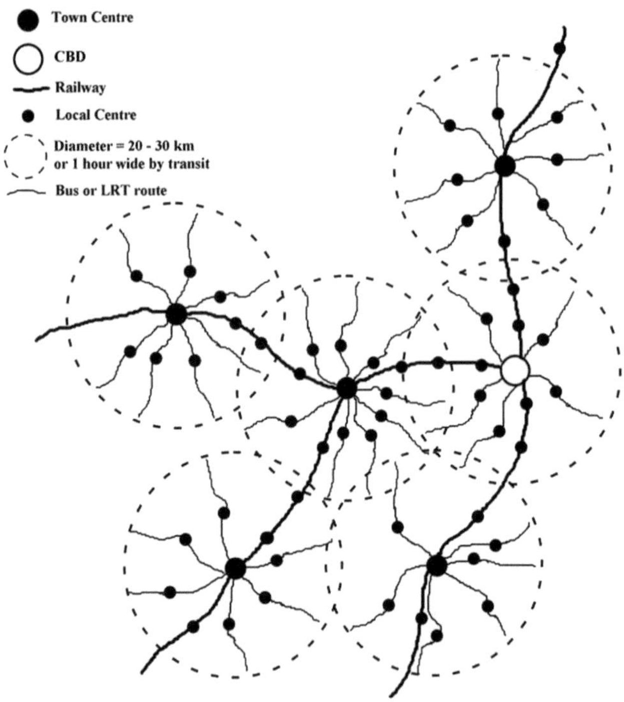

Fig. 9.7 A hub-and-spoke infrastructure network plan in Sydney (*Source* Newman and Kenworthy [34], p. 47)

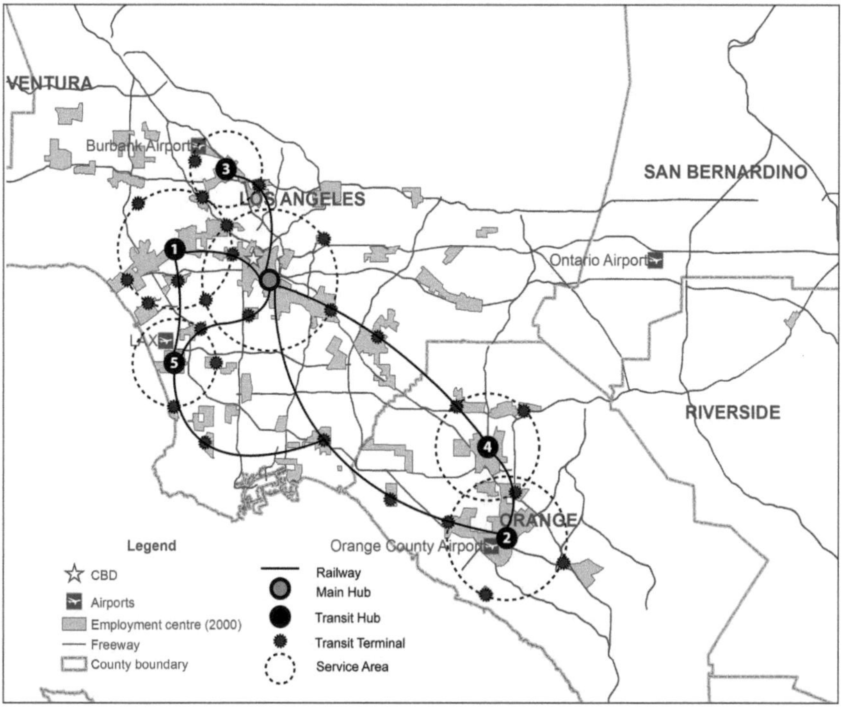

Fig. 9.8 A conceptual star network for the Los Angeles agglomeration economies (*Note* Main hub: LA Downtown—LA East. Transit hubs: *1* Santa Monica—Wilshire—Hollywood, *2* Santa Ana/Irvine/South Coast Plaza, *3* Burbank/Glendale/Universal City, *4* Anaheim, *5* LAX)

prioritises the five largest employment centres (LA Downtown–LA East, Santa Monica–Wilshire–Hollywood, Santa Ana/Irvine/South Coast Plaza, Burbank/Glendale/Universal City, and Anaheim). All of them have an employment level of over 100,000 (as of the year 2000). Furthermore, another transit hub could be co-located with (or near) the main airport (LAX) to improve the intermodal connection in the region. Compared to a point-to-point network, the hierarchical connection would inevitably cause different levels of accessibility across the region: the main hub and transit hubs would be served by public transport at a higher level of frequency and connectivity than anywhere else. Nevertheless, since the network structure and the accessibility level are determined by the underlying travel demand, as approximated by the size of the employment agglomeration, this type of network structure is more economically efficient and can better accommodate the traffic flow along the various corridors.

Given the political fragmentation of the Los Angeles region, and the potential regional impact of this conceptual public transport network, stakeholders from various sectors should be involved right from the planning stage. They might include the regional MPO (i.e. SCAG), local governments, state government agencies, transport commissions, resource agencies and conservation groups, the private sector, and the general public. In particular, transport authorities from both Los Angeles County and Orange County (i.e. Los Angeles County Metropolitan Transportation Authority and Orange County Transportation Authority) would ideally take more initiative to ensure the successful implementation of the plan. The star network should not be considered static or fixed; rather it is intended to be a starting point. Links connecting the nodes can be added and/or rearranged at a later stage, when node-to-node traffic has reached a sufficient level. When the network extends to other counties (i.e. Riverside, San Bernardino, and Ventura), involvement from their local governments and transport authorities as well as other local stakeholders might gradually increase.

Conclusion

Among other American cities, "Los Angeles was hardly unique in embracing the automobile as a democratic alternative to the railways" [7]. Easy access to the highway network and the convenience with which destinations can be reached are not features of a city that are in themselves blameworthy; rather they are a natural outcome of urbanisation and motorisation in many metropolitan cities. Critiques of Los Angeles' highway development and urban form stem from the worsening congestion and air pollution levels, as well as the time wasted in long-distance commuting. Despite these externalities, many people nevertheless still prefer driving to using public transport. This resistance to change in travel behaviour stems in part from people's residential location choice and the lack of a regional public transport network. Even those who live close to public transport routes may hesitate to use this mode of transport because of its old-fashioned image, which is related to many long-standing urban problems. Moreover, the abundant parking facilities in Los Angeles and the relatively low petrol prices in the US make driving inexpensive for the majority of households.

The challenges facing the development of public transport in Los Angeles would be best addressed with local knowledge and by means of customised solutions. From our mobility culture analysis, the importance of providing sufficient alternatives and of extending their coverage and service level can be seen. A regional rail network would be one strategy to move development a step forward. Owing to its geographic scale, the Los Angeles region needs a public transport 'backbone'. A regional rail network could serve this purpose, because rail systems are faster, more reliable, and less affected by traffic than bus systems. Existing car users, especially those who commute long-distance, might be more willing to switch to rail than to bus. Although there are many factors—such as the economic climate, oil supply and parking costs—that may affect the demand for public transport, discussion of these factors goes beyond the scope of this chapter. The traditional culture of public transport in the history of Los Angeles, and the recent enthusiasm shown by both the government and voters, mean that the political obstacles to rail development are relatively low.

In planning this backbone, we would need to factor in Los Angeles' unique urban form, which has been coined 'Los Angeles-style sprawl' [11]. As Ewing argued, "sprawl is a matter of degree. The line between scattered development, a type of sprawl, and multicentred development, a type of compact development by most people's reckoning, is a fine one" [11]. In view of the 48 employment centres presented above, Los Angeles is clearly a multicentred development. Identifying these local agglomeration economies and the underlying traffic demand between and within employment centres, and incorporating them into transport planning, could be a feasible tactic for making Los Angeles a transit metropolis. At these high-density locations, we can showcase how public transport development and land use may be creatively integrated to create more vibrant and liveable neighbourhoods. The conceptual star network, having several hubs aligned with the prioritised employment centres, has the ability to ensure satisfactory turnover and patronage, and hence improve the system's efficiency. When the urban core and other employment centres are better connected, new capital can be gradually infused into downtown, which could help to rejuvenate the urban core, reduce the region's spatial inequality, and support the effort to improve mobility.

References

1. ABC (2010) Los Angeles crime rates hit 50-year lows. http://abclocal.go.com/kabc/story?section=news/local/los_angeles&id=7204706
2. Adler S (1991) The transformation of the Pacific Electric Railway: Bradford Snell, Roger Rabbit, and the politics of transportation in Los Angeles. Urban Affairs Rev 27:51–86
3. Alonso W (1964) Location and land use. Harvard University Press, Cambridge
4. American Community Survey (2009) American community survey. http://www.census.gov/acs/www
5. American Public Transportation Association (2011) American public transportation association. http://www.apta.com
6. Banham R (1971) Los Angeles: the architecture of four ecologies. Penguin Books, London
7. Bottles SL (1987) Los Angeles and the automobile: the making of the modern city. University of California Press, California
8. BTS (2003) Research and innovative technology administration's bureau of transportation statistics. http://www.bts.gov
9. Cervero R (1989) Jobs-housing balance and regional mobility. J Am Planning Assoc 55:136–150
10. CRA (2006) Feasibility study for the resurrection of the red car trolley services in the Los Angeles downtown area (2006th edn), July 2006
11. Ewing R (1997) Is Los Angeles-style sprawl desirable? J Am Planning Assoc 63:107–126
12. Fishman R (2000) The American metropolis at century's end: past and future influences. Hous Policy Debate 11:199–213
13. Fung A, Wright EO (2003) Thinking about empowered participatory governance. Deepening democracy: institutional innovations in empowered participatory governance. Verso, New York, pp 3–42
14. Gerometta J, Häussermann H, Longo G (2005) Social innovation and civil society in urban governance: strategies for an inclusive city. Urban Stud 42:2007–2021
15. Giuliano G (2004) Where is the "region" in regional transportation planning? In: Wolch J, Pastor M, Dreier P (eds) Up against the sprawl: public policy and the making of Southern California. University of Minnesota Press, Minneapolis, pp 151–170

16. Giuliano G, Small KA (1991) Subcenters in the Los Angeles region. Regional Sci Urban Econ 21:163–182
17. Giuliano G, Small KA (1993) Is the journey to work explained by urban structure? Urban Stud 30:1458–1500
18. Giuliano G, Redfearn C, Agarwal A, Li C, Zhuang D (2007) Employment concentrations in Los Angeles, 1980–2000. Environ Planning A 39:2935–2957
19. Giuliano G, Redfearn C, Agarwal A, He S (2012) Network accessibility and employment centres. Urban Stud 48:77–95
20. Götz K, Deffner J (2009) Eine neue Mobilitätskultur in der Stadt—Praktische Schritte zur Veränderung. In: Bundesministerium für Verkehr (ed) Urbane Mobilität. Verkehrsforschung des Bundes für die kommunale Praxis: direkt, vol 65, pp 39–52. Bonn
21. Holzer H (1991) The spatial mismatch hypothesis: what has the evidence shown? Urban Stud 28:105–122
22. Ihlanfeldt KR, Sjoquist DL (1998) The spatial mismatch hypothesis: a review of recent studies and their implications for welfare reform. Hous Policy Debate 9:849–892
23. Innes JE, Booher DE (2003) Collaborative policymaking: governance through dialogue. In: Hajer MA, Wagenaar H (eds) Deliberative policy analysis. Cambridge University Press, Cambridge, pp 33–59
24. Institute for Mobility Research (2010) Mobility cultures in megacities: final report. Munich
25. Kenworthy JR, Laube FB (1999) Patterns of automobile dependence in cities: an international overview of key physical and economic dimensions with some implications for urban policy. Transp Res Part A 33:691–723
26. LAPD (2010) The Los Angeles Police Department. http://www.lapdonline.org
27. Los Angeles County Department of Regional Planning (2012) http://planning.lacounty.gov/luz/summary/category/residential_zones
28. Los Angeles Times (2008) Bottleneck blog: navigating Southern California traffic, transit and trends. http://latimesblogs.latimes.com/bottleneck/measure
29. Mees P (2010) Transport for suburbia: beyond the automobile age. Earthscan, London
30. Metro (2009) Long Range transportation plan. http://www.metro.ent/projects_studies/images/final-2009-LRTP.pdf
31. Metro (2011) Los Angeles county metropolitan transportation authority. http://www.metro.net
32. Neighborhood Scout (2012) Neighborhood scout. http./www.neighborhoodscout.com/ca/los-angeles/crime
33. Newman P, Kenworthy JR (1989) Cities and automobile dependence: an international sourcebook. Gower, Aldershot
34. Newman P, Kenworthy JR (2006) Urban design to reduce automobile dependence. Opolis Int J Suburban Metropolitan Stud 2: 35–52
35. Putnam R (1993) The prosperous community: social capital and public life. Am Prospect 13:35–42
36. Redfearn C (2007) The topography of metropolitan employment: identifying centers of employment in a polycentric urban area. J Urban Econ 61:519–541
37. Roberts I (1996) Children and sport: walking to school has future benefits. Br Med J 312:1229
38. Rodrigue J-P, Comtois C, Slack B (2009) The geography of transport systems. Routledge, New York
39. SCAG (2011) Profile of the City of Los Angeles. http://www.scag.ca.gov/resources/pdfs/2011LP/LosAngeles/LosAngeles.pdf
40. Shoup D (2005) The high cost of free parking. Planner's Press, Chicago
41. Snell C (1974) American ground transport: a proposal for restructuring the automobile, truck, bus and rail industries. U.S. Government Print Office, Washington: a report presented to the committee of the judiciary, subcommittee on antitrust and monopoly, United States Senate, 26 Feb 1974

42. Taylor B (2004) The geography of urban transportation finance. In: Hanson S, Giuliano G (eds) The geography of urban transportation. The Guilford Press, New York, pp 294–331
43. Texas Transportation Institute (2009) Urban mobility report. http://tti.tamu.edu/documents/ums/mobility_report_2009_wappx.pdf
44. Tudor-Locke C, Ainsworth BE, Popkin BM (2001) Active commuting to school: an overlooked source of children's physical activity. Sports Med 31:309–313
45. US Census Bureau (2010) US Census Bureau. http://www.census.gov
46. US Department of Transportation (USDOT) (2009) 2009 National household travel survey
47. Wilson R (1995) Suburban parking requirements: a tacit policy for automobile use and sprawl. J Am Planning Assoc 61:29–42

Chapter 10
Berlin: After the Growth: Planning Mobility Culture in an Environment of Dynamic Stagnation

Gunter Heinickel

Abstract After more than twenty years since the fall of the wall and the unification of Germany Berlin is still searching for its proper role and functions in Germany and Europe. For this, Berlin has not only to redefine and plan anew its traffic infrastructure in relation to long-distance links within Germany and into the European sphere, but this redefinition translates also in a complete overhaul of its inner-city transport networks and spatial relations in accordance with the general city development. Basically, this revision of the urban form can be interpreted as the quest for a new centrality for a city with a strong polycentric legacy. Yet, this reinvention and reorganisation of the city between the poles of polycentric and centralising forces reflects a repeated historical experience for Berlin. It is the aim of this chapter to show how in line with its own history, Berlin attempts to venture this reorganisation basically by a top-down approach, with the assistance of a number of master plans and regulatory devices, while at the same time rather unregulated processes from "below" redefine many of the contested and iconic inner city areas, which are at the focus of this city transformation. The consequential encounter of these top-down and bottom-up developments reveals new conflicts about deviant notions of urban life.

Introduction

Compared to the other metropolitan areas presented in this book, the Berlin example differs in several respects. Berlin is neither a megacity according to the usual definition (i.e. a conurbation consisting at least of 5 million inhabitants—and usually nearer 10 million; Berlin has a population of between 3 and 4 million), nor is the recent dramatic change of its urban exterior rooted in a vast growth of population and economic performance. However, much as Berlin may not seem to be a prime example of the present global shift towards intensified urbanisation,

G. Heinickel (✉)
Center for Technology and Society , Technical University Berlin,
Hardenbergstraße 16-18, 10623 Berlin, Germany
e-mail: Gunter-Heinickel@gmx.de; Heinickel@ztg.tu-berlin.de

Institute for Mobility Research (ifmo) (ed.), *Megacity Mobility Culture*,
Lecture Notes in Mobility, DOI: 10.1007/978-3-642-34735-1_10,
© Springer-Verlag Berlin Heidelberg 2013

probably nowhere else are the contradictory forces and trends of a European setting—and especially an Eastern European one—as visible. These express themselves in the form of present-day structural changes from an industrial to a service-based economy, reflected in a fast—even hasty—renewal of infrastructure and housing, while this complete redefinition takes place in a demographically stagnant, even receding environment. For that reason Berlin can probably serve as an example for future scenarios on a global scale: despite unabated population growth, birth rates are declining fast in most parts of the world. The current Berlin experience of sociopolitical changes in urban life in the context of demographic decline may have within it some pointers for the future of many megacities around the world in the coming decades.

Whereas London, the other European metropolis portrayed in this book, is characterised by longstanding political and social continuity, the immediate cause of the dramatic structural transitions that Berlin has undergone is the politically led changes in its positioning within Europe and Germany that have taken place over the last 20 years. That renders the Berlin experience somewhat similar to that of metropolitan regions outside Europe, such as Beijing, Shanghai and Gauteng (the South African province in which Johannesburg is situated). However, the planning strategies applied in Berlin do not have as their aim the governing or control of, or even dispute mediation between, the various strong self-sufficient forces pushing for social and economic change (be they from interest groups, entrepreneurs, independently minded citizens, or whoever) that are arising from a grassroots level and working, as it were, upwards; this is a profound distinction between this city's experience and that of these non-European metropolitan examples. In contrast to those cities, in Berlin these planning devices are implemented rather to *promote* such forces. That certainly affects any notion of mobility culture as applied to Berlin. Here, mobility culture is perceived by planners and political agents not as something which provides answers to longstanding, unavoidable limitations, originated and directed by the interests and activities of the citizens; but rather as something to be first envisaged, and then moulded by means of planning directives.

In view of the above, this chapter will cover the following points:

1. First, it will be shown how the desire to re-establish Berlin, the original German capital and now capital once more, as a key city in Europe has resulted in a top-down and centralised approach to development planning—and furthermore, that such a planning approach stands in line with the way that planning has happened historically in this city. The point will also be made that this continuity may be ascribed to Berlin's specific wider urban setting in Europe. As a result of this setting, the resources and potentials that are scattered throughout the region have to be consciously and repeatedly reallocated, by means of infrastructural and administrative interventions, to enable the city to attain to a level of growth and significance that extends beyond its own region.
2. Furthermore, it will be argued that this ambition and struggle for recentralisation and re-urbanisation has given rise to particular conceptions of applied land-use ideals, accompanied by a vision of a new mobility culture. In going

about this fresh start, and formulating an agenda for its development as a 'creative knowledge city',[1] the authorities in Berlin drew heavily on the metropolitan experiences of other major cities, especially when it came to lessons about the relationship between land use and mobility patterns. In recent times, researchers and planners have become increasingly aware of this interrelationship, translating mobility questions very much into land-use questions—that is to say questions about urban form: density, diversity and design [15]. In view of this relationship, and because of the relatively open planning context that has prevailed since 1989, solutions to mobility issues in Berlin have been conceived not merely as something which can help to resolve given limitations—concerning topography, infrastructures and economic diversifications—but, more than that, as something that can be actively deployed for the empowering of the city: the creation of a new mobility culture to initiate a virtuous circle of social and economic development.

3. The chapter also discusses how the way in which mobility concepts are received by the public helps to reveal different sets of constraints, which are often not even apparent at first sight. This is because the success or failure as regards the public acceptance of such concepts can uncover latent conflicts arising from different ideas about what mobility culture actually is. The latest conceptions of mobility culture include consideration not only of spatial forms, infrastructures, policy concepts and discourses, but also of the attitudes, behaviours and habits of transport users [11]. Such potentially hostile attitudes and behaviour towards the envisaged new urban and mobility ideals as applied to Berlin will be briefly highlighted by some examples of interviews. These interviews were collected in selected inner-city districts of Berlin during the ifmo programme on megacities. The neighbourhoods selected are highly representative, by virtue of the rapid social and infrastructural changes taking place in them, of Berlin's fashionable international image and reputation as a creative city. Moreover, the selected reference group of young urbanites so evidently visible in those districts represents the epitome of Berlin's hopes of rejuvenation and renewal. In these hot spots of development and change, the polished and well-meaning top-down implementation by city planners of transport and mobility schemes—sustainability-orientated though they may be—could encounter in unexpected ways a city that is incessantly improvising, and rather unruly.

[1] Creative knowledge cities are those whose culture, atmosphere and facilities are conducive to the settlement and development of creative industries (for example architecture, publishing) and knowledge-intensive industries (for example ICT, finance) and those who work in them—which is crucial to the future economic development and wealth-creation of the city. A 'talent pool' of well-educated and creative/technology-driven people are attracted to such cities. Transport systems are but one aspect of what constitutes a creative knowledge city; for example, university-city alignment and a business-friendly environment are two others.

The Past that Shapes the Present

The breakdown of the socialist states and economies of Eastern Europe during 1989/1990 and the subsequent unification of Germany brought Berlin abruptly back onto the political map in Europe. The opening of the formerly sealed borders between East and West assigned an unexpected new role to this city in eastern–central Europe. However, this new reality is sharply distinct from the situation extant before the Second World War, meaning that a simple return of Berlin to its pre-1939 significance is not possible. Instead, Berlin has to reposition itself in a manner similar to Warsaw, Kiev or Moscow, searching for and experimenting with new possibilities, trying to redefine a new centrality within Europe and within its own country. In fact, Berlin as both old and new capital of the German nation state never was clearly the social or economic centre of Germany [4].Traditionally, Berlin has, instead, played a part in a "German system of metropolitan regions" [17], competing in various spheres of specialisation for political, economic, social and cultural pre-eminence.

How Berlin Became a Metropolis and a European Transport Hub

The eventual rise of Berlin as a German political hub was based predominantly on political will—again comparable to the situation in Warsaw or Moscow—by the ruling monarchy and the developing Prussian state, and not on the basis of any outstanding strategic, geographical or economic advantages. And like Warsaw and Moscow, Berlin forms a distinct, prominent urban cluster in a geographical region of predominantly rural and small-town structures. These long-term characteristics of Berlin's geographical setting make it part of a European 'metropolitan corridor', stretching more than 2,000 km via Warsaw, Brest and Smolensk towards Moscow, a belt of significantly lower urbanisation than is found in western and central European regions. Berlin is somehow situated on the fringe of this corridor, in a region of Europe where two different tempos of development and densities of transport infrastructure come together [23, 24].

The fact that it is not situated on one of the major German rivers, but only on the modest River Spree, was another geographical drawback for Berlin that has hindered the development of its transregional significance in European commerce and traffic. Since the seventeenth century, the princes and kings of the Hohenzollern Dynasty, who had chosen Berlin as their seat of residence from around 1450, tried to overcome this obstacle by creating an extended system of canals. These linked the city to the not-too-distant Oder and Elbe rivers, and therefore to the global trade routes of the Baltic and the North Sea. The completion of the Finow Canal in the second half of the eighteenth century set Berlin at the heart of an extended system of waterways and inland harbours, at last elevating

Berlin to the status of largest inland port in the world in the 1920s, and enabling the city to rise in prominence as a manufacturing and industrial conurbation during the nineteenth century.

But the railway soon surpassed the waterways in significance. The tight, efficient railway system eventually secured the city's transformation into a major transport hub and industrial centre, and a major German metropolis. The elevated inner-city mainline railway is still the spine of this infrastructure, erected on 11 km of viaducts and bridges between 1875 and 1882, and confirming, in its east–west orientation, the primary orientation of Berlin's urban development and traffic routes thereafter. This railway line links all the major inner-city stations such as Ostbahnhof, Alexanderplatz, Friedrichstrasse and Zoo Station. It also serves as the backbone for the inner-city fast train system (S-Bahn), the first fully electrified railway system in the world (with work commencing in 1924 and electrification completed in 1929).

Aggregating Resources: Centralising the Polycentric

Top-down state planning and implementation of major infrastructures was therefore the repeated response of political leadership to the challenges posed by vast spatial resources and limited economic powers. The ambition to form an imposing and splendid Prusso-German capital led to the repeated aggregation and centring of dispersed and isolated economic clusters and pre-urban structures in the metropolitan region. To this end, strict planning and administrative devices were deployed. Over the last 200 years, a variety of schemes, frequently overambitious, have been implemented by government to draw these scattered regional potentials and functions towards the capital, and hence to structure and coordinate them.

To take an early example, the royal planning interventions of the seventeenth and eighteenth centuries could be cited, which turned the city's main course of development substantially to the west, reflecting the political will of the monarchy to compete with the most prestigious Western European capitals. This is most visibly manifest in the layout of the central boulevard Unter den Linden. This large and imposing avenue, starting from the former Stadtschloss royal palace, became the origin of an ever-extending east–west axis, which—like a spine—is the still most conspicuous arterial road. It eventually redirected large proportions of long-distance traffic from the former highways along the elevated plains around Berlin right through the heart of the city. Along this spine a number of the most prestigious new towns and later inner-city districts sprang up, sporting a 'modern' square pattern street design, which paid no regard to established traffic routes and principal directions, but tried to conform to the geometric ideals of the Baroque Age. Therefore not only was this new system of roads, squares and districts unsatisfactorily connected to the medieval centre, but it also created artificial barriers and hot spots for traffic, for example at Potsdamer Platz and Pariser Platz (by Brandenburg Gate), causing unnecessary congestion and traffic hold-ups from the twentieth century until the present day [14].

A later example of this kind of centralised restructuring attempt can be seen in the layout of the nineteenth-century bourgeois city according to the Hobrecht Plan. This attempted to manage the frenetic polycentric urban growth under one grand master scheme, which was intended to guide the city's development for the next 100 years. To achieve this it incorporated and realigned old village centres and country roads, without any regard for topographical features, into an imposing and splendid geometry of connections, junctions and regular housing blocks. But as is so typical of Berlin's planning history, the pioneering Hobrecht Plan was spoiled for practical as much as for financial reasons: the subsequent implementation of modern traffic infrastructure, especially train lines, and the decision to save public money by reducing the number of roads planned, destroyed the intended aesthetic flow and the proportions of the building blocks and streets. The reduction in number of roads increased the density of houses, creating less agreeable living conditions [25]. The disproportionately wide and open streets, on the other hand, proved to be most accommodating for the traffic requirements of the twentieth century, especially the car, but were often considered less appropriate for the contemporary pedestrian and occasional visitor, for whom the rigidly straight, regular and broad layout, bordered by strongly but indistinguishably decorated houses, appeared inhospitable and disorientating [2].

Nevertheless, Berlin grew under these eighteenth- and nineteenth-century planning provisions to attain genuine demographic and political significance in Germany—its population swelled from 57,000 in 1710 to 100,000 inhabitants in around 1750, catching up with Vienna in 1850, and exceeding one million in the 1870s and two million by around 1914. Finally, with the Greater Berlin Act of 1920, a further politico-administrative provision tried to reaffirm this centralising control over the ever-expanding metropolitan region: by incorporating over 90 village districts and newer towns—especially on its western and southern outskirts—into Greater Berlin, the Act added another two million inhabitants to the city. At this stage Berlin achieved its present size of 892 km^2, equivalent to an extension of 45 km east–west and 38 km north–south. Mark Twain, prompted by Berlin's relatively recent status as a national capital city, the lack of historically deep ingrained structures, and the precipitous growth in a space of only 200 years, compared it in the 1890s to contemporary Chicago [31].

The Lesions of Hot and Cold War

The outcome of the Second World War and the consequent political division of both Germany and Berlin eclipsed the dynamic and geopolitical position of the city at a stroke. Before 1945, Berlin's economic and cultural influence and pull had extended more deeply and strongly into the east of Germany, and into East Europe, and was not so pronounced to the west and south [4, 5]. The likelihood of any reprise of such a role is still uncertain: in contrast to the situation before 1945, Berlin is no longer the capital of the largest and most dominant German state,

Prussia. Instead, the city is located at the heart of one of the economically and demographically weakest German federal states, Brandenburg—and in fact only 100 km from the post-war border with Poland. Additionally, Berlin temporarily lost its role as a major national and international railway node for northern and eastern–central Europe. The former East German provinces of Prussia and Silesia, and other territories to which Berlin was the nearest German metropolis, now belong to Poland, and are mainly orientated towards Warsaw. The old transport links, economic relationships, and flows of both commodities and people were non-operational until 1990. As a consequence, Berlin lost its established role as an industrial centre.

Furthermore, the socialist experiment in East Germany and the resultant political partition which existed from 1945 to 1989 not only hampered industrial development, but in addition caused a major demographic downturn in the city and region as many people fled the socialist regime. As a result the population of Berlin was reduced from 4.5 million to its current figure of 3.5 million inhabitants—but under these conditions the would-be major suburbanisation by means of building detached housing, so typical in the development of post-war cities in the Western world, was curbed. The division of the city, and the building of the Berlin wall in 1961, led to separate development of infrastructure and transport systems within the conurbation for 40 years. In these circumstances the city developed two distinct main centres: as well as the historic centre (known as 'Mitte') in the eastern part, a new centre emerged around the Zoo area and Kurfürstendamm in the west. Duplicated communal facilities and public 'sister' institutions (state theatres, opera houses, museums, zoos, universities, libraries etc.) still bear witness to this duality.

However, the almost complete redefinition, from 1945 onward, of the urban and transport landscape of Berlin was occasioned not only by the war, the subsequent political partition, and the eventual building of the Berlin wall; facilitated by the profound destruction that the city suffered, the car-orientated doctrines of the Athens Charter could now be applied without reservation to the historic city [20]. A number of fairly radical central or master plans, commissioned by the capitalist western and the socialist eastern side alike—both competing for modernity—pursued the concept of the car-friendly city; these plans all had in common the ambition to completely transform the historical urbanity (i.e. the urban way of life, existence within a city), so as to separate out functions and establish the predominance of individual motorised transport [25]: widening many streets (in some places to the width of full-blown highways), reorientating inner-city roads and connections according to a grid-pattern ideal, and redefining neighbourhoods and residential environments [20]. Right up to the late 1970s, entire nineteenth-century boroughs were ultimately flattened to make way for open, scattered settlement structures. New passages were cut through the city's fabric, destined to become fast through routes. They were lined by wide green belts intended to shield the residential areas from noise and pollution, and also to offer recreational space for the occasional pedestrian. As a result the population density of many inner-city neighbourhoods dropped considerably, and shopping facilities often became concentrated and less variegated [25]. This reflects the city planning ideal at the time of segregated functions. All in all, access routes to shops and all-day services became more indirect, and modes of access less diversified, creating unattractive 'dead spots' and rather wearisome sections in a typical walk or stroll, in many of the inner-city neighbourhoods [7].

Berlin's Present-Day Dynamic Stagnation

The Legacy of History

These dramatic and brutal processes of ruptures to the city fabric and population shrinkage have resulted in a traffic infrastructure which is in many respects far too large—and outdated—for the existing population and the present-day economic capability of Berlin. Overly large roads, unused railway property, parallel structures of public transport lines—all these cause extra repair and maintenance costs. More importantly, it turns out to be difficult to introduce (directly or indirectly) new urban qualities into this superfluous space—or even to define what those qualities should be. As a result, large areas of brownfield and extended wasteland characterise many districts of the inner city even to this day, as isolated streets and segregated districts block junctions—or make them needlessly lengthy—and hinder opportunities for access within the historic neighbourhoods. Interim/stopgap solutions and cluttered improvisations scar many sites within the city [21]. The city still is fighting to overcome these divisions resulting from the great destructions of the twentieth century. However, a political initiative to form a common federal state (*Bundesland*) consisting of Berlin and Brandenburg, and thereby to reinstall the old political and administrative union of city and region, failed in a 1996 referendum.

Economic Stagnation

The same forces responsible for this rupturing lie behind the dramatic transformation of Berlin's economic setting from one of the foremost industrial and manufacturing centres of Europe (in the fields of steel, mechanics, electronics, chemistry and consumer goods, amongst others) into a would-be service metropolis. But as a legacy of the socialist economy in the East, and the formerly heavily subsidised economy in West Berlin, the public sector remains the dominant income sector, especially in the higher wage levels, illustrating the city's heavy dependency on public funds, and revealing the distorted income patterns, as compared with both German and international standards. Only in the sectors of biotechnology, medical technology and transport technology have some remnants of the old productive city survived, or been reborn. Gross Domestic Product was at a level of 101.4 billion Euros in 2011, and the metropolitan economy shows growth rates of only 1.8,

0.7 % lower than German average. Unemployment soars at 14 %, whereas net incomes stagnate at around 1,550 Euros per capita per month. The old economic elites left the city after 1945, and only a few returned after 1990. Over 80 % of enterprises in Berlin now belong to the service sector, which accounts for about 41 % of total employment, and of which the most important subsector is tourism.

Indeed, Berlin has witnessed a threefold growth in overnight tourism since 1990, which is unparalleled in its history. The city now ranks third among the most visited cities in Europe, with 17.8 million overnight stays, just surpassing Rome but coming behind Paris and London [21]. Paradoxically, the disjointed urban landscape, the all-too-visible tensions, and the spatial and architectural legacies of repeated shatterings of the past have become Berlin's most viable marketing asset: the sheer absence of many historic buildings renders Berlin a city full of history [20]. Because of the attractive power of this splintered urban landscape Berlin could—with some irony—be described as the first modern archaeological site in Europe.

Demographic Dynamic Stagnation

At first, Berlin may also seem to be a demographically rather stagnant place. Yet, behind this outward stability, massive changes have taken place. Indeed, the extent to which Berlin has been able to compensate for its economic shortcomings by attracting flocks of tourists and new residents since the 1990s is remarkable. Despite the fact that Berlin and Brandenburg each lost a million inhabitants after 1945, the city proper still comprises about 3.5 million inhabitants, and encompasses within the urbanised structures beyond the city limits 4.4 million inhabitants, while the so-called 'metropolitan region' of Berlin–Brandenburg totals around 6 million people (Statistik Berlin-Brandenburg. http://www.statistik-berlin-brandenburg.de/). This means that Berlin is still Germany's largest metropolis, being bigger than Hamburg and Munich combined. The population density of about 3,900 persons per km^2 is still the second highest in Germany after Munich, and stands in sharp contrast to that of the neighbouring state of Brandenburg, which has only 85 inhabitants per km^2 on average. Berlin, like Germany, has a fast-ageing society with a mean age of 45.

Yet, these unremarkable figures actually conceal the starker realities—there has in fact been a dramatic exchange of population; some have asserted that between 1991 and 2007 about one third of the city's population was completely exchanged, but that is probably an exaggeration. Be that as it may, Berlin certainly has a level of migration that far surpasses that of other states in Germany.[2] And whereas until 2004 the population was falling, there has been significant growth since 2006

[2] About 160,000 thousand persons moved into the city in 2011 (of whom around 70,000 were foreigners), while approximately 120,000 left it (of whom about 50 thousand were foreigners) (Statistik Berlin-Brandenburg. http://www.statistik-berlin-brandenburg.de/).

owing to this recent influx. In contrast to other more established megacities such as London, however, the city also simultaneously underwent a much less noticeable redefinition of its most socially and politically active and prominent social strata. The established socialist elites in former East Berlin were removed from public prominence, being rejected by society and severely weakened in influence, but the political and administrative elite of former West Berlin could not fill this gap and thus dominate the city that had now taken shape. This was due to their own lack of social and cultural capital, resulting from their strong dependence on the public money—the West Berlin establishment lived until the 1990s predominantly on a subsidised public-sector economy.

This politically caused void in terms of a well-defined elite opened up a sudden and unexpected social space for many in the incoming groups, who could then articulate and mould the transformation processes of the city in a way that is quite possibly more vivid and visible than in comparable ‘arrival cities’ [22] which are also experiencing an influx of immigrants having high social and professional expectations, whose elites are economically and politically more firmly established. Because of this, Berlin quickly acquired an international reputation for easy accessibility. In fact, the usual limiting factors such as living and housing costs, affordability, and the approachability of cultural and social arenas are less relevant than in all the other megacities addressed in this book. Segregating factors are also less obviously etched on the urban landscape, and tend to work more along sociocultural lines within boroughs and districts than between different parts of the city, especially when it comes to inner-city districts. While in London, for instance, the most costly and highly prized locations are to be found closest to the city centre, in Berlin it is the semi-suburban green belts, especially the established areas in the south-west of the city, that are traditionally the most valuable districts.

Sociocultural Dynamic

Furthermore, these new waves of immigration show significant differences when compared to the older patterns that Berlin had known since the late seventeenth century, when the population boost stemmed mainly from the east of Europe. They differ also from the situation in West Berlin after the building of the wall, when a low-skilled workforce was recruited mainly from south-eastern Europe and Turkey. Similar state contracts were made by East Berlin with their socialist brother states of Cuba and Vietnam, but never resulting in migration numbers comparable to those of West Germany. After 1990 the historical immigration pattern somehow re-emerged, as many Eastern Europeans rediscovered Berlin. But totally new, and unprecedented in terms of their social and cultural implications, are the arrivals from Western Europe—Spaniards and Italians, but also French and British—and from overseas, though it is still hard to come by exact numbers. Most of them arrive as students or long-term tourists, but eventually become residents. Often arriving with unclear intentions, they hope for serendipitous opportunities in the new and nebulous dynamic of the city.

These new people movements and migrations are inseparable from the recently intensified 'touristification' of metropolitan life—which occurs faster and is more prevalent than in other European cities—that has changed and reorganised the historic centre and neighbouring residential boroughs [21]. Low-cost tourism, in particular, frequently functions as an anticipatory factor, preceding migration—or even, for many, triggering personal decisions to move to and settle in Berlin, at least for a time.[3] These processes turned Berlin into the archetypal location for the 'creative class' as defined by Richard Florida [10], and recall some of the processes and associated problems that Charles Landry has described [16], cultivating new forms of low-cost urbanism [8]. Once this new dynamic had come about, the international perception of Berlin as a global city emerged, propagated especially by the cultural agents of the media and arts, although not backed up by reality in terms of economic importance or the presence of economic global players. This highly visible new citizenry now sustains the lion's share of newly founded enterprises and fresh expressions of business—predominantly in the sectors of tourism, communication and the media, IT, arts and other creative sectors—presenting Berlin with Germany's highest per-capita rate of newly established companies (124 per 10,000 inhabitants).

By the same token, these new inhabitants turned certain areas, which were alongside the former Berlin wall and consisted of the oldest existing settlement structures, into places of extremely elevated mobility, by comparison with other areas within Berlin, in terms of demographic regeneration, infrastructural renewal, economic and social activity, and, not least, in terms of physical movement, i.e. residential movement as well as leisure and professional travel. These neighbourhoods comprise, for the most part, long-neglected nineteenth-century housing structures of the Hobrecht era, threatened for a long time by demolition programmes in East and West alike. The specific conditions of the post-Cold War period—the unresolved legal status of many property rights (enabling cheap interim uses) and the subsidised or capped rents in East, as in West, Berlin—facilitated this sudden transformation of housing structures. In particular, districts that were formerly predominantly working class, such as Prenzlauer Berg (in the borough of Pankow), Friedrichshain and Kreuzberg, became largely or even completely redefined by the lifestyles of a left-liberal, ecologically orientated 'alternative' bourgeoisie and academic precariat (i.e. those who are well educated enough, but without any corresponding hope of job security, career path or reliable income source). Furthermore, these neighbourhoods promise to maintain the most outstanding growth rates until 2030, with Pankow (including Prenzlauer Berg) displaying an absolute dominance, having twice the rate of demographic growth as the next fastest growing borough [28].

---

[3] This was confirmed by several persons randomly interviewed in selected districts during summer 2011—see below for more details.

Struggling with Ambition

Striving for a New Mobility and Transport Centrality in Europe

This recent ambition of Berlin's to find a new centrality in Europe is most obviously expressed in the quest for a new transport infrastructure. As the original circle of rail termini around the historic centre—a design similar to most other European capital cities—was destroyed by war and partition, this east–west mainline was confirmed in its centrality after 1990: the reworking of the city's railway concept positioned the new central interchange station (opened in 2006) right on it, connecting for the first time the east–west and north–south train traffic of Berlin in one structure.

Yet, despite this latest railway design, intended to centralise and concentrate all mainline activities, Berlin's original role as an international railway hub is still hindered. Until 1990 all long-distance trains to and from East Berlin, and those connecting with West Berlin, were operated by the historical Deutsche Reichsbahn, the German state railway company given that name in 1924 but originally founded in 1920; from 1945 onwards this was, ironically (in view of the name, which was retained), under the jurisdiction of the socialist German Democratic Republic. Therefore, electrification on long-distance routes, especially in the direction of the capitalist Western neighbours, was not seriously pursued until after the fall of communism, and the rolling stock became completely outdated, with standards resembling those of the 1950s. Only in 1994 were the East German Reichsbahn and the West German Deutsche Bundesbahn (founded in 1949) fused under the name Deutsche Bahn AG ('German Rail'). Though there is a daily Berlin–Warsaw express, and (as from 2011) a direct Paris–Moscow service once more, the railway services connecting neighbouring regions of Poland remain insufficient, and are often inconvenient. In their place, a semi-legal—and at times clandestine—paratransit system of minibuses has supplanted these interregional routes. Originating at some train stations, and particularly airports, they serve destinations in neighbouring Poland, for example Szczecin (formerly Stettin, and today lying in the Polish part of the region of Pomerania) and Wroclaw (formerly Breslau, now in Polish Silesia). At Schönefeld airport alone, about 600 minibuses from Poland arrive during the course of a week. Only between 2 and 7 % of passengers travelling between Poland and Berlin do so by rail, as there are today fewer trains running on the mainlines between Germany and Poland than there were in the 1930s [9].

The political establishment, however, focused hopes of Berlin's return to its status as a hub of international travel on a project to construct a new airport, Berlin-Brandenburg Willy Brandt, at the outskirts of the city, to replace the two remaining operational airports, Tegel and Schönefeld. Berlin's other airport, Tempelhof, of all of them the most central and having supreme historic significance and reputation—being the focal point of the famous Berlin airlift and described as "mother of all modern airports" by Norman Foster—was closed in 2008. This aggregation of air traffic to one location serves as a textbook example of the general strategy behind the current city development plans: the accumulation and recentralisation of traffic and transport activities to foster the social and economic empowerment of the urban landscape. However, this ambition suffered an abrupt and most embarrassing hiccup in summer 2012 with the postponement—announced in a brusque and politically ill-advised manner—of the airport's opening date till 2013, owing to building delays. Once again, one of Berlin's highly zealous planning ambitions—to become a major transport hub for air traffic in central Europe—seems to have been obstructed by a lack of structural, or perhaps organisational, resources.

Ambitious Planning: Healing and Defining Urban Space

Behind this entire strategy of infrastructural reallocation and concentration stands the mission statement of the political class and the planning administration to redevelop Berlin into a "creative knowledge city" [1]. In fact, this mission statement can be read as an idiosyncratic reverse interpretation of recent studies on metropolitan development [17]: the observation about the positive correlation between manufacturing employment and service-sector employment led to the conclusion that strengthening the tertiary sector (service industries) could renew the production industries as well [1]. Furthermore, the authorities concluded that size and social diversity are genuine determinants of growth for cities—and not so much the results of an economically vibrant location: according to this way of thinking, the strengthening of centrality should somehow generate new economic life [10]. Against the background of this philosophy, the issues of land use, settlement structures and new designs for transport infrastructures became a major strategic vehicle for driving a clandestine sociocultural agenda. Various intra-urban and interurban approaches, analytical attempts, spatial agendas, and both integrated and cross-departmental strategies [1] were repeatedly amalgamated in many ways, to stimulate a kind of soft social engineering [13]. Taking the issue of the historic centre, not only is this agenda intended to overcome infrastructural tensions, but a painstaking process of economic transition is also supposed to be supported by it, empowering the metropolis by alluring a new stratum of professional high-flyers into the city [6]. In a way that recalls so many previous periods in Berlin's history, the idea of aggregating polycentric resources is once again being translated into an infrastructural planning agenda. Moreover, the creation of attractive quality housing is intended to inspire new residential and mobility preferences as well.

Most notable in this respect is the Planwerk Innenstadt ('master plan for the inner city', originally devised 1999 and revised in 2010), a massive—although not legally binding—scheme for rebuilding and socially re-engineering inner-city districts, incorporates this ambitious agenda: according to this general plan, the pre-war road and street layout is meant to reappear in at least some of its essential features—a compulsory block structure of architectural layout is specified, intended to bring back the historical blend of heterogeneous and socially mixed land use, i.e. living, shopping, working and possibly (clean) production [26]. The reconstruction of certain central squares and street layout with recognition value is supposed to strengthen the city's identity [3]. On an architectural level, this translates into bylaws about façade structuring, building materials (preferably stone coating) and proportions [12].This approach, with the aim of evoking the spatial qualities—and related mobility modes and rhythms—of historical urbanity, was labelled in the 1990s as the "ideal of the European City". Mobility was in fact explicitly listed among the guiding concepts of the Planwerk [29]. A typical architectural expression of this way of proceeding is the introduction of the multi-storey townhouse, a housing type hitherto unknown to Berlin. Nowhere else in Germany is this housing type so found in such numbers and so centrally located [19]. Property development and transport policy have therefore become intertwined, as they both aim for a new (actually, a traditional) quality of urban environment and use of space—the 'spatiality'—favouring short-travel-distance lifestyles by providing mixed land uses.

Regarding transport policy, the influence of this mission statement can be traced in various traffic and transport initiatives to promote the 'Umweltverbund' ('ecomobility', i.e. preference of green modes) according to the 2011 Stadtentwicklungsplan Verkehr ('city transport development plan'). In addition, the introduction of a low-emission zone in 2008 limits the access to the inner city to entitled vehicles [27]. Politically led models of coordination and cooperation of various transport suppliers and interest groups seek to promote non-motorised modes –pedestrian, cycling and public transport. The programmes thus concentrate on optimising existing infrastructure and transport supply in order to 'nudge' car users to use public transport, enhancing accessibility, safety and barrier-free mobility, while roadwork's are limited to renewal and maintenance.

Another prerequisite for the realisation of this transport strategy is the enhancement of intermodality in public transport. In achieving that goal, the main challenge after 1990 consisted of fusing the eastern and western rail services. This included not only the re-establishment of disconnected lines and the reopening of (hitherto non-serviced) 'ghost stations' of 'western' subway lines crossing eastern territory, but also the coordination of the S- and U-Bahn (subway) systems. Since West Berlin had extended the pre-war U-Bahn system as an alternative to the S-Bahn (which was under authority of the eastern Reichsbahn), the unsynchronised layout of stops and stations hampered intermodality between these systems after 1990. And even today, the tramway operates—with the exception of two extended lines—only in the eastern half of the city, as West Berlin had abandoned this system in the 1960s. The original intention of generally reintroducing the tramway to the western part of the city was scrapped for financial reasons. The decision to drop the plan was made easier by the prevailing and intractable prejudice on the part of the populace of the western districts against this 'alien' and potentially 'dangerous' rival for road space. Despite these discrepancies, the vehicles of the three systems operated by the Berlin administration—U-Bahn, trams and buses—have been decked out for some years now in matching bright yellow livery, citing historical models, but also imitating the renowned image of London Transport. It can thus be seen that the corporate identity of the transport system is intended to support the city's image by defining strong visual icons.

In accordance with these moves to strengthen public transport, and in order to promote a non-motorised mobility culture, the Berlin urban planning department (Senatsverwaltung Stadtentwicklung) adopted as mandatory the latest recommendations of the RASt 06[4] on the design of city roads. This new planning philosophy shuns efficient movement of traffic on roads as the criterion for planning traffic routes, replacing it by a consideration of the way that the roadsides are used, having regard to general amenity values, particularly for residents, pedestrians, cyclists and neighbouring roadside businesses. This implies a complete reorientation of city planning mentality by comparison with what has been the norm since the 1970s, when the idea of introducing popular roadside cafes and facilitating the general usefulness of streets for leisure purposes was rejected as completely alien and unviable ("Berlin will never be Italy" summed up the view of officials).[5]

Ambitious Governance

Because of these holistic ideals and the mutual interdependence of current land-use and transport planning in Berlin, one could probably speak of a governmentally developed and imposed neo-urban *concept of mobility culture*. And in contrast to London, Berlin's governance abilities are firmly anchored to a robust mentality of centralised institutional planning. These administrative structures acquired their fundamental modern form through the Greater Berlin Act of 1920, which defined 20 districts (*Bezirke*), incorporating 94 municipalities and localities. In 2001 the surviving 23 reunited districts were reduced to 12, with Friedrichshain-Kreuzberg as the only example of a new district comprising former eastern and western districts. The powers of these districts are limited, however, and especially in transport matters they are subordinate to the Senate of Berlin (the city government), with the 'reigning mayor' as head. The Senate commissions several agencies and counselling boards, such as the traffic control department, which

---

[4] RASt 06: **R**ichtlinien für die **A**nlage von **St**adtstraßen—these are noncommittal guidelines for the building of roads by the research institute Forschungsgesellschaft für Straßen- und Verkehrswesen.

[5] As stated in interviews with the Berlin planning department (Stadtentwicklung und Umwelt) in September 2011.

is responsible for the free flow and safety of road traffic. Remarkably powerful for such a big city is the commissioner and Cycling Council (the *FahrRat*), consisting of members of the Senate, the police, the districts and various transport and planning agencies, environmental and cycling organisations; and the presence of the lobbying organisation for pedestrians FUSS e.V. (rendered on their website as 'The Organization for Pedestrian Protection of Germany'), which is pressing for the relocation of cycle lanes away from the pavements onto the roads, and the introduction of bicycle 'fast-ways'[6] and pedestrianised 'green arterials' (carrying both pedestrians and cyclists directly through green spaces).

In contrast, the Center Nahverkehr Berlin (CNB—centre for public transport, Berlin) was set up in 2008 as an intermediary body standing between politicians and transport providers. Assisted by these powers and agencies, the Berlin Senate wants to achieve the envisaged reorientation of and decrease in travel activities, and a curbing of the suburbanisation processes—as confirmed in the newest city development plan from March 2011.

Utopian Ambitions for a New Mobility Culture

As outlined above, governmental planning devices such as Planwerk Innenstadt display an intention to use new settlement and architectural structures, not only for inducing social change and renewal by attracting a new citizenry of professional high-flyers, but also for promoting new styles of mobility in accordance with the ideals of ecomobility. In fact, the numbers plainly reveal, even now, a remarkably low car ownership in Berlin (324 per 1,000 compared to a German average of 570), and indicate a further receding share of car use in favour of bicycle and pedestrian modes, which have a combined share of approximately 30–40 %. Public transport modal share, by comparison, is estimated to be around 27 %. Against the backdrop of the fact that the car is not anything like as dominant in Berlin as in other cities, the official intentions to foster ecomobility seem to be enjoying a favourable reception by a compliant public. Recently, however, the fairly popular S-Bahn system, surprisingly enough, was plunged into a lengthy crisis: technical problems with a new generation of trains, together with financial issues and organisational problems concerned with maintenance and repair, resulted in painfully reduced and irregular services. This famed public transport system of Berlin, without warning, experienced a major blow to its image among residents.

On the other hand, the governmental notion of 'empowering' the metropolis by attracting a new stratum of qualified professionals and cultural elites by means of a revamped concept of the 'European city' is not something that is unanimously welcomed by the public, though. In particular, the current debate about 'gentrification' processes in Berlin—the rising rent levels, the transformation of flats into hotels/hostels as a result of the city's overwhelming new success as a tourist

---

[6] Cycle lanes designed to allow cyclists to travel at speed.

destination, and so on—is uncovering a certain amount of unease. It also raises a question: does this top-down definition of new urbanity work? As we have seen, Berlin has a record of failed, or reinterpreted, planning initiatives. So what are the ramifications of these comprehensive strategies, and to what extent is the social dynamic of the city in tune with these infrastructural parameters and their intended implications for a new mobility culture?

Competing Utopias of Urbanism: Mobility Culture as an Expression of Lifestyle

So how are the dynamic new citizenry of the 'creative class', as a central stratum of the hoped-for new urbanites and the most ardent agents of the new and unusual re-urbanisation processes, responding to these new planning provisions concerning transport preferences and urban spatiality? Are their expectations about a new (maybe even a 'low-cost') urbanism in tune with the governmental concepts of a new and modern mobility culture tailored to a new urban high-performing stratum of the populace? As part of the ifmo programme on mobility culture in megacities, a number of interviews were conducted among residents in selected locations of the above-mentioned neighbourhoods characterised by elevated mobility—these may give some indication of both actual and potential negative reactions centred on expectations about aesthetics, land use and mobility demands.[7] The items surveyed were their perception of: the existing mix of land use; street layout and use of road space; how welcoming the location feels to outsiders; visible evidences that area is undergoing transition/renewal; the nature of interactions between socially distinct groupings; and of tourists, in terms of numbers (too many?) and the effect that they have on the atmosphere.

Especially telling were statements made in reference to the planned reconstruction (Project K21) of the famous Kastanienallee in Prenzlauer Berg (revealingly nicknamed 'Casting Alley' among locals, in reference to the would-be starlets showing themselves, supposedly in the hope of being spotted by producers and photographers). Kastanienallee is an important access road running from the city centre to the borough of Prenzlauer Berg. It is a wide, busy avenue, large sections of which are not particularly smart, with a broad and open profile, not very well defined roadsides, and cobbled pavements that are bumpy and show signs of long-term neglect. This unlikely street, with its unremarkable features, has nonetheless undergone a transformation, becoming a popular shopping promenade for both locals and tourists.

The K21 Project serves, in fact, as a very good example of making the land-use and mobility concepts intended by the city administration a reality. The project was developed and proposed by politicians of the ecologically orientated Green Party,

[7] These interviews were carried out between June and August 2011 in the districts of Kreuzberg, Friedrichshain and Prenzlauer Berg.

supported by lobby groups for cycling such as the ADFC (Allgemeiner Deutscher Fahrrad-Club, the official German cycling federation)—for the reason, one assumes, that it is an attractive proposition for the new urbanites, who are (publicly, at least) so concerned about sustainability and the promotion of 'alternative' transport modes—i.e. public transport and bicycle. The road-building scheme in question does indeed stipulate a better sharing of modes in the thoroughfare, especially between the numerous bicycles and the tramway, by adding a cycle lane at the outer edge of each side of the road. Car parking is to be limited to a number of parking bays, using space taken from the pavements. The hitherto extremely spacious pavements, for their part, will be completely refurbished and at the same time narrowed.

The residents and regular visitors/shoppers interviewed voiced, however, a general sense of apprehension; in this regard, some of their statements seemed to express concerns that had a common theme: that the character of the street would be essentially altered and become too sleek, thereby forfeiting its quality of being somehow rough—even unruly—but at least an authentic urban environment, in which the 'regulation' of traffic is somewhat impromptu and a matter of negotiation. As an example of what they did not wish to see repeated, Kollwitzplatz, a square that was once the epicentre of the fashionable transformation of Prenzlauer Berg, was repeatedly cited. Now it is perceived by many longstanding residents as being too sleek and artificial, a place dominated by commercial interests for the sake of tourists. Comparable opinions were voiced in similarly structured districts such as Kreuzberg or Friedrichshain. In the latter, a young 'alternative' tourist couple from Amsterdam explained what they like about the place: "It's not perfect—which is good. Hopefully it will stay like that… By comparison, we live in a museum." And a young mother at Kastanienallee—coming from a not-too-distant neighbourhood as an occasional visitor and shopper—supported the refurbishment of the street in principle, but dreaded an associated shift towards a social monoculture in the area. The authenticity of the places in question is therefore very much associated with 'imperfect' and irregular structures, even if such 'authenticity' goes no deeper than the rippled cobblestone paving. Whatever the individual attitudes to the K21 design as a whole, resentment against the envisaged replacement of the existing—traditional—paving structure by more modern varieties and materials was unanimous among the interviewees. Additionally, local shopkeepers feared for their business turnover, and also were apprehensive about potential problems with delivery services.

Another concern related to Kastanienallee was that the street could become too 'fast'—especially the trams, which would no longer have to look out for bicycles, because of the segregated lanes. Currently many cyclists evade the competition for space with the tramway by simply using the pavements—which is easy because of their size, but increasingly generates conflicts with pedestrians.

In fact, in all locations of the survey the complaints about the latest 'cycling culture' were both unanimous and terse: even cyclists themselves are alarmed about ruthless riding styles—violating all traffic rules in a totally inconsiderate manner, paying no attention to children or the elderly. Cyclists are widely considered to be a growing nuisance, whereas complaints about cars—except for their noise and their consumption of public road space (by parking)—are surprisingly low-key and moderate.

Public debate in Berlin—as in Germany as a whole—about traffic planning and urban renewal tends to attribute all the virtues to cycling, with only negative statements being made about cars. Is it possible that this discourse—which, particularly in Berlin, so forcefully inserts itself into every new vision of urban mobility culture—has helped to create a new type of aggressive, self-righteous cyclist, with a mentality of privilege and a readiness to exhibit deviant behaviour? Moreover, there are now more bikes than ever, especially since the tourist industry discovered the suitability of this mode for offering guided tours—resulting in dozens of thronging cycling tourist groups. Unaccustomed to this mode of transport—especially in a city context—many of these tourists display an apparent insecurity, not to mention a lack of cycling ability, causing many minor accidents and hazardous situations. The interviewees frequently used the expression *Kampfradler* ('combat cyclist'), and reported many clearly dangerous encounters with cyclists. Concern about this phenomenon is certainly more than just a convenient media campaign on the part of car fanatics [18, 30].

These cursory impressions serve as little illustrations of the possible mismatch between the ambitious and encompassing mobility concepts commissioned by the Berlin city administration and the newly arriving—predominantly young—urbanites. Furthermore, they may give an indication as to the ways in which further closer investigation into the cultural aspects of mobility could be directed. These urbanites, however (and also others who feel so attracted to the habitats that have been shaped and defined by this reference group), would readily agree with the city planners about the goals of the new mobility concepts described: a mixed land use designed to reduce both the frequency and average distance of necessary trips, whether private or business; a preference for non-motorised modes; use of public transport wherever possible; and intelligent connections between different modes of transport.

However, the implementation of the infrastructural modifications needed for the achievement of these goals does reveal unexpected concerns pertaining to identity and lifestyle. These negative reactions may give some indications about widely diverging concepts of urban 'utopia' as pertaining to mobility styles. Such conflicts are certainly not so serious as to impede the wider planning strategies of the administration. However, these utopian ideas are obviously of sufficient relevance to generate interest, and some sympathy, among a significant proportion of the influx of visitors associated with the new international mass tourism in those parts of the city. It is evident that even the most well-meaning planning intentions orientated towards the sharing of road space and the promotion of multimodal activities can find itself fundamentally in conflict with the image of splintered—i.e. imperfect—urbanism, which may have found its expression within a mobility culture specific to Berlin.

Discussion and Conclusion

In most cities, especially megacities, factors that constrain the ability of individuals to move around in the way that they wish to (whether by necessity of for pleasure) are pretty self-evident, directly influencing them and featuring in everyday life.

Such limitations are to a very large extent the consequence of well-established interest groups which favour certain economic and social powers and activities.

Not so with Berlin. Neither in its international nor even its national setting, nor in its internal structuring, has this city ever developed a definite focal point that could give it a sense of orientation. The chief causes of Berlin's limitations are not any predetermined structuring of land uses—a relatively uninfluential factor—nor ownership rights or other legal instruments that determine use of space and the character of locations, but rather the *absence* of any such structures, and the continuing *presence* of voids.

This is only in part a legacy of the self-destructive history of Berlin in the twentieth century. In fact, the reasons are more profound, though hidden under the surface: the current situation of a splintered urbanism is in many ways a repeat—a déjà vu—of earlier periods of the city's history. This recurring lack of aggregating powers (other than those exercised by the state) is the cause of Berlin's impressive record of developing and implementing master plans, whereas the conditions prevailing in London and other megacities, on the other hand, prohibit such comprehensive planning.

Against this background, the themes of the resulting mobility culture in Berlin are not shaped by a mechanism of conflict and negotiation among individual citizens, or between social groups, with their vested social and economic interests; rather, they come from the reaction of the citizens to centrally drafted outlines and idealised plans originating from the administrative authorities.

However, Berlin, because of the unresolved and open breaches in its physical and social landscape, its zones that are in transition from 'old' to 'new', and its incomplete structures, has quickly become an utopia for various imagined ideals of new urban lifestyles. These urban visions consist not so much in hopes of social advancement or of high incomes, as is the case in so many other cities, as in aspirations to a lifestyle involving specific forms and expressions of mobility that are not nearly so constrained, externally, by cost or by boundaries of social interaction (whether legal or more informal ones). In other words, we are talking about aspects specific to a local mobility culture. But it is precisely this idiosyncratic and unplanned notion of local mobility culture that is, perhaps, coming under pressure from concepts of holistic mobility.

References

1. Adelhof K, Bontje M (2008) Looking for the "Holy Grail": how amsterdam and berlin want to become creative knowledge cities. In: Adelhof K, Glock B, Losseau J, Schulz M (eds) Berliner Geographische Arbeiten—Urban Trends in Berlin and Amsterdam vol 110. Berlin, pp 1–11
2. Bab J, Handl W (1918) Wien und Berlin: Vergleichendes zur Kulturgeschichte der beiden Hauptstädte Mitteleuropas. Oesterheld and Co, Berlin
3. BMVBS (Bundesministerium für Verkehr, Bau und Stadtentwicklung) (2008) Identität durch Rekonstruktion? Positionen zum Wiederaufbau verlorener Bauten und Räume.

Dokumentation der Baukulturwerkstatt vom 16. Oktober 2008 im Bärensaal des Alten Stadthauses in Berlin. Berlin
4. Briesen D (1992) Berlin—die überschätzte Metropole. Über das System deutscher Hauptstädte zwischen 1850 und 1940. In: Brunn G, Reuleke J (eds) Metropolis Berlin: Berlin als deutsche Hauptstadt im Vergleich europäischer Hauptstädte *1870–1993* Berlin, pp 39–77
5. Brunn G, Reuleke J (eds) (1992) Metropolis Berlin: Berlin als deutsche Hauptstadt im Vergleich europäischer Hauptstädte *1870–1993*. Berlin
6. Colomb C (2012) Staging the new Berlin: place marketing and the politics of urban reinvention post-1989. Taylor and Francis, London
7. Eberling M (2011) Schönheitsoperation oder Kahlschlag? Brunnen (1/4). Das Magazin aus dem Brunnenviertel
8. Färber A, Vetter A (2011) Low-Cost Urbanism: Ein Forschungsprogramm zur Untersuchung des Städtischen unter dem Eindruck von intensivierter Mobilität in der EU. In: Johler V, Malter M (eds) Mobilitäten. Europa in Bewegung als Herausforderung. Kulturanalytische Forschungen Göttingen, pp 109–216
9. Filter M (2010) Neue Wege in der Mitte Europas? Berlin und seine verkehrlichen Verbindungen nach Osten. Ein diachronischer Vergleich: 1930–2010. Master-Thesis. European University Viadrina, Frankfurt (Oder)
10. Florida R (2005) Cities and the creative class. Routledge, London
11. Götz K, Deffner J (2009) Eine neue Mobilitätskultur in der Stadt—praktische Schritte zur Veränderung. In BMVBS (Bundesministerium für Verkehr, Bau und Stadtentwicklung) (Ed) Urbane Mobilität. Verkehrsforschung des Bundes für die kommunale Praxis: direkt 65 vol 2009. Bonn, pp 39–52
12. Hertweck F (2010) Der Berliner Architekturstreit: Architektur, Stadtbau, Geschichte und Identität in der Berliner Republik 1989–1999. Berlin
13. Hoffmann-Axthelm D (2001) Warum sind die Bündnisgrünen gegen das Planwerk Innenstadt? In: Stimmann H (ed) Von der Architektur- zur Stadtdebatte: Die Diskussion um das Planwerk Innenstadt, pp 83–93
14. Kiaulehn W (1981) Berlin—Schicksal einer Weltstadt. C.H. Beck, München
15. Klinger T, Kenworthy JR, Lanzendorf M (2010) What shapes urban mobility cultures? A comparison of german cities. European transport conference, 2010, Glasgow, Scotland, UK. 11 Oct 2010–13 Oct 2010
16. Landry C (2006) The art of city making. Earthscan, London
17. Läpple D (2007) The German system. In: Burdett R, Sudjic D (eds) The endless city. The urban age project by the london school of economics and deutsche bank society, pp 226–243
18. Matthies B (2011) Ein Pfarrer macht mobil: Mit Gottes Segen gegen die Kampfradler. Der Tagesspiegel. http://www.tagesspiegel.de/berlin/ein-pfarrer-macht-mobil-mit-gottes-segen-gegen-die-kampfradler/4677908.html. Accessed 30 Sep 2011
19. Neisen V (2008) Townhouse-living in the centre of Berlin. In: Adelhof K, Glock B, Losseau J, Schulz M (eds) Berliner Geographische Arbeiten—Urban Trends in Berlin and Amsterdam Vol 110. Berlin, pp 24–38
20. Oswalt P (2000) Berlin. Stadt ohne Form. Strategien einer anderen Architektur. München, London, New York
21. Richter J (2011) Low-Cost Urbanism. Die räumlichen Auswirkungen des "Low-Cost Tourism" auf die Berliner Innenstadt. In: Johler VR, Malter M (eds) Mobilitäten. Europa in Bewegung als Herausforderung. Kulturanalytische Forschungen pp 199–208
22. Saunders D (2010) Arrival city. How the largest migration in history is reshaping our world. Karl Blessing Verlag, Munich
23. Schlögel K (1999) Kiosk Europa. Oder Berlin als Perspektive und Zentrum. In: Berlin. Metropole: Kursbuch Kursbuch, Heft 137), pp 161ff
24. Schlögel K (2001) Promenade in Jalta und andere. Hanser, München, pp 262–271
25. Schmidt AK (2008) Vom steinernen Berlin zum Freilichtmuseum der Stadterneuerung. Die Geschichte des größten innerstädtischen Sanierungsgebietes der Bundesrepublik: Wedding-Brunnenstraße 1963–1989/95. Hamburg

26. Schwedler HU (2008) Overcoming spatial and cultural barriers: urban planning and management in berlin since reunification. In: Williams JA (ed) Berlin since the wall's end: shaping society and memory in the german metropolis since 1989 Newcastle, pp 33–58
27. Senatsverwaltung für Stadtentwicklung, Abteilung VII Verkehr, Land Berlin (2011) Stadtentwicklungsplan Verkehr. Berlin
28. Senatsverwaltung für Stadtentwicklung und Umwelt, Ref. IA (2009) Bevölkerungsentwicklung für Berlin und die Bezirke 2007–2030. Stadtentwicklungsplanung in Zusammenarbeit mit dem Amt für Statistik Berlin-Brandenburg
29. Senatsverwaltung für Stadtentwicklung, Umweltschutz und Technologie (1997) Planwerk Innenstadt Berlin: Ein erster Entwurf. Berlin
30. Stresing L (2011) "Kampf den Kampfradlern": Schluss mit den Kleinkriegen! *Der Tagesspiegel.* http://www.tagesspiegel.de/meinung/kampf-den-kampfradlern-schluss-mit-den-kleinkriegen/4384276.html. (12 July 2011)
31. Twain M (1892) Berlin: The Chicago of Europe. *Chicago Tribune.* http://twainquotes.com/Travel1891/April1892.html. (3 Apr 1892)

Chapter 11
London: Culture, Fashion, and the Electric Vehicle

Ivo Wengraf

Abstract In this chapter, I outline the mobility culture of London through the example of the G-Wiz EV. London, as a global megacity, has a mobility culture that must function under a wide range of transport, environmental, political, economic, land use and architectural constraints. In short, a modern global city squeezed into a historical shell, with a mobility culture marked by little negotiation room for large scale shifts through built environment alterations or behavioural change. This "make do and mend" mobility culture limits the conceptualisation of the issue of mobility in London. This chapter considers the G-Wiz, a small EV which experienced a remarkable and seemingly sudden increase in sales. Through a confluence of social, economic and technical conditions, EV sales increased sharply, following running cost incentives (significantly the Congestion Charge) and the fashion of green conspicuous consumption dominant in UK public discourse at the time. This chapter argues that the G-Wiz was used as a mobility *tool* to adapt superficially to mobility constraints without a change to either the ways of moving in the city or the ways of understanding those movements. Thus, the G-Wiz serves as a useful example of London's mobility culture at work.

Introduction

The mobility culture of London must function under a wide range of transport, environmental, political, economic, land-use and architectural constraints. While London is a lively, modern global megacity, it is also a patchwork of settlements going back to the Romans, with a built environment that is now essentially fixed, partly by choice through planning regulation, and partly by necessity because of the cost and upheaval of introducing change. London is a modern global megacity squeezed into a historical shell, with a mobility culture marked by little room for negotiation when it comes to large-scale shifts through alterations to the built

I. Wengraf (✉)
Royal Automobile Club Foundation for Motoring, London, UK
e-mail: Ivo.Wengraf@racfoundation.org

Institute for Mobility Research (ifmo) (ed.), *Megacity Mobility Culture*,
Lecture Notes in Mobility, DOI: 10.1007/978-3-642-34735-1_11,

© Springer-Verlag Berlin Heidelberg 2013

environment or behavioural change. This restriction is a product of political, economic and social conditions. This chapter will argue that the city's mobility culture is one of 'make do and mend'. In London's mobility culture, the conceptualisation of mobility revolves around the type of mobility, the source of funding for it, and the sign value of any mobility technology, rather than, say, the need for mobility itself, its location or its role.

This chapter reviews the rapid uptake of the G-Wiz electric vehicle (EV) at a certain point in London's mobility history. It sets out a number of key elements of the underlying mobility culture of London, and uses the example of the G-Wiz to demonstrate how mobility in London is constrained. The G-Wiz, is, in fact, a *direct response* to mobility constraints in London. The city's mobility culture restricts the possibilities for fundamental changes in infrastructure or behaviours that would improve the costs, availability or environmental impacts of mobility. Instead, shifts in ways to travel in London are more often comparatively small modifications—economically and politically straightforward. They are without major impact on the shape or appearance of the metropolis, and, especially more recently, have a particular focus on the symbolic, sign values of mobility technologies. This chapter argues that the G-Wiz EV was used as a mobility tool to superficially adapt to mobility constraints without a change to either the ways of moving in the city or the ways of understanding those movements. The G-Wiz serves as an example of London's mobility culture at work: limiting infrastructural and historical conditions, a need to be seen at the forefront of progressive and fashionable behaviour, and citizens who are unwilling—or unable—to make the fundamental changes required for more sustainable transport in the city.

Mobility Culture

To explore fully the mobility culture of a city as large and diverse as London is, of course, well beyond the scope of this chapter. However, there are fundamental differences—key constraints—that frame mobility culture in London and mark it as distinct from those cultures seen in cities to which one may reasonably compare London (such as New York or Berlin). These differences can be considered in a number of ways—in this chapter, I examine London mobility culture through three of these frames: (1) development and governance, (2) economics and employment and (3) infrastructure and change.

Development and Governance

Although one may speak of the 'city of London' to mean the modern metropolis, London is in fact a patchwork of settlements from the urban, through the suburban, to the rural, and encompasses development from across the history of modern settlement design. Essentially, the modern megacity of London was created by means of a collection of cities, towns and boroughs growing into one another. Three points emphasise the nature of this complex and varied development. Firstly, the term 'London' was not used officially to describe beyond the small Roman/medieval core until 1889, with the creation of the County of London. Secondly, the governance of London is through a number of independent borough and city councils (32 London Borough Councils and the City of London Corporation) within a citywide government of historically varied structure. Thirdly, there is no functioning distinct universal boundary for London as a megacity.

Despite a long history as a global city, London has not always had a mayor; in fact, London has only recently created a well-established, citywide, elected mayoral government. Originally administered as a group of individual, historical cities and boroughs, London was reorganised to be run through a county council (excluding the City of London) in 1889. In 1965, the Greater London Council was created, to run the city along with the 32 London councils and the City of London Corporation. The Local Government Act 1985 abolished the Greater London Council and the metropolitan county councils, so that London did not have a citywide political leader. In 1999, as part of a process of devolving governmental responsibility to London, Wales, Scotland and Northern Ireland, the Greater London Authority was created, with an Assembly and an elected mayor who had specific responsibilities for citywide policies, perhaps most notably transport. The Greater London Authority boundary does not entirely reflect the size of London as a megacity, and four other 'boundaries' are worthy of note:

1. the greenbelt, an area surrounding London with strict development controls to limit London's expansion;
2. the M25 peripheral motorway;
3. the ill-defined London 'commuter belt', and, for the purposes of data collection;
4. the administrative areas of Transport for London (TfL) and southeast commuter rail.

The London Mayor's role now puts him in control of the public transport body, TfL, to manage the transport network with a strong, centralised and multimodal remit.

Since the establishment of the office of the Mayor, and the development and acceptance of the Congestion Charge to control entry to motor vehicles into the centre of the city and at the same time raise hypothecated (ring-fenced) revenues for transport improvements, London now has the governance structures, the access to capital, and the governmental and public interest to enable involvement in new developments in transport technologies, schemes and policies. Since the Congestion Charge, there have been a range of efforts (amongst which are smart ticketing with the Oyster card, cycle hire, new cycle routes, new bus development and EV promotion) which suggest a more proactive and progressive approach to mobility—although, as argued below, limited by the scope of London's mobility culture.

Economics and Employment

Greater London—modern, colloquially understood London—is around 1,590 km^2 in area and in 2010 was estimated to have a resident population of 7.8 million people—under 15 % of the UK's population. Its economic and political influence, however, extends throughout the UK. It is a Hybrid City (see "Chap. 2"). In other words, it has a dense urban core with high levels of public transport use, and a more automobilised low-density suburban area surrounding it. In addition, in a way rather particular to London, it has a large peripheral commuter belt, extending far beyond the city. London has a pre-eminent position in the economy, society and geography of both England and the United Kingdom. As noted in the London Plan [10], London is one of the top three global centres for finance, and the largest city within the European Union. It is, of course, also the centre of UK government, although some parts of the civil service have been relocated to cheaper, less economically developed areas of the UK in an effort to reduce costs and bolster local employment in depressed areas. London is also a major cultural centre, with a number of buildings, theatres, galleries and museums of global significance. It has 43 universities and institutes of higher education—the highest density in Europe—and it is has a significant knowledge economy. These knowledge and cultural economies are vital to London. London is also a major transport hub internationally (Heathrow, one of the five international airports for London, is the busiest airport in the world by international passenger numbers, and third after the major airports of Atlanta and Beijing in terms of numbers of all passengers including domestic), and nationally (the UK road and rail networks are set out radially from London).

London, being a global city and international air hub, and also the home of English, the de facto international language of business, experiences significant international immigration, both to the city itself and as a gateway to other parts of the UK. For many sectors of the economy, working in London is important as part of career progression (a fact supported by the fertility rate statistics for women in London, which show a larger proportion of mothers having children later, relative to the UK average). As manufacturing in the other large cities of the UK (such as those in the North West and West Midlands regions) has declined, London's economic importance relative to them has increased—and London's large financial, cultural, academic and scientific sectors have accentuated this prominence. In addition, as economic ties with Europe have strengthened, London and its related satellite towns in the southeast (e.g. those along the M4 corridor, or Ashford in Kent) play an increasingly important role in international trade. London has been left to drive the UK economy, and this has put pressure on its housing, transport and infrastructure.

In view of all of these factors, one should view London's economic situation at two different scales: firstly, the importance of—and perhaps overemphasis on—london for the UK economy, with resultant pressures on mobility; and, secondly, the way that it relates to and competes not so much against other UK cities as with an elite group of global cities, which gives rise to demands and expected standards not applicable elsewhere in the UK (consider, for example, the issue of airport expansion) and to the political and financial problems concerning the funding for its infrastructure (for example, the long-planned Crossrail project involving the construction of major new railway connections under central London, linking neighbouring counties to both east and west).

The pre-eminent economic position of the city has created a robust market for employment, putting pressures on housing prices and extending the commuter belt far beyond the region administered as part of London government. Accessibility to jobs is measured by TfL in terms of those within a 45-minute commute time, but many thousands have a far longer commute than this. While outer Londoners have between 250,000 and 500,000 jobs available to them, within a 45-minute commute time, central Londoners have 2.5 million. Economic opportunities have helped to raise property values and housing demand in London relative to the rest of the UK. Affordable housing is in short supply, and this helps to drive the increase and expansion of the London commuter belt (the surrounding area from which it is realistic to make a daily rail trip into London to work). The housing crisis endemic to London, the demands on further development in the southeast commuter area and the mobility culture in this city are all related—it is common for employers to provide additional bonuses (called the 'London weighting'[1]) or an interest-free loan for the purchase of a railway season ticket, so that staff can afford to travel to work from an area in which they can afford to live. Approximately 700,000 people commute from outside London, mostly by train; the price of the service, and the pressure on it, are significant. The centre of London, therefore, is congested and expensive, but critical for employment and the national economy.

Infrastructure and Change

London was the first city to develop a metro rail network[2] and was the largest network by length until Shanghai overtook it in 2007. London was a leading city in the development of the railways. As one would expect from a city with such a public transport history, London has significant infrastructure, much of which is Victorian, Edwardian or Georgian, and constrained in development by the city's early leadership role in infrastructure creation. In fact, there are a great number of historically listed Underground and mainline railway stations and rail loading gauges are limited by many low Victorian rail tunnels and bridges. While there has more recently been an increase in light rail (for example, the Croydon Tram and the Docklands Light Railway), and the introduction of the high-speed commuter line and service to the Channel Tunnel, much public transport infrastructure is still based on these early efforts. So, while London may well have been a leader in transport infrastructure development, the Tube and rail networks have not been significantly expanded in the city in recent times, and there have been problems regarding its

[1] Generally approximately £3,000 to £4,000 per year.

[2] Hereafter referred to as either 'the Underground' or 'the Tube', and first opened in 1863.

modernisation, with systems often constrained by cost, complexity and historical infrastructure. Further to this, transport in London is a complex animal, as it is a city with 33 local administrations and the centre of the national transport network, including a convolutedly privatised and expensive rail network running passengers from outside London into and through 12 separate main railway stations.

In addition to the problems of small tunnels too expensive to make larger, and historically protected transport infrastructure, there is also the fact that London's transport is a fundamental part of the city's identity. Harry Beck's classic topological Tube map streamlines the information required for negotiating London, and a strong brand identity, simplifying things for travellers negotiating the system, is formed by this map, the London Transport (LT) typeface, and the Underground's world-famous logo (the roundel consisting of a red ring with "UNDERGROUND" in white letters on a blue banner across the centre). They are carefully managed symbols of the city. The black cab and the Routemaster bus are also symbols of London (see Fig. 11.1), and the London Transport Museum has a major focus on art and design in public transport that acknowledges the importance of the particular appearance of certain transport infrastructure in the making of the city. The best example of the importance of transport to the city's public identity is the campaign to save the Routemaster. These buses are still used on routes which pass key landmarks, meaning that the photo opportunities are maintained. When they were replaced with 'bendy buses', there was significant complaint, in spite of compelling arguments from a point of view of transport provision and planning. The current Mayor has promoted the design of a new bus which echoes the design of the Routemaster, at great expense and with a necessary reduction in capacity.

Fig. 11.1 Transport as a city symbol

London is the site of the textbook example of urban congestion charging. Although the scheme has fluctuated over time in terms of cost, exemption details and structure, it basically covers the area of London within the innermost central ring road (i.e. the centre of the centre of London). Cars pay a fee of £10 per day to enter the charging area, which is monitored by number plate recognition cameras. There are exemptions for low-emissions vehicles, public transport, government vehicles and powered two-wheelers. London also has a Low Emission Zone, which covers almost all of Greater London with a similar camera network. Emissions limits are set for all vehicles larger than cars, and one can check with TfL to see whether a particular vehicle meets these criteria (which have recently been made more stringent).

The Congestion Charge marked a significant shift in the funding of transport in London, and also in the stomach for and interest in new transport projects. London now has a number of projects that are based on new mobility concepts. The Barclays Cycle Hire Scheme (renting out what are commonly called 'Boris Bikes' after the present Mayor of London, Boris Johnson) has proved highly visible, although the success of the scheme in terms of transport provision is debatable. Expansion of this scheme will be supported by the related expansion of cycle superhighways—these are wider cycle routes which have proved contentious, not only with those who have no interest in cycling infrastructure provision, but also among committed cyclists.

What is critical to London's mobility culture is the related importance of housing supply and cost, access to jobs, and the necessity for transport. London is a city constrained by its own historical development. The greenbelt limits further growth, and the city and related infrastructure have often been developed:

1. for smaller, pre-London 'towns';
2. independently, by Victorian engineers, private firms or local councils; and
3. early in terms of technological development (recall, for example, what has been said about the tunnel loading gauge).

As other cities in the UK, more directly dependent than London upon manufacturing, have declined since the 1970s; as trade has increasingly focused on Europe; and as the UK economy has shifted towards tertiary industries, London has become ever more important as a location for jobs—a change that has been helped by the fore mentioned role of English in business, and the creativity and knowledge base of London. This has brought people to the city, and then pushed them out to travel back in, as they seek to balance employment with quality of life and affordability of housing. Much of inner London contains housing not originally designed for its current use. Many houses have been converted into flats. These flats are not purpose-built, either as flats or for UK lifestyles and values. London is, therefore, a city of travellers, who are trying to balance these related issues: the good jobs of London, the costs of travel, and the costs and quality of housing.

London's travellers include 700,000 commuters entering the city from *outside* Greater London—travelling significant distances at significant cost. Those who work in London must balance the economic strength of the city and resultant access to good jobs, with the costs of living, shortage of housing and problems with quality of life. Many are pushed further and further out of the city. The housing crisis and the north/south economic divide play a profound role in shaping mobility culture. People find themselves in positions that dictate a mobility pattern, with little scope for change. Of the megacities, London is perhaps the one with the greatest prioritisation of history, in that the built environment is best considered as a fixed entity, and the greatest acceptance of anachronism; the mobility culture is indicative of a broader UK cultural tendency to modify or adjust systems rather than enact root-and-branch reform.

The G-Wiz

The REVA G-Wiz (see Fig. 11.2) is the most successful EV in UK history, representing about half of the EVs on the road in London today. Although there is nothing prohibiting the use of the G-Wiz EV beyond London, it was marketed and sold, and was and still is serviced solely within London. One cannot consider the G-Wiz without reference to the Congestion Charge, which was one of the most prominent developments in transport policy in the recent past. The Charge had a surprising outcome: the high cost of driving, taken with the various breaks and benefits given to EV users, made London a competitive market for the EV, whereas otherwise it would not be. The G-Wiz sold remarkably well, and London became both a major market for alternative vehicle technologies and, arguably, the most prominent exemplar for EV use amongst the cities of the world. This conducive climate resulted not only from the Congestion Charge scheme, but also a dynamic and fortuitous combination of local council parking and charging strategies, popular discourse about the environment at the time, and innovative marketing strategies for a new product.

Storage of electricity in batteries is the main barrier to wider use of EVs. The energy density of batteries is very poor compared to that of fossil fuels. An EV must balance range, speed, maintenance requirements, charge time, safety, comfort, features and costs. Many of these elements are essentially in opposition to one another: safety features, and features in general, require additional weight, thereby limiting range and speed [3, 14]. The main way in which EVs became commercially feasible in the UK—the way designers squared the circle, as it were—was through the exploitation of the quadricycle[3] category of automobile. This allows for a road-legal design of an ultralight, low-power EV without any of the safety or design restrictions applying to standard vehicles. The G-Wiz is, therefore, *not a car*, but rather a quadricycle, because the quadricycle category of road vehicle absolves the manufacturer

[3] The term 'quadricycle' as applied to road-legal vehicles is defined precisely in European law (as a motor vehicle with four wheels and specific weight, speed and engine capacity characteristics which diverge from standard cars), but not in UK registration and licensing legislation, under which such vehicles can fall into a number of different registration categories, depending mainly on their intended use [5].

Fig. 11.2 The G-Wiz: a common sight on the streets of some of central London's wealthy neighbourhoods

from the obligation to incorporate 'car-standard' levels of safety equipment and design (elements which limit range and speed through added weight).

Environmental Discourse at the Time of the G-Wiz's Appearance

Along with the Congestion Charge, central to the development of the G-Wiz was the growth of public interest in climate change [2]. This moral concern about the environment led public discourse on environment, policy decisions, fashions in consumption and use, business behaviour and, of course, debates and developments in mobility decision-making. The growth in the G-Wiz's popularity was set in a public discourse of the environment where green products, policies and behaviours—both genuine and 'green wash'—was fashionable and dominant. Green consumerism began to develop hugely, and included 'carbon-offsetting' programmes; a fashion for reusable shopping bags; and a range of environment-related books and films, including *An Inconvenient Truth* [8, 9] (which echoed the book *Silent Spring* in its impact) and *Who Killed the Electric Car*? [4] Climate change found itself a place in new parts of the media, such as fashion magazines and newspaper Sunday supplements. The environment, and buying something to save it, was cool.

In addition, the Stern Report [13] also set out an economic argument for major socioeconomic changes to negate the potential impacts of climate change. This report received significant media coverage and was at the centre of political discourse. Not only was environmental concern cool, it was beginning to feature in normal economic and governmental policy.

In the first review of the G-Wiz in 2004, the Daily Telegraph pointed out the 2,000 colours available, and the financially advantageous grants and discounts associated with it. "With electricity costing less than 1 p per mile—90 % less than petrol—[that means an estimated] annual savings of £9,000 on an average commute from a London suburb to the City or Canary Wharf" [6]. The value for urban commuters was stressed, along with tax rebates that were particularly associated with wealthy Londoners—as it highlighted savings for drivers to central, high-density financial districts. Later on that year, Scarlett [12] wrote of the financial benefits:

> Electric cars might have a clunky image, but they are exempt from the congestion charge, benefit from free meter parking and—apart from an annual administration fee—enjoy complimentary access to 26 central London car parks, some of which offer free battery recharging. Selected boroughs also allow up to 83 % discounts on residential parking permits.
>
> Even outside the capital, electric cars are exempt from road tax and, with running costs as low as one penny per mile, they are up to 90 % cheaper to run than the average petrol car. Even going slowly has its advantages, such as the lowest insurance group and company car tax rate.

Early reviews of the G-Wiz were almost exclusively focused on the financial benefits in comparison to the internal combustion engine (ICE) car, rather than, say, the convenience or the environmental benefits. Yet it was not promoted as a permanent, sensible long-term financial choice. Instead, early reviews implicitly suggested that the G-Wiz could function as an *avoidance* tactic for circumventing the Congestion Charge and related motoring expenses. These reviews put forward the idea that such a tactic would be *the* reason to purchase it. The G-Wiz was presented as a highly visible purchase in line with then-current fashions for conspicuous green consumption. So, while the G-Wiz could have been a new form of mobility, or part of a significant shift in mobility culture, it was not being discussed as such, but rather as part of a car mobility culture which promotes *more* car ownership.

Interviews with users suggested a particular engagement with the G-Wiz as a vehicle [15]. Brenda, a wealthy lawyer who drives to the centre of the city, saw a photo of one in the Evening Standard (London's main daily paper) and was drawn to the diminutive, non-threatening design: "They just looked so sweet to me. I loved the fact that there were a mass of different colours, it made somehow driving potentially fun again". Brenda saw two reasons to buy a G-Wiz: "[the] green side of it all, and the fun". What is this fun, though? The fun is that she:

> …can park it anywhere; clearly, not paying Congestion Charge is a key feature, and of course Westminster and the City of London… ha[ve] special policies enabling you to park on the street… you can often find you can fit into spaces where there isn't any other space generally for cars, so that is a great convenience.

Here, the design, the tax breaks and exemptions are attractive because they allow the G-Wiz to function as the 'un-car'. It is marked out as distinct, an exception to the drudgery of driving in a congested city. The price of motoring in it is advantageous, and the symbolic value of the car as a green gesture (but perhaps not actually green in itself) was noted by many.

The G-Wiz was marketed through upmarket magazines and web-based methods, with a resulting class bias in clientele. That said, despite the high costs of EV technology, the product still had to be sold cheaply. At the time, an EV could cost three times what an internal combustion equivalent might. The business model of GoinGreen, the firm that sold the G-Wiz, functioned under the assumption that there were a significant number of motorists that would choose to 'green' their motoring, but who would not pay any premium to do so. Therefore, the G-Wiz was sold via an operating model designed specifically to reduce costs as much as possible. It was sold via the Internet, rather than through dealers. There were no showrooms. Instead, test drives were arranged online, and customers were met at railway stations. There was no advertising budget, and they made no brochures. The G-Wiz was a 'car' sold without car salesmen.

This unique operating model not only lowered costs, but fitted into the Internet-enabled and time-constrained lifestyles of the target market, which—because charging required off-street domestic parking (a comparative rarity in London)—was predominantly the older and more affluent consumer. It was bought as a second or third vehicle to service commuting distances of 5–20 miles. So while it was *marketed* as a green car, it was being *purchased* and *used* only for a particular, narrow set of mobility requirements—essentially commuting short distances within and across the Congestion Charge zone.

The targeting of affluent urban car users tapped into a significant and influential section of central Londoners: wealthy individuals in finance and cultural industries keen to avoid new congestion charging, maintain urban mobility as they had been doing, and enjoy being part of fashionable efforts in support of greener motoring. GoinGreen had a specific strategy to attract such customers—refusing to let motoring journalists review the vehicle, the company instead promoted it to lifestyle, environment and fashion journalists. These reviews generated a high profile for the G-Wiz, but, critically, not a profile which would lead to the vehicle being judged against the ICE benchmarks employed by experienced motoring journalists. Fashion and lifestyle media coverage tended to give technical information that was not strictly accurate. These (unsurprisingly) superficial analyses were explicitly sought by GoinGreen, whilst all other forms of coverage were avoided.

It is important to note here that the financial incentives for the G-Wiz were both critical for the success of the vehicle and also dependent upon a range of actors collectively supporting such an effort. There was no centralised strategy for supporting EV development in this way, and the conditions key to the success of the marketing and sales strategy were momentary. Even though the Congestion Charge is relatively popular, it has been a point of contention for some Londoners, not least for its connection to Ken Livingstone, the first Mayor of London, who came before Boris Johnson and introduced the Charge. Thus the financial benefits were not only accruing to a largely wealthy clientele, but also inevitably connected with London's politics, whether as part of an a green agenda, or as a defiant response to congestion charging and/or a personal challenge to Ken Livingstone, in that the G-Wiz could be seen as something of a 'cheat' against the Charge. A similar point is noted by Beth, a G-Wiz owner:

> [With a child and dog to transport] cycling is just not really an option any more… and I got fed up because I was… paying about £1,500 or £1,600 a year—after I'd been taxed—just to Ken Livingstone, in order to get to work, to drive half a mile across the zone.

Beth purchases a G-Wiz to avoid what she considers to be a tax that is unfair given her situation, but she personifies the tax itself—labelling the tax as something paid to Ken Livingstone personally. The conditions which encouraged the adoption of the G-Wiz are highly political and subject to fluctuation. The financial argument for G-Wiz purchase was rooted in the Congestion Charge. The terms of the Congestion Charge are subject to change, especially as transport policy is such an important part of the role of the London Mayor and such a key element of debate in mayoral elections.

The G-Wiz as the 'It Car'

The G-Wiz began to be popular amongst a set of wealthy, socially conscious London commuters. Guardian editor Rusbridger [11] wrote of his own G-Wiz experience, trading upon key points of the G-Wiz story. He bought a bicycle to commute with, and to make a small and largely symbolic contribution to reducing carbon emissions. He maintained a large ICE car, and his bike purchase has much more sign value than use value in terms of environmental impacts, but it helps to maintain his extremely mobile habits—for him, an additional 27 min on public transport is too much. He then gave up on the bicycle and bought a G-Wiz. He loves it, but still rubbishes it after a fashion, saying that it is "designed with a crayon". Rusbridger is not the only one. Other celebrities also bought and endorsed the G-Wiz as an urban London runabout, and it became (in part because of the manner in which the G-Wiz 'message' was spread) a symbol of green mobility and something of a fashion icon.

By November 2008 GoinGreen had sold approximately 950 vehicles, making London home to "the largest privately owned electric fleet by a single manufacturer in one city in the world" [7] and "the most successful electric car of all time" [1]. However, it retained its position as the butt of jokes in the popular motoring press. Perhaps more surprisingly, it did so also within the media *sympathetic* to the environmentalist ideals upon which it relied, and the types of fashion (or fashion-conscious) media sought out by GoinGreen as part of its campaign. In other words, even keen owners disparaged the vehicle.

The Fall of the G-Wiz

Problems arose for the G-Wiz after both the Department for Transport and the BBC's *Top Gear* magazine purchased G-Wiz vehicles for crash testing in or around April 2007. *Top Gear* magazine, which had been avoided by the GoinGreen marketing team as typifying the 'petrolheads' who are regarded by many EV enthusiasts as the enemy, put a car through a Euro NCAP-style test. It failed badly. Although the safety test was a genuine concern to the Department for Transport and the Transport Research Laboratory, as well as to *Top Gear*, some saw this concern as misleading. Beth noted:

> But if you read, there's an article online which I'm sure you've probably seen about how the *Top Gear* thing was totally taken out of context and actually it was all a bit of a stitch up, because obviously *Top Gear* are a bunch of petrolheads and don't like, you know, any kind of [EV].

This crash test, the remarkably emotive photographs of it, and the publicising of the test via BBC News and the newspapers, created a PR crisis for GoinGreen and, more interestingly, forced some G-Wiz owners to re-examine the process by which they made their mobility decisions, and the factors and values that they used in arriving at those decisions. The shock of the test results shifted the discourse of green motoring in London, and compelled users to reassess their behaviour. Whatever the merits of the G-Wiz, for very many G-Wiz owners and those considering buying one, travel had to be perceived as safe. Boris Johnson, who was at the time a Conservative MP and journalist, and who later succeeded Ken Livingstone as Mayor of London, was a member of the developing London celebrity G-Wiz-owning crowd, and defended it publicly. Perhaps it is not surprising that Johnson, as a Conservative politician, did so on libertarian grounds, nor is it surprising that he emphasised its trendiness and its London-centric convenience in terms of mobility. Even as a promoter of the car, he resorted to humour when talking about its appearance and origin. The engagement with the G-Wiz in discussion highlights the way in which it was considered as a fashion accessory to London's mobility culture, rather than a potentially serious agent for change.

Questions arose as to the degree to which owners were aware of the safety information, and to what extent GoinGreen may have been misleading about safety (an issue about which compromises *have* to be made, given current technology, in order to balance range, weight, safety, cost and comfort). In addition, it raised questions about the extent to which safety was or was not part of the decision-making process in buying a G-Wiz (relative to the dominating role of the environment in popular discourse, and the consumption culture, of the time). It was clear from the public reaction of owners and prospective owners that the graphic depiction of a hideous crash caused people to see the G-Wiz from a different perspective—one in which safety featured far more prominently.

The fundamental problem here is that the G-Wiz design is that of a light, small vehicle with a battery and motor combination which can, by *automobile* standards, barely power the vehicle, and which offers only basic range. Any EV design must balance the diametrically opposed elements of the design brief. For an EV to be safe, it must almost certainly be heavy—and this comes at a cost. For it to be efficient, it must be light, and advanced technology costs money. To reconcile this, the G-Wiz was designed as quadricycle. This allowed for a design where environmental responsibility and cost could balance at a point which was affordable. However, this entailed sacrificing the element of safety, at least in comparison to the modern automotive norms of the 'car'. At this point, what had originally been introduced to the public as Indo-Californian, was now almost exclusively referred to as dangerous and Indian-made—thus crassly playing on racist notions of Indian engineering. The fashion for the G-Wiz was clearly in steep decline.

These safety-related stumbling blocks, although damaging enough to the brand and to sales, were only one element in the decline of the G-Wiz as a fashionable icon of green motoring. The London Mayoral election and the Congestion Charge tax changes proposed as part of manifestos were highly damaging to the sales of EVs in London. Firstly, Ken Livingstone proposed altering the Congestion Charge to an emissions-based scheme, so that low-emissions vehicles were allowed free entry to the city centre, while larger vehicles, often disparagingly called 'Chelsea Tractors', would pay a £25 charge. This would obviously move the Charge away from being a congestion-alleviating measure and turn it into a climate-conscious tax with an express interest in taxing a symbol of wealthy urban polluters. This created an impetus for major manufacturers to market certain redesigned vehicles that complied with the free category, as the proposed barrier for exemption was not placed extremely low. Therefore, under the new scheme, a G-Wiz would be afforded the same financial assistance as small hatchbacks and even a handful of specially modified medium-size vehicles. All these alternatives:

1. were from established and reliable manufacturers;
2. were built to Euro NCAP safety standards;
3. did not involve compromises or adaptations; and
4. could be the 'second car' purchase of a commuting family, rather than occupying the G-Wiz's typical position of 'third car'.

This proposal made a competing EV dealer, the NICE Car Company, go bust.[4] In addition, the nature of the Congestion Charge became a matter of debate in the election campaign, meaning that the crucial issue of the future structure of the charges remained unresolved, and therefore that the purchase of a G-Wiz would be based, at least in part (a very important part for most would-be owners), on a set of as-yet-undecided criteria. The proposals to charge 4 × 4 s (in fact, all vehicles in carbon emissions band G, in other words those emitting over 225 gCO_2/km) at a higher rate (£25), and to lower the restrictions on smaller, moderate-CO_2 vehicles, hardened into a political division during the campaign, involving drivers, non-drivers and people with various environmental concerns. The factors that made the G-Wiz a viable choice for a particular type of London mobility (generous tax breaks, low costs, and high sign value in terms of fashions of consumption) shifted, and the G-Wiz's sales collapsed. In the event, Ken Livingstone, whose proposals were part of his manifesto for re-election in 2008, lost the mayoral race, and his successor Boris Johnson scrapped the higher-rate Congestion Charge plans which would have also allowed low- (not just zero-) emitting vehicles exemption. Nevertheless, a modification to the proposed changes at the lower end of the scale was implemented on 4 January 2011: cars which emit 100 gCO_2/km or less, and that meet the Euro 5 standard for air quality, now qualify for a 100 % discount.

[4] NICE sold electric quadricycles, along with EV bikes and vans. It had a more traditional business model and dealership setup, with higher overheads, and a less fashionable image. It has since resumed trading, but simply as the London sales arm of AIXAM-MEGA.

Conclusion

Various technical, social, economic and legal elements work together to form mobility cultures, and in particular combinations these can work to create conditions conducive to the adoption of vehicles that might otherwise seem outlandish. That a cheap vehicle so counter to the prevailing ideas of automobility still remains a common sight on some of the most expensive streets in Europe (in the City of Westminster and in London's financial district), speaks volumes about the combination of a unique mobility culture, favourable cost structures and clever marketing that created the G-Wiz's remarkable sales growth.

Many factors worked together to make London a major city for EV adoption: local parking regulations, a city-level congestion charging scheme, a public discourse about environmental issues and related conspicuous green consumption, savvy marketing and sales, a mobility culture focused on the sign values of transport, and a vehicle designed to exploit a technical loophole.

However, these elements were fluctuating. The public discourse on environment was in reality only momentary. A shift in that discourse towards safety priorities (brought about by a set of disturbing visual images) destroyed the basis for purchasing the vehicle in question. Policy, and political engagements with transport, change frequently. Elements beyond the control of those keen to promote EVs (in this instance, the *Top Gear* test) can force reassessment of individual behaviour that is adverse to the market for those vehicles. None of these issues are either stable or settled, and the dialectics of these debates are fluid and complex. Conditions and policies fluctuate and work independently from one another, not necessarily to good effect, and rarely coming together to bear the fruit of clearly positive progress. A move towards EVs may well be temporary, and can be superficial, not resulting in the hoped-for environmental changes. Any effort to encourage significant changes to mobility in London needs to be mindful of the 'small steps' of London's 'make-do-and-mend' mobility culture. In such cases, small steps could well actually be backward steps. If there is an effort to create conditions conducive to the adoption of EVs, it is critical to ensure that such adoption is meaningful, and an awareness of the fluid nature of the components necessary to implement such change is essential, so as to ensure that they are managed in such a way that they work concurrently.

Within the subset of megacities marked by prominence before automobilisation (e.g. London, Paris, New York), London is distinct because of the particularly intense constraints on mobility and mobility culture. London has a particularly strong concentration of jobs, wealth and economic and political significance, both nationally and internationally. Although this is arguably also the case in a city such as New York, the USA has other competing cities offering comparable opportunities. London does not. Furthermore, in contrast to many of the major cities of Europe (e.g. Berlin), London conducts the work of a global megacity within a rigidly preserved historic shell. London stands out within Europe as a city in which the historic centre is a seat of employment and business that is critical to the national economy.

Within the constraints obtaining in London, a 'make do and mend' mobility culture exists. Transport infrastructure is an asset to be 'sweated'—its value to the city should be exploited to the maximum: in this way, the most is made of the city for the least amount of economic, visual, infrastructural, political or cultural change. The G-Wiz, therefore, is both an example and a product of this culture, allowing private mobility of a particular class of affluent urbanites to continue unhindered despite the Congestion Charge (which is in itself an example of sweating an asset—the central London road network). However, the conditions that caused the G-Wiz to come into being, and—for a period—thrive, were not sufficiently strong or stable to bring about a more significant cultural change—namely, a large-scale shift to electric drivetrains. Rather, the G-Wiz was a high sign value (and relatively low use value) product of the mobility culture of London and the green discourse of the time. In the long run, it was not capable of competing with other, more fully featured vehicles, nor with the competing discourse of safety.

References

1. Boxwell B (2008). Website of the G-Wiz owners club. http://www.g-wiz.org.uk/information-electric-car.html
2. Boykoff M (2007) Flogging a dead norm? Newspaper coverage of anthropogenic climate chance in the United States from 2003 to 2006. Area 39(4):470
3. Brant B (1994). Build your own electric vehicle, Vol 1. Blue Ride Summit, TAB books
4. Davies S, Paine C, Kirsch DA (2006) Who killed the electric car? Sight Sound 314(5798):424
5. Department of the Environment (2012) Proposal for mandatory wearing of helmets on quadricycles. http://www.Applications.doeni.gov.uk/publications/document.asp?docid=18703
6. Foster P (2004) G-wiz, it's India's new electric runabout. Telegraph. http://www.telegraph.co.uk/motoring/news/2729562/G-wiz-its-Indias-new-electric-runabout.html
7. GoinGreen Website. http://www.goingreen.co.uk/store/content/gwiz_faq/%20
8. Gore A (2006) An inconvenient truth: the planetary emergency of global warming and what we can do about it. Bloomsbury, London
9. Guggenheim D (2006) An inconvenient truth
10. Mayor of London (2008) The London plan. Greater London Authority, London. http://www.london.gov.uk/thelondonplan/docs/londonplan08.pdf
11. Rusbridger A (2006) He's electric. The Guardian. http://www.guardian.co.uk/environment/2006/apr/25/ethicalliving.lifeandhealth
12. Scarlett M (2004) Fun, up to a (power) point. The Telegraph. http://www.telegraph.co.uk/motoring/2730010/Fun-up-to-a-power-point.html
13. Stern NH (2007) The economics of climate change. Cambridge University, Cambridge
14. Wakefield E (1977) The consumer's electric car. Ann Arbour. Ann Arbor Science, Michigan
15. Wengraf I (2009) Places and practices of automobilities: case studies of british motoring Department of Geography. University of Liverpool, Liverpool

Part III
Perspectives for Megacities on the Move

Chapter 12
Singapore's Mobility Model: Time for an Update?

Paul A. Barter

Abstract Singapore's urban transport system is well known as a policy success story, and its key elements have been widely recounted. However, while acknowledging the strengths of the Singapore approach, this chapter discusses some of its shortcomings. It reviews the key elements of Singapore's approach to its urban transport challenges, starting in the early 1970s. This approach established a prominent role for public transport and avoided the locking in of significant automobile dependence. However, a source of new tensions is an overemphasis on high traffic speeds in too many contexts. This emphasis is increasingly in danger of undermining other key elements of Singapore's approach to transport issues, such as the requirement for space efficiency and the need to raise the status of alternatives to the car. Finally, the chapter points towards possible adjustments to Singapore's model that might allow it to overcome its dilemmas.

Introduction

Singapore's urban transport system is well known as a policy success story, and its key elements have been widely recounted (for example [9, 19, 28]). However, while acknowledging the strengths of the Singapore approach, I want to also discuss some of its shortcomings in this chapter, thus joining a small critical stream in the literature on this fascinating case (such as [21, 29]).

I will first quickly review the key elements of Singapore's approach to its urban transport challenges, starting in the early 1970s. This is the approach that has attracted praise from many quarters for establishing a prominent role for public transport and avoiding the locking in of significant automobile dependence.

P. A. Barter (✉)
LKY School of Public Policy, National University of Singapore,
469C Bukit Timah Road, 259772, Singapore
e-mail: paulbarter@nus.edu.sg

Institute for Mobility Research (ifmo) (ed.), *Megacity Mobility Culture*,
Lecture Notes in Mobility, DOI: 10.1007/978-3-642-34735-1_12,

© Springer-Verlag Berlin Heidelberg 2013

However, I will then argue that various tendencies in transport policymaking and debate have been exposing inadequacies in the old approach. One source of new tensions is an overemphasis on high traffic speeds in too many contexts, in both rhetoric and practice. Although widely seen as evidence of success, this emphasis is increasingly in danger of undermining other key elements of Singapore's approach to transport issues, such as the requirement for space efficiency and the need to raise the status of alternatives to the car. Such tensions are, of course, not unique to Singapore. They are, however, more keenly felt here.

Finally, I draw on recent urban transport policy literature and experiments elsewhere to point towards possible adjustments to Singapore's model that might allow it to overcome its dilemmas.

The Singapore Approach and Its Strengths: 1970 to 2000s

From the early 1970s, the great strength of Singapore's urban transport policymaking was its unusual willingness to face up to the need for difficult trade-offs [2]. Singapore's circumstances prompted it to 'do better with less' by focusing on moving people and goods space-efficiently and by reducing the need for travel, especially car travel, via metropolitan planning orientated to public transport.

The impossibility of meeting unconstrained demand for private motor traffic through road expansion was recognised in Singapore's State and City Planning study of the late 1960s, which resulted in the 1971 Concept Plan [7]. This adopted a long-term perspective to ensure adequate space for development over many decades ahead. At about 100 persons per urban hectare, Singapore's urban population density is high by Western standards, but only moderate relative to most large cities in Eastern Asia [13].

In the early 1970s, this sense of spatial limitation—together with increasing traffic congestion, a low-quality bus system, and a lack of funds for infrastructure—prompted a 'hard-nosed bargain'. This involved the promised benefit of gradual improvements in space-efficient public transport, and the development of an urban fabric to suit, while the price to be paid for that was making ownership of space-wasting cars unattainable for the majority of the population by means of unpopular vehicle taxes and fees. A key aim of this approach was to maintain efficiency of movement for the sake of commerce and trade [2].

Although Singapore's political system is relatively authoritarian, allowing a famously technocratic approach to policymaking, politics is never absent, and elections still necessitate considerable responsiveness to public opinion. So, although Singapore's transport policies are hard-nosed in facing the need for trade-offs, they also need to some extent to be politically savvy.

This approach, presaged in the 1971 Concept Plan and pursued from the mid-1970s, was initially both space-efficient and thrifty, focusing initially on low-cost improvements and efficient use of existing infrastructure, while allowing large road and public transport investments to wait until the mid-1980s and beyond.

Let's look at key elements of Singapore's approach in a little more detail, beginning with the spatial strategy, before turning to travel demand management and then public transport improvements.

Compact Spatial Strategy

The 1971 Concept Plan and its successors (strategic plans giving high-level guidance rather than detailed master planning) call for a strong-centred urban structure with clear corridors, with the centres of high-density new towns located along mass transit spines. Residential uses occupy only about 15 % of Singapore's land surface, and a very high proportion of the population lives in high-density housing, both in private apartments and in Housing and Development Board (HDB) estates and new towns, which are built and managed by the public-sector HDB (Fig. 12.1). This approach is in keeping with Singapore's inability to sprawl.

This strategy involved a transport vision as much as an urban structure vision. The Concept Plans have thus also provided the framework for transport strategies, especially as relating to their spatial aspects. The planning agency, the Urban Redevelopment Authority (URA), and the main urban transport agency, the

Fig. 12.1 Dense HDB (Housing and Development Board) housing in the new Singapore town of Punggol, with an elevated people-mover system which provides feeder service to the Mass Rapid Transit (*MRT*) station

Land Transport Authority (LTA), coordinate closely, a process which is aided by Singapore's single-level governance structure (in which town councils lack any significant planning powers).

Effective integration of transport and land-use planning is also assisted by the large percentage of land in government ownership, a legacy of both the colonial era and of the land acquisition law, strengthened (by the Land Acquisition Act) in 1966, and used with vigour in the development of public housing, industrial areas and city-centre redevelopment. The URA also uses land releases as a means to guide development. Furthermore, widespread state ownership of land has meant that the government automatically captures much of the land value increment from mass rapid transit (MRT) development.

Vigorous Travel Demand Management

The study team behind the 1971 plan predicted that traffic trends would quickly become incompatible with this compact spatial strategy and with any feasible programme of road construction [19]. It therefore recommended travel demand management (TDM). This led in 1975 to the Area Licensing Scheme. This was the world's first cordon road pricing policy to manage car travel to a city centre, and required the purchase of a ticket to drive a motor vehicle into the central area. This was replaced in the late 1990s with the more extensive Electronic Road Pricing (ERP) system.

The State and City Planning study also called for restraint of vehicle ownership, and these policies have in fact proved much more significant than the more widely known road pricing. In 1972 a step-by-step set of increases in purchase taxes and ownership taxes was begun. In 1990, these purchase and ownership taxes were supplemented by the Vehicle Quota System (VQS) with its auctions of Certificates of Entitlement (COEs), which are required for the registration of each vehicle and last for 10 years. Under the VQS, the growth of the total Singapore fleet of vehicles was limited to 3 % per year, although in practice annual increases have fluctuated around this level. This has now been reduced to 1.5 % and is scheduled to reduce further to 0.5 %. Currently, the main purchase taxes on a new car are: the Additional Registration Fee (ARF) at 100 % of the open market value of the car, Excise Duty at 20 %, and the COE price (an auction price)[18]. As of early June 2012, the COE price for cars below 1,600 cc engine size is S$59,003 (about US$46,000).[1] As a result, a new mid-sized non-luxury sedan, such as the Toyota Corolla Axio 1.5 costs just over S$100,000 (more than US$78,000). Despite the fact that a proportion of ARF and the COE can be redeemed if the car is deregistered within 10 years, cars in Singapore are clearly very pricey.

The annual cost of car ownership is also substantial, with the annual 'road tax' for a modest sedan being typically between S$600 and 800 per year

[1] In early June 2012, S$1 was worth US$0.783 (i.e., US$1 = S$1.28).

(approximately US$470–630). Moreover, residential parking is charged separately from the cost of public housing flats. Around 80 % of the resident population lives in this kind of HDB-developed housing estates, where car owners pay between S$65 and 90 per month (roughly US$50–70) for a parking permit for their locality [10].

The VQS has not always been precise in hitting its vehicle growth targets. For example, between 1998 and 2004 car numbers increased at only 1.2 % per year, well below the 3 % target. Then from 2004 to 2008 car numbers increased at 7 % per annum, well above target [6]. The problem has been the use of forward estimates for the number of vehicles to be deregistered, which turned out to be systematically high or low in each period. Unusually low COE prices resulted in the late 2000, bottoming below S$10,000 (a little under US$8,000) for small cars in 2009. However, with recent reductions in the quota growth rate and a planned clawback of the excess, COE premium prices have soared again.

Are motorcycles an easy option to avoid these high car costs? Motorcycles do account for about 16 % of registered vehicles [17], which is higher than in most high-income cities. But it is much lower than in nearby Malaysia or Indonesia. The VQS system does apply to motorcycles, but their modest COE prices (currently about S$1,000, or nearly US$800, as of mid-2012—and this is unusually high) suggest that motorcycle demand may be inherently limited in such an affluent society.

Despite such hiccups, the system has been broadly effective at limiting growth in the vehicle fleet. Only 35.4 % of resident households owned a private car in 2002/3. This rose to 38.3 % by 2007/8 [24]. Car ownership in 2011 was 100 per 1,000 persons [17].

Before leaving this topic, it is important to be clearer about ERP. Prices for each gantry for each 30 min period of the week can be adjusted separately. This adjustment is performed every three months, with a price change for each time and place only if the average speed there and then was outside the target range. The target speed ranges are chosen to maximise the efficient flow of traffic. ERP prices are thus, in one sense, market responsive. It is therefore important to note that, unlike London's congestion charge for example, Singapore's ERP does not actually aim at limiting traffic. Its objective is to optimise speeds in order to maximise traffic flow at congested times. According to LTA sources, the main short-run motorist responses to Singapore's ERP are shifts in the timing of car trips and (to some extent) their routes, rather than mode shifts to public transport or other alternatives.

Steadily Improving Public Transport

Public transport improvements have also been significant but, again, there are some common misconceptions about the timing of the investments. The rail-based MRT is the linchpin. However, it is important to note that the decision to build the MRT had not yet been finalised in the 1970s. The early improvements to public transport focused on buses, starting with forced mergers of bus companies and the imposition in 1973 of professional management on the unified company, together with service obligations. Other early steps included two major reorganisations of routes, the banning of pirate taxis, allowing non-corporate 'supplementary' services to help with peak demands, and the creation of a significant network of bus lanes.

The gradual nature of motorisation in the city gave Singapore time to build up bus usage until mass transit became affordable [5]. The initial MRT system was built in the mid-1980s. The first segments opened in 1987. Rapid expansion is continuing, with an MRT network of 278 km expected by 2020, up from 138 km in 2008 [15].

There were strong efforts in the 1990s to improve the level of network integration, including common ticketing (although not with free transfers, which were only achieved in 2010) and improved passenger information by means of a new entity called Transit Link, co-owned by the two major public transport operators. Integration efforts were also intensive on the physical side, by means of the newly formed Land Transport Authority (LTA), created by a merging of several relevant agencies in 1995. One prominent result was a greater effort to provide bus interchanges at many MRT stations.

As you would expect, a key result of these policy efforts is a high modal share for public transport. Public transport's share of total daily motorised trips was 56 % in 2008 [6]. This does not include the significant use of subscription chartered buses for commuting.

Synergy and Consistency Over Time

There are obvious synergies among these policies. The deliberate slowing of traffic growth and the successful improvement of public transport helped the high-density corridors and strong city centre model to succeed. In turn, this urban structure helps public transport to thrive. The public transport strategy offers steadily improving mobility for the masses who do not own cars. Efforts to contain car ownership and usage has helped to retain much of the growing middle class as a solid customer base for public transport.

Transport and urban planning policies since the 1970s have been remarkably consistent, with only gradual changes in emphasis and some additional innovation, but with few clear breaks from the approach adopted in the first half of the 1970s. The updated Concept Plans of 1991 and 2001, the LTA's 1996 White Paper, and the 2008 Land Transport Master Plan all remain broadly consistent with the policies in place by the mid-1970s. One gradual change has been a trend towards a higher capital cost approach over time, as wealth has increased and as easy, lower-cost infrastructure opportunities were exhausted [23].

Challenges and Tensions in the Singapore Approach

Without wanting to minimise the strengths and successes of the city's land transport policies, I turn now to a brief discussion of some shortcomings and tensions in Singapore's approach.

In several ways, Singapore's key urban transport strategies, which have done so well since the 1970s, may now be reaching a point where fundamental change will be needed rather than just reinforcing and adjusting the same approach. Singapore does somewhat worse than you might expect (given its firm TDM policies) on indicators such as vehicle kilometres travelled per capita, transport energy use per capita, non-motorised transport use, and even its public transport modal share [6, 12]. What is going on then?

Let us look at examples from several key urban transport policy arenas. We will see that, ironically, some of the difficulties arise from tensions that are inherent in the hard-nosed bargain itself.

Travel Demand Management Dilemmas and Side Effects

Longstanding dilemmas over TDM seem to be worsening. The need to contain the spatial demands of cars is, of course, as pressing as ever. This has always been politically difficult, but the difficulty seems to be increasing.

The range of households aspiring to car ownership has broadened during the last decade. This may be due to various factors. Obvious possibilities include growing affluence, and government statements sympathetic to aspirations to car ownership. Cars also enjoy a large speed advantage over public transport, arising from road design priorities and from the success of the TDM system. And, as mentioned earlier, for a period in the late 2000 the car fleet was inadvertently allowed to increase more rapidly than intended, resulting in low COE prices and car ownership that was newly affordable for households of median income or slightly below.

Yet the recent tightening of car ownership restraint has again sent car prices soaring from a low point in 2009. So many see their car-focused aspirations receding again. This contradicts past government statements that encouraged the expectation of a gradually expanding car ownership, and comes at a time when Singapore politics is becoming more vigorously contested, along with a strengthening of civil society and energetic online debates. Restraining car ownership in the old tough-minded way is getting harder, yet public expectations for excellent mobility remain largely focused on the aspiration to car ownership. The main alternative to car ownership that has been put forward, the MRT, is still not widely seen as a serious competitor in terms of either prestige or convenience.

The current approach to TDM, which focuses on car ownership containment and very high fixed costs for cars, also has some negative side effects. For example, a high annual distance travelled per car figure of about 19,000 km is in part a result of these high fixed costs. Since the mid-1990s, Singapore's authorities have been concerned that high fixed costs might be creating a sunk cost [2] effect in which motorists seek to get good use of their expensive capital asset, and that high fixed costs undermine the effectiveness of usage restraints, which in cost terms seem relatively insignificant by comparison [8, 14]. The high cost of cars also inadvertently puts them on a pedestal in terms of social prestige and signalling of status. In addition, a practical consequence for public transport policy and practice is an apparent tendency for the public transport operators to assume that most of their passengers cannot afford cars, and therefore have few travel alternatives to the public transport system.

For some time during the 1990 and 2000s, Singapore seemed poised to address these side effects via a gradual shift towards usage restraint and away from ownership restraint [14, 20]. However, recent developments reveal a change of heart on the part of the government. As mentioned above, the rate of growth of vehicle ownership is being drastically reduced, even if this means dramatically higher purchase costs. This will again magnify the side effects above.

Singapore's TDM policies also seem to have unfortunate side effects for road policies and priorities, to which I now turn.

Road Priorities That Conflict with Space Efficiency and with Place-Making

Singapore's road development policies have some surprising features that sit uneasily with its commitments to containing traffic growth and to improving the more space-efficient alternatives.

The predominance of a Corbusian 'towers in the park' style in most of Singapore's public housing estates actually leaves rather generous spaces for large roads. A relatively dense network of six-lane arterial roads has been created. Speed and capacity are high priorities in arterial road intersection design, with almost all having generous slip lane treatments, often including multiple turning lanes. This is an example of how prioritising road capacity undermines the space efficiency of the transport system.

To be fair, HDB town centres and enclaves along the Singapore River have a very high-quality pedestrian realm, and there are laudable efforts to make the area agreeable, such as very generous tree planning for shade and a programme of covered walkways throughout the new towns. Nevertheless, the road priorities mentioned above still tend to create an urban environment that is more hostile for walking and cycling than one might expect.

[2] A sunk cost is one that, having been incurred, is impossible then to recover. In this context, a large part of the high fixed cost of purchasing a car in Singapore is a sunk cost.

Probably no city in the world has really solved the design issues of 'arterial streets', with their inherent tensions between the traffic purpose and the access and place purposes of these established streets that also serve significant traffic flows [26]. However, it is surprising that a paragon of sustainable transport like Singapore is so determined to apply 'road' or highway-like designs, with firm access control efforts to limit friction, to such a large proportion of its arteries. It generally avoids traditional street designs, even on the former traditional streets. Even the central area's streets are designed for relatively high traffic speeds, and to give priority to traffic flow (Fig. 12.2). An extensive system of one-way streets, especially in the central area and within and near major sub centres, encourages high traffic speeds and creates detours for bus and bicycle users. To confirm these points for yourself, I suggest a 'virtual tour' with the help of Google Street View.

Moreover, regulations give pedestrians surprisingly low priority at minor intersections. For example, pedestrians only have right of way at signalised crossings and marked zebra crossings and not at unmarked crossings, even those at intersections. Road designs have ignored bicycles, although some effort to provide off-road bicycle infrastructure has commenced recently [3].

Singapore's road expansion and building programme seems to have a marked tendency towards a predict-and-provide approach. Major intersection revamps, such as the recently completed Woodsville intersection, are held up as

Fig. 12.2 Marina Boulevard with its multiple lanes of one-way traffic is an illustration of the high priority to traffic at relatively high speeds and volumes

promising to 'ease traffic congestion'. Several key arterials have been converted to 'semi-expressway' status in which flyovers, viaducts or underpasses eliminate most traffic light delays. Urban expressway construction also continues. For example, the 12 km Kallang–Paya Lebar Expressway was completed in 2008 and the 21 km North–South Expressway is now under construction for completion in about 2020 [16]. Both are radial expressways which include long underground sections. The ambitious 5 km underground Marina Coastal Expressway is another project under construction. It will, however, allow an existing expressway link to be converted to an ordinary arterial within the expanding central business district at Marina South.

Some of this road expansion threatens much-loved places. There were earlier controversies over an expressway bisecting the Bukit Timah Nature Reserve and over the demolition of the old National Library, for example. Agitation over a road project's impact on the historic Bukit Brown Cemetery and protests over evictions associated with the planned North–South Expressway seem to indicate new levels of intensity of feeling about such issues in Singapore. Moreover, large numbers of homes are located very close to heavy traffic flows [19].

The priority given to roads is reflected, too, in a reluctance to lower speed limits, and in the speed enforcement priorities of the traffic police. Enforcement focuses primarily on high-speed, high-capacity roads, rather than along ordinary arterials with 50 km/h speed limits. There is also relatively little awareness among motorists that 50 km/h *is* the general speed limit, with 74 % believing it to be 60 km/h or higher, according to a survey by the insurance industry [27].

Singapore's road deaths (at 3.8 per 100,000 people in 2010) are relatively low by world standards, reflecting the prominent role of public transport, but are high compared with world leaders such as Sweden. And vulnerable road users are strongly over-represented in the road death figures—in the three-year period 2008–2010, pedestrians accounted for 27 % of the road traffic fatalities, cyclists 9 %, and motorcycle riders and pillion passengers 48 %, while only 8 % were car occupants [25]. To put these in perspective, bicycles are estimated to account for fewer than 1 % of trips [3] and, in 2010, motorcycles accounted for roughly 14 % of vehicle kilometres by Singapore private passenger vehicles[3] (estimated from data in [17].

On-road priority for public transport is a mixed but generally more positive story. The bus lane network, with 23 km of all-day bus lanes and 155 km of peak-only bus lanes (as of 2008), is being further expanded. On the other hand, most Singapore bus stops have bus bays, which make returning to the traffic stream a significant source of bus delays. A new effort to overcome this involves signage and markings to require motorists to give way to buses exiting certain bus stops.

---

[3] The source does not account for Malaysians commuting into Singapore each day. They are roughly 30,000 in number, and a high proportion of them use motorcycles. However, even with generous assumptions, these could boost the motorcycle share of vehicle kilometres by only 1–2 %.

Disappointment with Alternatives

Meanwhile, disappointment with public transport and the relative lack of attention to vulnerable modes like walking and cycling is also making the politics of TDM worse.

Although public transport in Singapore is by most standards both comprehensive and of a good standard, it is still popularly seen as a distant second best for those who can afford a car. The city-state has long sought to improve public transport, aiming in the 1970 and 1980s for an efficient, cost-effective and functional system. From the mid-1990s more ambitious goals emerged, using the slogan "world class" [14]. Since 2008, the goal has been "making public transport a choice mode" [15].

However, both the system itself, and explicit policy ambitions for its future, seem to be lagging behind Singaporeans' growing aspirations. Public transport's share of total daily motorised trips actually dropped from 63 % in 1997 to 58 % in 2004, falling further to 56 % in 2008, according to Singapore's Household Interview Travel Surveys [6]. Rising car numbers may be part of the explanation for this, but there must also be other factors, since the rise in car ownership only took off after 2004.

The late 2000, especially, saw vocal dissatisfaction with public transport, as traffic congestion became more widespread[4] and crowding rose to unprecedented levels during a period of rapid increase in the immigrant population. The steep rise in COE prices since 2009 has also focused renewed attention on the inadequacies of Singapore's public transport. A spate of MRT breakdowns in 2011 and 2012 has intensified a growing perception that the promise of an excellent alternative to cars is not being delivered, even as the possibility of car ownership slips increasingly out of the reach of many.

The centrepiece of efforts to improve the alternatives to cars has been rapid expansion of the rail system. This could be called a 'World City'-inspired strategy, looking to London, New York, Tokyo and Paris as cities where large percentages of people, including affluent individuals, rely on urban rail for mobility. However, the bus system, which remains important, has been relatively neglected. Although the network and its services are comprehensive, the typical waits of between 12 and 20 min throughout the day, without posted schedules on most lines, render buses rather unappealing for people who have a choice. It is true that the well-regulated franchise approach to regulating Singapore's buses was a dramatic improvement on the poorly regulated route license one which preceded it in the early 1970s. There are now two major private operators who take the revenue risk. Unfortunately, this prompts rivalry that undermines

[4] In the late 2000, as the car population rose more quickly than intended, congestion has become more widespread across the network and across times of the day and week. However, the ERP system prevents congestion becoming more acute, since ERP prices rise if speeds on the key priced links drop below the target range.

integration, especially in information and marketing. Government policy has for a long time firmly ruled out operation subsidies, but supports taxpayer funding for capital costs. With bus purchases considered to be an operating expense (as opposed to a capital cost), this arrangement tends to underfund buses relative to rail in Singapore [21]. All of these features of the current regulatory arrangements make it difficult for government to set more ambitious goals for the bus system [1].

New efforts to improve the bus system are now underway, but progress is slow and the ambitions are still relatively limited. A future shift to competitive tendering for route bundles has been declared as policy, but with no date attached. Priority is at present focused on peak-hour frequencies, with all-day service levels still being neglected. Since 2009, the LTA has taken over bus route planning from the operators (which previously undertook this within their respective areas of monopoly, with approval needed from the regulator, the Public Transport Council). The February 2012 Budget included a one-off injection of funds to expand bus system capacity as a stop-gap until new MRT lines open. However, this required messy and ad hoc arrangements, since the existing regulatory system lacks mechanisms to implement such operational funding injections.

The LTA was far-sighted in the 1990s in realising the value of complementing public transport with various other modes, including taxis, car sharing, bicycles and walking [14]. However, there has so far been little action in this area. For example, a car sharing industry was fostered in the 1990s but then given little material help. Official views on the potential place of bicycles in the transport system have been ambivalent. There are only modest signs in the Land Transport Master Plan 2008 that this may be changing.

The road priorities discussed earlier also undermine the alternatives, as already discussed in the case of pedestrians and cyclists. Public transport is also affected. High levels of service for motorists fuel the perception that first-class mobility requires car ownership. Even for radial peak-hour trips, public transport barely competes with the car for speed. Singapore bus stops are located far from intersections for the sake of traffic efficiency, thereby hindering connections and making it very difficult to plan an integrated grid of bus routes.

Diagnosis of the Challenges and Possible Ways Forward

We have seen that Singapore's successful approach to urban transport policy of the last four decades is now running into difficulties that threaten to undermine continued improvements. This section identifies some of the sources of the problems, and suggests possible refinements to Singapore's central urban transport 'bargain'.

Sources of the Difficulties

Are there common origins of the various difficulties facing Singapore's urban transport policy that are discussed above? Recall that the hard-nosed bargain that has been driving policy since the 1970s sought to keep the arteries moving by containing car ownership and use, while promising in return steadily improving public transport and a good level of service on the roads. Unfortunately, key elements of this bargain seem to be increasingly in conflict with each other, and with other important objectives. I would like to suggest that many of the difficulties discussed in the previous section arise from three tendencies:

- An overly narrow conception of the payoffs to be gained from containing cars;
- A somewhat narrow conception of the alternatives to cars; and
- A tendency to overemphasise traffic speeds and a high level of service for private motorised traffic.

Let me elaborate briefly on these.

Singapore's approach seems to be undergirded by a comparatively narrow notion of the benefits of its strong policies for constraint of car growth. This has been consistently justified in terms of the efficiency benefits of avoiding congestion by putting in place a space-efficient transport system [2]. The official rhetoric on travel demand management has focused on keeping the arteries moving for the sake of commerce. Other benefits such as environmental sustainability and energy efficiency were only in the background. Furthermore, benefits such as preserving or enhancing the liveability of places have never been prominent in Singapore. Traffic limitation has not been justified in terms of rescuing much-loved places or spaces in the city. This contrasts with various Western cities, where saving loved places from the ravages of traffic, road expansion or parking has often played a key motivating role in urban transport reform.

Another source of tensions is the tendency to frame the alternatives to cars overly narrowly, with a focus especially on the urban rail system. As the only alternative mode that can even remotely compete with the car on speed for long urban trips, the MRT is certainly important. Household survey data show that, in contrast to the buses, the MRT is well used across all income groups in Singapore [6]. However, various complementary elements in the transport system—car sharing, cycling and walking, and even the buses themselves—have been relatively neglected.

Singapore also seems to overpromise on traffic speed. In effect it is seeking too much of a good thing. As mentioned above, the goal of maintaining mobility for the sake of commerce and economic efficiency is central to the hard-nosed bargain in Singapore's land transport policy. Given the spatial limitations, this has for the most part been understood to imply prioritisation of public transport. However, with goods transport especially in mind, this goal has also been interpreted as requiring keeping the traffic moving. Unfortunately, the road priorities discussed above suggest that the promise of high traffic speeds has been overemphasised to the point where it undermines the overall thrust of the hard-nosed bargain.

This tendency to overemphasise speed (in road designs, for example) is reinforced by the relatively narrow conceptions of the objectives of the transport strategy, and by narrow thinking when it comes to alternatives to cars. More emphasis on place-making and on the vulnerable modes that complement mass transit might have prompted a more moderate approach to traffic speeds.

The insistence on maintaining high traffic speeds as a goal in Singapore's TDM approach threatens to create a vicious circle, whereby the high costs of purchasing a car oblige the authorities to keep delivering on high traffic speeds, which in turn is only possible by limiting the number of cars and making them expensive to buy. In other words, a high level of service for motorists on the road has come to be seen as a quid pro quo for high car fees and taxes, and there seems to be a consequent reluctance to further inconvenience motorists, who have already paid a great deal for their cars [19].

Ways Forward: An Upgrade for the Singapore Model?

Here I would like to suggest a number of changes which might help overcome the growing tensions within the existing approach. The need is for a change of emphasis, not a wholesale reversal of approach. Much in the existing strategy needs to stand. More than ever, mobility in Singapore needs to be space-efficient, so private car traffic still needs to be firmly contained.

The points below are merely my suggestions as a resident and student of Singapore's transport policy trends. It is for Singaporeans themselves to debate and shape a new strategic bargain. Nevertheless, my suggestions are made in the full awareness of the emergence of a new political mood in Singapore. The old bargain was a technocratic one. Any successful update will need more widespread appeal and can only emerge from a much wider debate.

The Latest Strategic Thrusts

The Land Transport Master Plan 2008 identified three 'strategic thrusts', with the following labels:

1. Making Public Transport a Choice Mode
2. Managing Road Usage
3. Meeting the Diverse Needs of the People [15].

Points 1 and 2 represent the current version of what I have been calling the 'hard-nosed bargain'. Point 1 is ambitious and promising, but retains the narrow conception of car alternatives that was discussed above. The discussion of Point 2 in the plan reaffirms the desire to constrain car growth, but still also emphasises improving and expanding roads in order to provide a high level of service. Point 3 softens the bargain a little by emphasising diversity, differing abilities, and more attention to short trips. But this seems framed as 'welfare' for needy groups without cars, rather than as a positive mainstream agenda.

So this latest set of strategic thrusts has not yet overcome the difficulties discussed above. Singapore needs an update which poses a more positive vision of success, rather than being so Spartan in its focus on efficiency.

Hold up a More Positive Alternative

The unpopular elements of Singapore's TDM policies might be more palatable if making the non-car-owning lifestyle an attractive and high-status option becomes the focus of much more strenuous and integrated efforts. This could involve promoting a broad vision of a 'combined mobility package' or 'mobility mix' that can compete head-on with car ownership (see for example, [11]). Various places around the world are already experimenting with such efforts to be more ambitious with public transport, and to integrate it with its various natural allies (allies in the sense that, although they may compete with one another, they all work together to lessen the pressure to own a private car), which include bicycles, bicycle sharing, taxis, car sharing, peer-to-peer car sharing, delivery services, real-time mobility information and payments platforms, and so on. This approach would also help to make owning a car less desirable and prestigious.

Emphasise Place-Making Payoffs from the Bargain

Greater emphasis on the positive payoffs from hard-nosed mobility policies could involve more focus on liveable places, including streets, as central parts of the public realm. Singapore currently gives most of the 'dividend' from its strict TDM policies to existing motorists by keeping traffic speeds high and delays to a minimum. It might be better to give more of this dividend to place-making.

In a dense city which its residents increasingly perceive as crowded, an expanded public realm, as a payoff from changed road design priorities, should be valued very highly. Saving loved places and creating new loved places could be linked much more closely with the sacrifices of the TDM system. Singapore urban transport policy could thus be much more positive about the 'soft connective tissue' that forms the heart and soul of the city—i.e., its streets as places, which if redesigned could do much better at serving short trips, slow modes, and local activities.

Transcend the Obsession with High Traffic Speeds

These changes would require Singapore to overcome its obsession with high traffic speeds. This would need to involve changes in road design priorities for most streets except the high-level arterials and expressways. A key measure that would reclaim space for the public realm, without imposing overly harsh restrictions on private vehicles, would be taking a smarter attitude to speed in road design priorities [4].

This could involve adopting street design innovations from northern Europe. These often make a clearer distinction between roads that are 'traffic space' (designed with traditional traffic engineering) and streets that are part of the 'public realm', designed for the peaceful coexistence of public place activities with slow movement by vulnerable modes as well as motor vehicles [22].

Singapore is rightly praised for avoiding widespread congestion on roads whose central purpose is mobility. But its emphasis on speed goes too far in its aversion to the slowing of traffic on multi-purpose streets where lower speeds and an acceptance of delays for the sake of access movements, pedestrians, buses and cyclists would be entirely appropriate.

Conclusion

Singapore is a useful case for the study of mobility models and public sector decision-making because of its relative coherence in policymaking and its unusual determination to accept sharp trade-offs. Since the 1970s, Singapore's transport and urban policies have been firm in facing up to the tight spatial imperatives of this island city-state. This has meant taking a long-term perspective to ensure space for urban development over many decades. In a hard-nosed bargain, unpopular vehicle taxes and fees have greatly slowed the growth of the vehicle fleet and managed its use. The more popular payoffs of the policy were steadily improving public transport and efficient traffic movement. The resulting urban transport achievements have been widely praised.

However, the old version of this hard-nosed bargain contains some contradictions. The high level of service on the roads is justified primarily for the sake of commerce, but has gradually come to be seen also as a quid pro quo for motorists in return for the high costs they experience. This has encouraged an overemphasis on high traffic speeds as an end in itself in road design, undermining the more important goal of space-efficiency. The bargain also involves a rather narrow framing of its objectives, focused on efficiency of movement but ignoring potentially popular place-making and liveability payoffs that would result from containing traffic. Finally, with such a premium having been put on speed over the years, it has proved difficult for policymakers to see the importance of the slower modes that are necessary in order to complement urban rail, which is understandably seen as the only alternative which has serious potential to compete with cars.

It might seem churlish to complain about a set of policies that have achieved so much. Nevertheless, it seems clear that Singapore may need a change of emphasis in its transport planning in order to build on its achievements, continue to face unavoidable constraints, and yet at the same time offer more appealing mobility choices to an increasingly demanding and informed population.

References

1. Barter P (2007) Wanted—An Ambitious Vision for Public Transport. Opinion Piece for Ethos Magazine (Magazine of the Civil Service College), 2/2007, pp 50–57
2. Barter P (2008a) Singapore's Urban Transport: Sustainability by Design or Necessity? In: Wong TC, Yuan B, Goldblum C (eds) Spatial Planning for a Sustainable Singapore. Springer, Dordrecht, pp 95–114
3. Barter P (2008) The Status of Bicycles in Singapore. In: Tiwari G, Arora A, Jain H. (eds) Bicycling in Asia. Interface for Cycling Expertise (I-CE). Utrecht, pp 49–66
4. Barter P (2009) Earning a Public Space Dividend in the Streets. Journeys (Magaz LTA Acad) 2:32–39
5. Barter PA, Kenworthy J, Laube F (2003) Lessons from Asia on Sustainable Urban Transport. In: Low NP, Gleeson BJ (eds) Making Urban Transport Sustainable. Palgrave-McMillan, Basingstoke, pp 252–270
6. Choi C, Toh R (2010) Household Interview Surveys from 1997 to 2008—a Decade of Changing Travel Behaviours. Journeys (Mag LTA Acad) 2:52–61
7. Dale O (1999) Urban Planning in Singapore: the Transformation of a City. Oxford University Press, Shah Alam
8. Economic Review Committee (ERC) (2002). Restructuring the Tax System for Growth and Job Creation: Recommendations by the Sub-Committee on Policies Related to Taxation, the CPF System, Wages and Land of the Economic Review Committee (ERC): Report to the Singapore Parliament, 11 Apr 2002
9. Han SS (2010) Managing Motorization in Sustainable Transport Planning: The Singapore Experience. J Transp Geogr 18(2):314–321
10. Housing and Development Board (HDB) (2012) Car Park Facilities: Apply New Season Parking Ticket. http://www.hdb.gov.sg/fi10/fi10327p.nsf/w/w/CarPark1NewSP?OpenDocument#Charges
11. International Union of Public Transport (UITP) (2011). Becoming a Real Mobility Provider—Combined Mobility: Public Transport in Synergy with other Modes like Car-Sharing, Taxi and Cycling: Position Paper, Brussels Apr 2011
12. Kenworthy J (2012) Moving to the Dark Side or Will the Force be with us?—A Sustainability Report Card on Trends in Transport and Urban Development in Thirty-Three Global Cities, 1995-6–2005-6: Pre-publication chapter for book on Ecocities from the International Ecocities Conference, Montreal, Aug 2011
13. Kenworthy J, Laube F (2001) UITP Millennium Cities Database for Sustainable Transport: (CDROM Database). International Union (Association) of Public Transport (UITP). Brussels
14. Land Transport Authority (LTA) (1996). White Paper: A World Class Land Transport System. Singapore
15. Land Transport Authority (LTA) (2008). Land Transport Master Plan. Singapore
16. Land Transport Authority (LTA) (2011). LTA news release: 'Full Alignment of North–South Expressway Unveiled', 15 Nov 2011
17. Land Transport Authority (LTA) (2011). Singapore Land Transport Statistics in Brief. Singapore
18. Land Transport Authority (LTA) (2012). Vehicle Tax Structure. http://www.lta.gov.sg/content/lta/en/motoring/vehicle_ownership/vehicle_tax_straucture.html
19. May A (2004) Singapore: The Development of a World Class Transport System. Transp Rev 24(1):79–101
20. Ministry of Finance Singapore (2002). Fiscal Year 2002 Budget: Part III: Tax and Fee Changes
21. Richmond J (2008) Transporting Singapore—The Air-Conditioned Nation. Transp Rev 28(3):357–390
22. Shared Space Project (2005). Room for Everyone: A New Vision for Public Spaces: Report of the European Union Iterreg IIIB Project "Shared Space"

23. Sharp I (2005) The journey: Singapore's Land Transport Story. SNP Editions for the Land Transport Authority, Singapore
24. Singapore Department of Statistics (2012). Household Expenditure Survey 1997/98-2007/08. http://www.singstat.gov.sg/stats/themes/people/hhldincome.htm
25. Singapore Traffic Police (2010). TP Annual 2010. http://www.spf.gov.sg/prints/tp_annual/2010/index_tp_10.htm
26. Svensson A, Marshall S (2007) Arterial streets: Towards an Integrated Approach. In: Marshall S, Banister D (eds), Land Use and Transport, Elsevier, pp 22–292
27. The Straits Times Singapore (2 July 2010). 75% of Singapore Motorists Admit to Speeding: Survey. The Straits Times Singapore. http://motoring.asiaone.com/Motoring/News/Story/A1Story20100702-224976.html
28. Toh R, Phang S (1997) Curbing Urban Traffic Congestion in Singapore: A Comprehensive Review. Transp J 37(2):24–33
29. Willoughby C (2000) Singapore's experience in Managing Motorization, and its Relevance to Other Countries. Discussion Paper TWU-43. The World Bank. Washington http://www.worldbank.org.transport/publicat/twu_43.pdf

Chapter 13
Perspectives on Mobility Cultures in Megacities

Gebhard Wulfhorst, Jeff Kenworthy, Sven Kesselring and Martin Lanzendorf

Abstract Megacities are facing multiple challenges in urban mobility, linked to energy scarcity and climate change, unprecedented urbanisation and suburbanisation, as well as local issues of social and spatial inequalities, traffic impacts on health, severe congestion, conflicts over urban space and complex regional governance tasks. This chapter explores how megacities can address these issues to create well-functioning mobility systems, while simultaneously enhancing their liveability, economic performance and sustainability. Every city is unique and complex, so there is no one simple solution. It is argued here, however, that the mobility culture concept helps us to navigate a path through this complexity and find suitable mobility solutions in each city. Some key outcomes of the mobility culture research and workshop exchanges are discussed in terms of local policies for challenges, urban structure and transport supply, the critical value of urban space, travel demand management and creative processes in urban mobility development. Appropriate local strategies have to be developed by communities in a bottom-up and top-down approach.

Introduction

Megacities around the world are evolving in many different contexts. Some of the key characteristics of these cities have already been identified in the preliminary study of the fellowship programme on megacity mobility culture (see "Chap. 3"

G. Wulfhorst (✉)
Chair of urban structure and transport planning, Technische Universität München, Munich, Germany
e-mail: gebhard.wulfhorst@tum.de

J. Kenworthy
Curtin University Sustainability Policy Institute (CUSP), Curtin University, Perth, Western Australia

S. Kesselring
Technische Universität München, Munich, Germany

J. Kenworthy · M. Lanzendorf
Goethe University Frankfurt a.M., Frankfurt, Germany

Institute for Mobility Research (ifmo) (ed.), *Megacity Mobility Culture*, Lecture Notes in Mobility, DOI: 10.1007/978-3-642-34735-1_13,
© Springer-Verlag Berlin Heidelberg 2013

in this book). The major finding has been confirmed by the city stories on selected megacities by the research fellows: diversity.

The framing conditions of mobility in every city geographical setting, climatic conditions, transport networks, urban form, policymaking, stakeholder involvement, cultural habits and individual lifestyles and preferences are manifold and varied. Against this background, the megacities studied show a broad picture of mobility behaviour and transport conditions. In consequence, there is no single simple solution to respond to the major challenges of sustainable mobility.

This chapter sums up the findings from two workshops on challenges and strategies for the future of mobility in megacities with invited experts. It will be shown how the approach of mobility culture could help to develop successful local solutions to global challenges in a bottom-up/top-down approach.

Multiple Challenges for a Common Future

There are multiple challenges to be faced when considering sustainable mobility in megacities. These include global factors such as energy scarcity and climate change, demographic trends, economic globalisation, accessibility, and spatial and social inequalities. These common global challenges interact with the cultural, social, historical and economic conditions unique to each urban context and are demanding specific local solutions.

Travel Demand

Business experts foresee a further global growth in travel demand [29]. The volumes are expected to grow continuously and exponentially, more than doubling within the next 40 years, today's approximately 35 trillion annual passenger-kilometres. This represents a major challenge for both the global environment and also affected local environments. Average annual growth rates are estimated at between 0.7 and 1.2 % in the Organisation for Economic Co-operation and Development (OECD) member states, and up to 3 % in Latin America and China.

There are, however, countervailing opinions, backed up with data, which suggest that at least in highly developed economies, mobility has reached a peak and is declining [14]. There is some evidence that this peak applies specifically to car use within cities [17].

Whether such 'peaking' in mobility might occur also in the rapidly motorising countries, and therefore globally, is a moot point. Physical restraints such as density, road capacity, energy costs and regulation triggered by greenhouse issues may suppress private motorised mobility in cities. This is especially pertinent in the case of megacities, where such constraints can multiply.

What seems clear, though, is that in the near to medium future at least, the world as a whole faces something of a mobility explosion which must be accommodated in some way. Such growth in travel demand is already causing significant congestion, local air pollution, noise impacts, safety concerns (two-wheelers in particular), public health issues, degradation of public spaces, and deteriorating quality of urban life.

This research programme, however, has revealed that there is considerable diversity amongst cities. It has shown that the urban development pathways and resulting mobility patterns are highly dependent on the decisions taken by governments, private stakeholders, individual citizens and civil society groups. There is no inevitability about the direction of change.

Climate Change

Transport systems account for about one-third of global primary energy consumption. Up to 95 % of this transport energy demand has, thus far, been satisfied with fossil fuels, so that a direct link to carbon emissions, global warming and climate change is clear [10]. Fast-growing urban agglomerations are responsible for a major share of these phenomena even if their per-capita consumption of energy might be relatively low [27]. Until now, rising motorisation, together with increasing daily travel distances, can significantly negate lower fuel consumption per vehicle-kilometre from more fuel-efficient cars [15].

In consequence, the urban transport sector must contribute to mitigation strategies for limiting greenhouse gases and tackling the critical challenge of global warming. This problem necessitates clear policies for climate protection in urban transportation [19].

Energy Scarcity

The increase in energy demand and the finite nature and instability of fossil fuel supply, will in all likelihood lead to ongoing volatility in energy price levels, and potentially sharp price increases in crude oil, (as well as alternative fuels and electricity). The costs of mobility are already greatly affected by these pressures, and are likely to be more seriously so in the near future. Lower-income households are likely to suffer the most in a scenario of more expensive transport energy, because they can't afford less car-dependent inner area locations or the escalating cost of running multiple cars in car dependent suburbs [4]. In some cities in the USA and Australia, properties in distant low-density suburbs, which offer only poor transport alternatives to the car, have already been subject to depreciation.

Demographic Trends in Urban Development

From a global perspective, the ever-increasing human population will become more and more 'urban'. On a local level, this megatrend deserves detailed consideration and calls for great care in handling the consequences. 'Old' economies, such as Japan, the United Kingdom and Germany, have ceased population growth. Many of their conurbations are facing shrinking population and ageing societies, thus putting pressure on public services, including the viability of public transport. However, in different local areas, confounding trends of compact re-urbanisation and low-density suburbanisation can occur simultaneously.

Other parts of the world seem to show exponential trends in population, GDP and motorisation. In the rapidly developing economies, such as China, India and some Eastern European Countries, urbanisation due to migration from rural areas to urban hubs is very strong. However, this urbanisation often also leads to uncontrollable urban sprawl. This puts immense pressure on land demand and undermines the mobility advantages of high density, mixed-land uses and urban design amenable to non-motorised modes [11]. It also threatens the capacity of public infrastructure supply to keep up and can reinforce social segregation. The historically high urban densities of rapidly growing megacities may collide with the desire for more space per citizen and increasing motorisation rates, fuelled by higher income levels. In different contexts, high-income population groups may also seek central places and urban locations that offer more opportunities such as cultural amenities [18]. These diverging trends can create social inequalities between different parts of the city.

While such growth creates problems, it also provides opportunities that cities with stagnating populations don't have. BRIC countries (Brazil, Russia, India and China) and the second-tier economic powerhouses, or 'CIVETS' countries (Colombia, Indonesia, Vietnam, Egypt, Turkey, South Africa), contain over half the population of the world, and most of the fast growing megacities. This growth can be used to forge new directions in city development and mobility, provided that the right governance arrangements and policies can be implemented.

Processes of Globalisation

Megacities are important economic hubs in a globalised world. A worldwide network of knowledge, communication and trade relies on distinct nodes in which these commodities can be exchanged. Successful megacities will continue to act as strong economic centres of a country or region within a global network.

Added value in today's economies depends to a great extent on knowledge. However, knowledge economies and innovation processes are fundamentally rooted in face-to-face contacts and cannot survive by virtual communication alone. Physical proximity is needed for innovation. Cities offer abundant opportunities to meet and exchange, either by means of the geographical proximity of diverse urban functions, or by the relational proximity to other hubs provided by the connectivity of transport networks and mobility services.

Accessibility

In this sense, accessibility becomes a key factor in the mobility systems of megacities. Accessibility is a function of:

(1) The urban qualities of the locations in question such as physical extent, population size and density, availability of job opportunities, attractiveness of urban amenities, presence of points of interest, as well as,
(2) Transport supply defined by travel time, travel cost and comfort of the transport modes connecting these locations.

It describes the opportunities for participation in economic processes such as job markets, shopping activities and social life. It may well show major differences between one location and another in trip purpose, transport mode, moment in time or social group [7].

Accessibility changes can therefore lead to manifold effects on the level of economic, social and environmental development. Increasing travel speeds and physical spread might imply longer travel distances, social segregation and environmental damage. Increasing urban qualities might foster proximity, reduced travel need and support social inclusion. Accessibility planning can support a change in paradigm in urban mobility from speed to proximity [3].

Two levels of spatial scale seem today to be of great significance. The first is the immediate surroundings of home or job locations. Local neighbourhoods can provide a context where a great variety of activities take place in a small area, improving the viability of short, non-motorised trips. Locations with a poor urban structure, such as mono-functional residential areas or peripheral business parks quickly become car-dependent. Secondly, there is the wider functional urban area that represents the arena in which our daily activities and trips take place. Therefore, the location of housing, jobs, shopping, service and leisure facilities and how they are linked to one another is of great importance. These characteristics within a context of personal preferences are influencing to a large extent our individual travel options and mobility decisions.

Spatial and Social Inequalities

Urban structure and transport networks also have a bearing on the socio-spatial inequalities in megacity regions. Distinct urban functions are separated from one another according to economic agglomeration effects, regulatory policies and social clusters within megacities. Historically, the provision of areas with transport infrastructure has evolved with the social status of these areas. There is often a huge disparity in developments on a local scale (e.g. between east and west Ahmedabad, or within Hybrid Cities in the study, which have highly car-dependent fringe areas and highly accessible and diverse inner cities). These have important consequences for social segregation (as in Johannesburg, with its densely populated townships at the edge of the agglomerations). They result in major social and spatial inequalities in different neighbourhoods (conflicts and crime, safety issues etc.), which governance of megacities must address.

The social inequalities within megacities frequently correlate with the unequal distribution of transport infrastructure and networks. There are at least three local contexts that encapsulate respective mobility options.

(1) In some megacity areas, people depend on the private car as the only socially acceptable mode. The 'car orientation' of travel patterns and related housing types, is found, for example, in the car-dependent suburbs in North America, Australia or Europe. However, in European cities, there are numerous suburbs where public transport still provides a viable alternative for many, in contrast to the situation in the suburbs of American or Australian cities (see the case studies of Los Angeles [Chap. 9] or Atlanta [Chap. 8] in this book).
(2) The context of 'public transport for the poor' is found in those frequently low-income and densely populated areas where people rely on public transport for accessing other areas of the megacity. This applies, for example, to the *favelas* within Latin American megacities, where informal bus systems provide a mobility safety net for a captive clientele. Often public transport provision is of low quality in its vehicles, comfort levels, and frequency (for example in Sao Paulo—see "Chap. 7" in this book). In the case of Atlanta, there are the twin problems of low density and very poor public transport service levels resulting from sprawl (see "Chap. 8" in this book).
(3) The 'multimodal' context is characterised by a mixture of private cars, attractive public transport, and even highly attractive non-motorised modes. 'Multimodal' areas are thus attractive residences for a diversity of social groups. Ideally, these neighbourhoods display a diversity of social groups and travel patterns. Examples of these neighbourhoods are to be found in most cities, e.g. Berlin (see "Chap. 10" in this book) and Ahmedabad (see "Chap. 4" in this book).

Individual Preferences and Habits

Individual and household travel patterns differ between and within megacities. Travel patterns are driven by a broad variety of social and technological factors. Some of these are 'hard' or 'objective' characteristics of the spatial and transport infrastructure, such as the accessibility of functions within megacities, the provision of attractive public transport, walking or cycling infrastructure, the price of transport and parking, and so on. Others, frequently called 'soft' or 'subjective' characteristics for travel choices, are often less easy to assess for the purpose of comparative analyses.

It is widely acknowledged that subjective preferences are key factors affecting the travel choices of individuals and households. Over time, each person develops travel routines stemming from an interaction of individual and household choices with social norms, attitudes and preferences. The term 'lifestyle' is used to assess a latent construct comprising the attitudes, values and orientations of individuals [8]. It is interesting to see how residents' lifestyles evolve, and how these interact with individual travel patterns. While the 'car-orientation' lifestyle archetype has long served as a model for the ambitious middle classes in evolving megacities, the case studies show a broad diversity of lifestyles and ambitions. Individual preferences or habits can be caused by in-built physical factors such as a sheer lack of choice, by cultural factors, or a combination of both.

Resilient Policies and Programmes

Megacities may appear to be self-organising urban and semi-urban structures or organisms. However, many conscious decisions continuously shape the forms and the urban, social, ecological and economic conditions within these built environments. The network structures of megacities and the highly diverse conditions within them represent an accumulation of these decisions. It follows that one of the key challenges for mobility in megacities, especially those with weak governance structures and poor public accountability, is to make consistently better and more transparent decisions involving cooperation between stakeholders at every decision-making level.

The case studies within the fellowship programme tell us that conditions governing living, working, and mobility in each of these cities are culturally highly diverse, almost to the point of being incomparable. Each city's own specific social situation and unique history of how relations have formed between stakeholders rule out a single governance model applicable to all of them. These analyses of mobility culture, including the political culture of a city, suggest that governance models need to be found and developed which bring together the most powerful stakeholders, as well as the most vulnerable actors in the field.

It becomes obvious that the design of sustainable mobility for megacities is a highly cultural topic. Processes of governance, and the mode of organisation of actors for generating consensus and solutions, depend on the way that relationships have historically grown up between political, economic and civil society stakeholders. As many case studies have shown, the development of rational policies to address the above-mentioned mobility challenges facing megacities is a complex task, which requires new political, cultural and social skills to guide them [5]. Megacities are under pressure to harmonise, or even sometimes just to cope with, the demands and needs of global, regional and local stakeholders and citizens [1]. Case-sensitive research and analysis are needed to develop appropriate and flexible political cultures in megacities.

Overall, some keys to more sustainable mobility are the development of consistent policies and models for the transfer of knowledge, the sharing of power and responsibility, and the implementation of feasible, step-by-step development paths. Otherwise any business model, new mobility service or product will tend to fail in an environment that has not been sufficiently understood.

The Mobility Culture Approach as Key to Consistent Strategies

Mobility culture, as introduced in Chap. 3, is the theoretical framework underlying the case studies and the analyses developed in this book. For the analysis of megacities, a theoretical approach is needed which can deal with the enormous complexities of these places. Although there are still some open questions in defining the elements of the mobility culture concept, we found that for implementing sustainable transport policies in urban areas, understanding the city's mobility culture is critical. Transport planning and related policies have for a long time focused on the built environment and provision of transport infrastructure. Using the mobility culture framework, light is also shed on other elements of megacity governance, such as preferences, lifestyles and discourses (see Fig. 3.1 in the Reader's Guide).

Mobility culture could be considered as a weak or 'soft' factor in transport policies, planning and engineering. In fact, these cultures are extremely hard, stable and resilient when it comes to the question of how they can be changed, for instance towards more sustainable mobility. Of course, there is always change in society. New lifestyles and products emerge, and every day in every place new cultural practices of social interaction and community develop. But there is no simple or linear way of generating a specific mobility culture. Many different actors, stakeholders and individuals working at many different social, political and economic scales contribute to the mobility culture of nations, regions, cities and social groups.

Mobility cultures are based upon societal principles, values, norms and regulations, which are deeply embedded in social practice. They are also etched into political routines, the design of products and services, and the everyday mobility of citizens. The association of mobility with autonomy, freedom and individuality originates from ancient times [13]. The practices and logic people and institutions apply in their everyday decision-making constantly refer to these general assumptions and interpretations of mobility [6, 12, 20, 23]. Political and economic actors apply these when they are active in debates about the present and decisions about the future of mobility in megacities.

Easy and simple transformation of mobility cultures is impossible. City differs from city when it comes to their specific opportunities and their political programmes, the cultural and consumption preferences of consumers in them, and their mobility behaviour. They also display great variety in national and regional transport systems, the availability of infrastructure, transport modes, parking spaces, and ways in which both public transport and public spaces are used. However, from the case studies and the mobility cultures framework, we have been able to derive five important factors crucial to developing more sustainable mobility in megacities:

(1) accessibility,
(2) the business sector,
(3) emotional geography,
(4) education, and
(5) governance.

In the case studies we found great differences in accessibility, which concern land-use and transport developments, and how human and natural environments can be accessed and occupied in megacities. Questions relating to the public realm, the liveability of urban spaces and the sustainability of mobility provision and infrastructure, play an important role in how people can appropriate and 'embody' these spaces, and how mobility cultures can be influenced and transformed. So-called 'rational' decisions in transport and mobility depend to a very large extent on peoples' needs, wishes, values and 'mobility potentials', expressed in their ability and readiness to pay for transport and mobility.

When developing a comprehensive approach we need to consider the business sector: the companies themselves, their corporate mobility management, and their motivations for and openness to develop new, more sustainable services and products. There is a need to introduce the political framework in which a transition towards sustainable mobility services and products can be effected. This can be achieved by 'conditioning' the markets. Setting the right conditions and constraints, from a social and environmental perspective, is one of the key tasks for mobility governance and regulation within global megacities.

The emotional geographies of megacities also need to be considered as a key factor in shaping the conditions for mobility within megacities. Economies, communication, media and interaction today are global and local—at the same time—glocal [21, 26]. Mobility cultures are local and global by nature. The emotions and everyday cultures of citizens and customers need to be included, not only for improving the acceptance of public and private policies, but for understanding the socially innovative potentials of mobility in everyday life. Mobility concepts that take for granted the rational decisions of individuals will generally miss the point and frequently fail. Innovative policies in mobility and product development need to aim at emotional and not just rational behaviour [24].

Besides these already mentioned factors, education and the communication of mobility ethics, norms and values play an important role in establishing and influencing mobility cultures. Education is part of the communication between individuals and society and it affects the individual's mobility needs and wishes. 'Learning mobility' occurs in families, but also in kindergartens, schools, universities and professional education. Therefore, it makes a big difference how ethics and responsibilities about ecological, social and economic aspects of mobility are embodied in education and life-long learning.

Finally, the most important success element for sustainable mobility in megacities is governance. It needs innovation in cooperation between private and public, individual and collective, state, economy and civil society. We opt for a bottom-up—top-down approach in setting up innovative solutions, products and services. This should also take into account the integration of new policies into society, meaning that neither the state nor the economy alone can establish sustainable mobility cultures and solve the challenges of mobility. In other words, instead of relying on hierarchical concepts of regulating and governing mobility, the future of mobility in megacities needs to be conceived as a collective and societal task of mobilizing all institutional resources available for making better places [9].

The Strategic Task: Finding Diverse Solutions for Unique Cities

Strategic approaches to the above-mentioned challenges and specific framing conditions have been discussed in the expert workshops of the fellowship programme, against the backdrop of the mobility culture concept. Even though every local situation requires the development of individual solutions, some general components of these solutions have been considered. In what follows, selected aspects of this discussion are developed.

Perspectives to be Addressed

Megacities across the world do not only face global challenges that are common to all; their development also has to take into account local conditions that are unique. Cities contribute significantly to these global challenges and must be part of their solution. Mobility Culture we have argued, is a suitable framework to develop appropriate strategies for sustainable urban mobility.

How can the megacities of today handle their responsibility to contribute to the development of tomorrow's world? The Brundtland Commission's declaration on sustainable development [28], states that our individual needs should be fulfilled. However, this must not jeopardise our common resources or limit the options available to future generations, or other regions, for fulfilling their own needs. So, looking beyond our individual, local desires and actions, we have also to consider and respect our common constraints at the global level.

It seems that one of the key problems in sustainable development is that we can never be sure to end up with a system-wide optimum by combining individual behaviour that is motivated by user-optimising criteria. In that case, how can we strive for sustainable urban mobility, while our daily individual mobility choices may be heading in the opposite direction?

In this conflict between individual decisions and common targets, the system boundaries used are crucial. Top-down and bottom-up strategies have to meet.

Efficient internal market mechanisms have to be framed by clear boundary conditions, set by the public. In consequence, we have to develop consistent and coordinated new policies that will assist in putting these well-intentioned approaches into practice.

Be Visionary: And Pragmatic

The general objective is clear: We urgently need to develop a more sustainable mobility system. But this is not an easy, linear optimisation problem. It is more complex than just modelling optimal prices. Besides, there might be no 'optimal solution', no equilibrium state for a dynamic, non-linear, complex system like this. The following questions need to be answered in problem-solving strategies: What, precisely, is the objective function? Which are the multiple targets and interests, and by whom are they addressed? What are the framing conditions? Which specific constraints apply?

We have to take up the challenge, which involves setting a common vision and clear targets, and developing shared strategies that enable realistic steps to be taken. Cities should never let mechanistic solutions lead the future through self-fulfilling prophecies. Cities need to have a vision of their future, and mark their steps in policy and strategy terms by yearly increments with a view to accomplishing those visions. Any successful visioning exercise requires both bottom-up, community driven activities, and top-down government involvement.

In summary, we should be less illusionary in our thinking, but more visionary in our ambitions, and more pragmatic in the realisation of our projects.

Think Globally, Act Locally

The slogan "Think Globally, Act Locally", adopted by the Agenda 21 movement, has a strong application to mobility in megacities. The spirit of this idea, derived from the UN Conference on Environment and Development that took place in Rio de Janeiro in 1992, should be clearly maintained and even reinforced according to today's challenges of globalisation. Every city is unique, so there cannot be one single answer, one simple solution to the challenges that present themselves. We need new, creative solutions for 'our city', which requires investment in an understanding of the local context. It is for this reason that we study how different cities try to tackle the global challenges at a local level. From these local, case-study-based observations, we can try to develop a general understanding of processes, and thereby generate knowledge and experiences that are transferable to more general situations.

Understanding the complexity of different scales, the local, regional, national and global levels, is needed. We need to consider the interactions between these multiple and overlapping scales. What kind of bottom-up strategies, or top-down mechanisms, can we apply? Can we develop joint strategies for urban solidarity and the governance of sustainable mobility? What kind of formal or informal instruments support the integration of local knowledge and interests into common societal solutions?

On the local scale, new cultures and instruments of participation need to be found that link citizens to policy decisions that take place at higher levels. This will avoid political frustration, and improve acceptance of policies. Decisions and solutions agreed upon at the level of a metropolitan region must then be sufficiently reliable and resilient to be implemented and secured at a lower level.

Consider all Modes of Transport

When looking at transport systems, the above cooperative approach would have to consider all modes of transport, weighing up their various potentials and challenges.

Walking and cycling will probably gain greater stature in sustainable urban mobility concepts. Historically, urban mobility has always been based upon non-motorised travel. In many urban places, more than 50 % of all trips are still non-motorised. These modes are likely to become central to urban regeneration and city-friendly mobility. Non-motorised travel can satisfy individual needs locally while helping address societal challenges at a global scale. With the appropriate urban structure embracing density, diversity and design and attractive networks, walking and cycling could attain a high modal share [22].

Public transport certainly has a vital role to play as a backbone for structuring large-scale urban agglomerations. It can support the development of a polycentric urban system through better access to and attractiveness of centres [2]. Its competitiveness against the private car will, however, have to be demonstrated, not only in travel times, travel costs and individual comfort, but also energy efficiency, local pollution and global greenhouse gas emissions.

Private motorised travel will no doubt continue to be a strong mobility option, and a driver of local development, but it remains an illusion to imagine building attractive and resilient cities that are based predominantly upon the car. Rapid motorisation, regardless of the car technology, will always conflict with one of the most important urban resources: space.

Make Space a Value

Urban space is a scarce and precious resource, especially in growing megacities. Conflicts over space should therefore be reflected upon carefully. To which user groups is public space being allocated? Engaging citizens as local experts would likely help.

Space allocation problems can be partly overcome through subway systems, elevated roads or tunnels. However, new challenges can emerge, such as financing investment and maintenance, integration of urban design, and arguments about aesthetics.

Congestion might be a serious problem for many car users in peak hours, but capacity constraints can also be considered as "limits to growth", offering a form of system self-regulation. Rather than combating congestion by providing additional urban roadway, mobility policies can improve mobility options, allocating more dedicated space to non-motorised modes and public transport and helping to attract people from cars. Fair and equitable solutions to urban space allocation probably need to include altering long-term spatial travel patterns by changing urban structure, providing alternative modes and traffic management schemes which organise vehicle flow more efficiently.

Some very effective measures to overcome the mismatch of travel demand and road space supply rely on mobility pricing schemes including parking management. Certainly, charging for congestion is a rarely practised, but potentially effective route to follow. The local and global effects of such pricing policies, synergies (for example, climate protection plans), and conflicts (involving, for example, social factors) must be studied within the local context.

Generally speaking, making the scarce resource of public urban space into a value certainly can assist in achieving decisions that treat all modes more equitably.

Manage Transportation Demand

With many megacities having both growing populations and economies, motorisation and vehicle mileage have increased accordingly. As soon as certain thresholds of disposable income and household budgets are surpassed, the demand for individual motorised transport in many cities rises to a point where infrastructure provision cannot keep pace. Coherent urban development, together with network design, will not of itself be able to satisfy the demand, although the provision of, for example, excellent public transport infrastructure integrated with dense, mixed-use development can help a great deal [16]. A lack of space, compounded by the absence of sound, integrated land-use/transport planning in many dense, fast growing cities, often leads to rapid suburbanisation.

Mobility pricing schemes, therefore, are more and more frequently being considered as the only strategy for the relief of congestion. They also are a very much-welcomed resource for financing local transport. Transport for London, for example, is obliged to spend the revenues of the charging zone on multimodal transport policies. Stockholm, with its electronic urban toll system, not only reacts to peak hour congestion levels by adjusting time-dependent tariffs, but also allocates the money collected to implementing strong improvements to the supply of public transport. Because of this clear public benefit, the pricing scheme is widely accepted among the citizens.

The attractiveness and amount of driving in cities are also highly influenced by parking regulations. Cutting down the supply of on-street parking space will provide scope for reorganising public space to create attractive urban qualities. Cheap and abundant parking, which extends even to the central business district (CBD) of a city, as it does in Los Angeles, Houston, Phoenix and other car-orientated cities worldwide, certainly won't contribute to urban attractiveness. On the other hand, a lack of parking space, especially for residents, might reinforce the suburbanisation processes. New urban development does mostly need parking space, but the question is how much?

Not only does the overall amount of parking have to be considered, but, by use of balanced parking management schemes, the parking pricing and policies applying to different user groups (residents, commuters, visitors) should be regulated. This can motivate them to use alternatives to the car, thereby reducing the amount of traffic simply searching for parking. Strategies for tackling the problem of commuting to work, such as payment for a parking space are particularly needed. When combined with the subscription to a monthly public transport ticket, or even 'cash-out' regulations, whereby employers are required by law to give employees who have a parking space the choice of using that free parking or being paid its equivalent cash value, could function as strong incentives to modify behaviours [25].

Stay Active in the Policy Process

Public bodies have to face up to their responsibility to ensure societal goals and respect environmental constraints, while at the same time supporting development of economic activity. Institutions have to be trustworthy in their areas of responsibility, and must be held accountable in their fields of competence. A clear policy should be formulated, based on a comprehensive participation mechanism, which increases awareness and acceptance of policy. Consequently, political strategies, once they have been decided at a higher level, are more likely to be respected and implemented at the local level.

Working on open systems such as the transport network and mobility patterns of a megacity represents a challenge to institutions whose areas of authority and competency are clearly restricted to limited spatial areas. Functional territories and organisational territories never can perfectly be in line with each other, so we have to look for cooperation and coordination amongst stakeholders.

These policy processes, of course, form part of a cultural setting and occur within a continuous history, not in an instant of time. They are influenced not only by the current 'power game' of stakeholders, but also by political systems, physical geography and local habits. Increasingly, we have to consider, in a pluralistic society, the manifold concerns and attitudes that exist, and both individual and common values, as these relate, for example, to consumption. There is no single public interest, but rather many 'public interests'.

Develop Creative Solutions Together

For substantial innovation to take place in the mobility sector, creative solutions are needed. Competing partners (for example municipalities within an urban metropolitan area) normally can only be led to cooperate when the gain in terms of common benefits exceeds the individual investment. However, it becomes more and more clear that real success in sustainable urban mobility cannot be found alone, but only in a collaborative way.

The Mobility Culture Approach might help to de-code the hidden interwoven relationships. It will make clear that overall success has to be based upon bottom-up processes that attend to individual needs and interests, translated into top-down integrated policies that address not only those individual needs, but also the societal needs on a system level. It becomes obvious that new solutions which address multiple crises will have to be developed from such bottom-up/top-down interfaces.

Acknowledgments We would like to thank all the fellows and experts within the fellowship programme 'Mobility Culture in Megacities' for the rich exchange and discussion that took place both within the group and in the workshops. This has been a fruitful experience for all of us.

References

1. Brenner N (2004) New state spaces: urban and the rescaling of statehood. Oxford University Press, Oxford
2. Cervero R (1998) The transit metropolis: a global inquiry. Island Press, Washington
3. Crozet Y, Wulfhorst G (2010) Urban mobility and public policies at a crossroad: 50 years after Hansen W, the paradoxical come-back of accessibility. Paper presented at world conference on transport research WCTR 2010. Lisbon
4. Dodson J, Sipe N (2006) Shocking the surburbs: urban location, housing debt and oil vulnerability in the Australian city. Urban research program, research paper 8. Griffith University, Brisbane
5. Flyvbjerg B (1998) Rationality and power: democracy in practice. University of Chicago Press, Chicago
6. Freudendal-Pedersen M (2009) Mobility in daily life. Between freedom and unfreedom. Ashgate, Burlington
7. Geurs KT, van Wee B (2004) Accessibility evaluation of land-use and transport strategies: review and research directions. J Transp Geogr 13(12):127–140
8. Götz K, Ohnmacht T (2012) Research on mobility and lifestyle—what are the results? In: Grieco M, Urry J (eds) Mobilities: new perspectives on transport and society. Ashgate, Farnham, pp 91–108
9. Healey P (2010) Making better places. The planning project in the twenty-first century. Houndmills Basingstoke Hampshire, Palgrave Macmillan, New York
10. Intergovernmental Panel on Climate Change (IPCC) (2007) Contribution of working group III to the fourth assessment report of the intergovernmental panel on climate change
11. Kenworthy J (2010) An international comparative perspective on fast-rising motorisation and automobile dependence in developing cities. In: Dimitriou HT, Gakenheimer R (eds) Transport policy making and planning for cities of the developing world. Edward Elgar, London, pp 74–112

12. Kesselring S (2008) The mobile risk society. In: Canzler W, Kaufmann V, Kesselring S (eds) Tracing mobilities: towards a cosmopolitan perspective. Ashgate, Burlington, pp 77–102
13. Leed EJ (1991) The mind of the traveller: from Gilgamesh to global tourism. Basic Books, New York
14. Millard-Ball A, Schipper L (2011) Are we reaching peak travel? Trends in passenger transport in eight industrialized countries. Transport Reviews 2010 pp 1–22)
15. Newman P, Kenworthy J (1988) The transport energy trade-off: fuel-efficient traffic versus fuel-efficient cities. Transportation Research 22A(3):163–174
16. Newman P, Kenworthy J (2006) Urban design to reduce automobile dependence. Opolis: Int J Suburban Metropolitan Stud 2(1):35–52)
17. Newman P, Kenworthy J (2011) Peak car use: understanding the demise of automobile dependence. World Transport Policy and Practice 17(2):31–42
18. Newman P, Kenworthy J (1999) Sustainability and cities: overcoming automobile dependence. Island Press, Washington
19. Newman P Kenworthy J (2011) Evaluating the transport sector's contribution to greenhouse gas emissions and energy consumption. In: Salter R, Dhar S, Newman P (eds.) Technologies for climate change mitigation-transport sector. UNEP RISO center on energy, climate and sustainable development and riso DTU National laboratory for sustainable energy
20. Rammler S (2001) Mobilität in der Moderne: Geschichte und Theorie der Verkehrssoziologie. Edition Sigma, Berlin
21. Ritzer G (2010) Globalization. Wiley, Malden
22. Schiller PL, Bruun EC, Kenworthy JR (2010) An introduction to sustainable transportation. policy, planning and implementation. Earthscan, London
23. Sennett R (1996) Flesh and stone. The body and the city in Western civilization. W.W. Norton & Company, New York
24. Sheller M (2005) Automotive emotions: feeling the car. In: Featherstone M, Thrift N, Urry J (eds.) Automobilities. Thousand Oaks, SAGE, New Delhi, London, pp 221–242
25. Shoup D (2005) The high cost of free parking. The American Planning Association, Washington
26. Swyngedouw E (1997) Neither global nor local: 'globalization' and the politics of scale. In: Cox K (ed.) Spaces and globalization. Reasserting the power of the local Guildford, New York, pp 137–166
27. Trubka RL (2011) Agglomeration economies in Australian cities: productivity benefits of increasing urban density and accessibility. Ph.d thesis. Curtin University, Sustainability Policy Institute
28. United Nations (1987) Our common future. Report of the world commission on environment and development, published as annex to general assembly document A/42/427, Development and international co-operation: Environment 2
29. World Business Council for Sustainable Development (2004) Mobility 2030: meeting the challenges to sustainability. The sustainable mobility project. Full report 2004 http://www.wbcsd.org/web/publications/mobility/mobility-full.pdf

Epilogue: The Seven Mobility Culture Temperaments of Cities

Tobias Kuhnimhof, Johan Joubert, Laurel Paget-Seekins Ivo Wengraf, Swapna Wilson, Gunter Heinickel, Ziqi Song, Sylvia He and Marcela da Silva Costa

The opening chapters of Part One of this book introduced us to the world of megacities and the numerous challenges and pressures which they face. With regard to transport, there are many factors driving its path of progression, such as demographic developments and increasing motorisation, which constantly force cities to move on. The second chapter of Part One presented an approach to categorising cities worldwide on the basis of quantitative indicators. This categorisation aided us in navigating the overwhelming complexity of global megacity mobility culture, as it provided for the identification of certain groups of cities and stages of development.

Our 'Reader's Guide to Mobility Culture', which links Part One of the book to the rest, began by introducing four key dimensions that shape and constitute mobility culture:

(1) spatial structure and transport supply;
(2) policymaking and governance;
(3) perceptions and lifestyle orientations; and
(4) mobility behaviour.

This four-dimensional model represented our first step in making mobility culture something more tangible after the relatively abstract discussion of it in the introduction of this book. As a second method of approaching mobility culture, the Reader's Guide also reinterpreted the results of the preceding chapter's cluster study analysis, presenting the clusters as a tool for understanding and interpreting *change* in urban systems: seen like this, they permit us to map the way in which cities move on and to identify the challenges which cities face as they journey into the future.

However, if we really want to understand mobility culture, we ultimately have to ask the question: "What determines how cities move on?" It is through the close scrutiny of changes in transport systems and travel behaviour that we can

Institute for Mobility Research (ifmo) (ed.), *Megacity Mobility Culture*,
Lecture Notes in Mobility, DOI: 10.1007/978-3-642-34735-1,
© Springer-Verlag Berlin Heidelberg 2013

understand the underlying patterns of mobility culture—the values, conventions and social practices associated with mobility. Hence, as a third approach to analysing mobility culture in the Reader's Guide, we set out a number of research questions which are key to understanding processes inherent to the mobility system, and to preparing for the future:

(1) Where have we come from?
(2) Where are we today?
(3) What does the future hold?
(4) What action needs to be taken?

By using these first three approaches as tools, we were in a position to begin our voyage of discovery into the mobility culture of megacities worldwide in Part Two of the book: the four-dimensional model of mobility culture provided a breakdown of the entities that we should look out for; the cluster analysis provided a framework for positioning a city in terms of its evolutionary path into the future; the four research questions allow us to look 'behind the curtain' and understand the nature of changes taking place.

Equipped with this toolbox, we moved on to the 'Stories from the Megacity' in Part Two of the book. These stories take a closer look at specific cities, and at particular facets of mobility culture that these cities represent, together with the associated changes that have taken place in them over time. We found that each of these city stories in fact consists of many interwoven storylines describing how the city's transport system and travel behaviour have been and continue to be shaped. These stories thus shed light on fundamental facets of mobility culture in each of these cities. They do make use of quantifiable indicators—but even more importantly, they give us an insight into the characteristics of the different actors who are shaping mobility culture. Moreover, they describe mechanisms of interaction between these different actors, or between them and the pertaining framework conditions.

As the fourth and final approach to mobility culture, our Reader's Guide encouraged readers to look out for common denominators in the city stories found in Part Two of the book: even though the city stories confirm that each of the cities presented is unique, they also show that some of the characteristics of relevant actors, and some of the interaction mechanisms also, are so fundamental that they appear in almost every story. These common denominators form common threads running through all the city stories, and thus represent, as it were, generic character traits of cities. Of course, these character traits are not equally pronounced in all study cities, and they impact on the development of the transport system differently in different locales. Therefore, we feel justified in labelling them 'mobility culture temperaments'.

In the following pages we present seven important mobility culture temperaments. Even though each of these threads runs through almost every one of the megacity stories, their relevance differs from city to city. It is the individual mix of temperaments that constitutes the mobility culture character of a specific city.

Heritage and Inertia

The present urban form is shaped by history—and sometimes stands in the way of improvement.

Cities and their transport systems have been shaped in a certain era. In the case of many cities, it helps to identify the time period or periods which were characterised by the strongest population growth, in order to understand the principal characteristics of the local urban transport system: European metropolises such as London and Berlin experienced their most rapid growth in the late nineteenth and early twentieth centuries—an era characterised by the emergence of modern public rail transport, before mass motorisation set in. On the other hand, many US cities—and particularly cities in the Sunbelt such as Atlanta—experienced rapid population growth during the era of mass motorisation, and at the same time as suburbanisation and the development of the interstate highway system. In the urban fabric of today's European and American cities, the way in which these eras of technological development and urban/transport planning role models have shaped them can still be seen. However, there are also other phenomena belonging to the zeitgeist of cities that leave their imprint on them, such as the political system—as can be observed in South Africa and China.

As a result, cities carry a historic heritage: their present-day urban form has been shaped by their history. This is often a positive thing, in that it makes cities unique and gives the populace something to identify with. Nevertheless, this is often also a restriction, because the given built environment cannot easily be changed, giving the cities an inertia that obstructs further development of the transport system.

The megacity stories in this book have provided ample examples of this mobility culture temperament as representing a fundamental characteristic of each city's transport system.

Firstly, history has left its mark in the spatial distribution of the population. Johannesburg and Atlanta, both places where racial segregation is ingrained in settlement patterns and has made obvious impacts on the prevailing mobility culture, provide examples of this phenomenon. Since such settlement patterns cannot be changed easily, they are likely to continue shaping the transport systems of these cities in the future.

Secondly, the transport system itself and its various components have also been shaped by history. This is epitomised by the Victorian, Edwardian, and Georgian infrastructure of the London Tube: many historic tunnels are too small by today's standards, but owing to cost and complexity issues are hard to modernise—not to mention that parts of the Tube infrastructure are listed as historic and cannot be altered. In contrast, some of Berlin's infrastructure is oversized, reflecting the megalomaniac visions of a bygone era. However, the heritage characterising an urban transport system concerns more than simply the built infrastructure. The iconic double-decker Routemaster bus, which has recently been reincarnated in new modernised form in London, despite the cost advantages of alternatives, is an example of the relevance of tradition in transport systems.

Finally, heritage can also be seen in travel behaviour. Travel habits of entire populations cannot be changed quickly, and a certain inertia characterises the development of travel behaviour. However, the city stories also provide ample evidence for the rediscovery of that time-honoured mode of travel, cycling, and its far more ancient companion, walking, in places as diverse as São Paulo and London. Likewise there are new perspectives on a renaissance of once-dominant public transport modes such as streetcars (trams) in American cities. Such developments can be partly interpreted as the resurfacing of mobility culture heritage.

Overall, we have been able to show that in some cases historical heritage shapes the transport system because it is ingrained in the infrastructure, but sometimes because it is written into the traditions and habits of the citizens also. Respecting a city's heritage—whether willingly or because obliged to—enforces a certain inertia on decision-makers and travellers; this is a fundamental characteristic of urban transport systems, and one which shapes all of the study cities to a greater or lesser extent.

Progressiveness and Prematurity

Cities can make change happen fast—sometimes too fast.

Despite this resistance to change—such as that which originates from the inherited built environment—cities are nevertheless great agents of change. Transport systems are constantly being developed, and travel behaviour also adapts and evolves as time goes on. As this process continues, those who shape the transport system, and those who travel on it, are not merely forced to react—they design their transport systems proactively, and in many cases embrace progress with enthusiasm. And this attitude is not restricted to the spheres of policy and planning: the populace, particularly the travelling public, often supports progress with open-mindedness, and displays a willingness to adopt new technologies with alacrity. In some cases, however, it may be that this embracing of perceived progress is actually happening too fast: valuable elements of a city's past are being thrown overboard, and the new technologies or travel options that have supplanted them have brought new problems with them.

Among our study cities, Beijing is probably the most notable in its determination to be progressive—and in terms of the downsides resulting from this modernism. In the Chinese capital, the progressive mindset of the many players in the transport arena, ranging from administrators through industrial policymakers to travellers, has recast the prevailing mobility culture repeatedly in less than two decades—from that of a Non-Motorised City (as Beijing was categorised in the cluster study based on 1995 data) to an almost Traffic-Saturated City, which is now heading towards becoming a Transit City: with one major focus of the economic agenda being on the car industry, and fuelled by a growth in personal income, Beijing has been in an era of rapid motorisation since the 1990s. Elements of the inherited built environment that stood in the way of this 'progress', such as small alleys and old neighbourhoods particularly suitable for non-motorised traffic, were demolished—irreversibly reshaping the urban fabric. Increasing numbers of the city's population are letting their bikes rust, forgetting about inherited mobility patterns and embracing a car-dependent lifestyle. Recognising the downsides of too heavy an emphasis on the car, however, Beijing has recently again reacted with an amazing determination to effect rapid change: whereas two thirds of total transport infrastructure funding was allocated to roads in 2002, by 2009 priorities had changed to the extent that 70 % of that year's budget was spent on public transport.

Such quick turnarounds in mobility culture—motivated by an urge to modernise—are not unprecedented in the history of transport. The fate of public transport in American cities can serve as an example here: it took Atlanta 35 years, beginning in the 1890s, to develop an amazingly dense streetcar system. Twenty-five years after its heyday in the 1920s it was gone, destroyed in the name of progress. The fate of the Los Angeles streetcar system, which was once the largest in the world, was similar. Today, decades after the streetcars have disappeared, their revitalisation is being conceived as a sign of progress: Atlanta's first new streetcar line is scheduled to open in 2013.

In this book there are various examples of how the determination to make progress in the arena of transport shapes mobility culture. It can be seen in the open-mindedness with which the population of Ahmedabad embraces urban renewal projects, in the changing perception of cycling from a mode belonging to the poor to an expression of a healthy lifestyle in São Paulo, and in many more facets of the stories presented. All of these make it clear that it is the progressiveness—sometimes premature—of actors in the transport system that enables the overcoming of inherited inertia.

Diversity and Contrast

Transport can bring people closer together—and it can drive them further apart.

Cities not only differ one from another: the diversity *within* cities can be even greater than that *between* cities. This diversity can be an asset—but only if it makes available a variety of options to the travellers. In many cases, however, the diversity of cities does not give rise to any options for its population—at least for a segment of its population. Instead, diversity in these cases is associated with significant contrasts—often along racial or class lines. Moreover, the choices which travellers and decision-makers make often reinforce these contrasts.

Firstly, the megacity stories show that in places with a history of racial segregation such as Atlanta or Johannesburg, this heritage still shapes travel in the present day. The South African paratransit is locally often referred to as 'Black taxi'; the acronym for the Atlanta mass transit system, MARTA, is sometimes disparagingly reinterpreted as standing for 'Moving Africans Rapidly Through Atlanta'. In both cases, these modes serve parts of the population that do not have any alternative travel options. Even though this is today primarily due to economic conditions, there is obviously a strong racial component involved.

Secondly, the transport system not only mirrors contrasts in society, it often actually reinforces such contrast. This is because decisions have been taken, often many years ago, which still today reinforce segregation. In Atlanta, the mass transit system is limited in scope and extent as a result of the racism prevailing in the 1970s when it was created. A lack of access to public transport continues to restrict economic opportunities for low-income, primarily African-American, residents.

The chapter on São Paulo presents examples of how spatial segregation between rich and poor has been aggravated not only by housing policies, but also by the fact that key links in the transport network were until recently being designed exclusively for the wealthy parts of the population. Restricted to use by cars and no other form of transport (motorised or non-motorised), they thus provide yet more options for the haves and deny access for the have-nots who cannot afford a car, further reinforcing the contrasts between different socioeconomic groups within society.

Contrast and diversity don't only pertain to different socioeconomic classes or races. We find contrast and diversity also where medieval modes meet modern mobility, as on the streets of Ahmedabad; or where East meets West, as in Berlin. It is in the hands of the local decision-makers to determine whether such challenging diversity leads to clashes of various kinds, or to a reconciliation between the differing groups that yields an enriching diversity for the whole population. Thus diversity and contrast, as well as the way in which they are dealt with, represent a basic temperament that defines local mobility culture.

Creativity and Confusion

Necessity is the mother of invention—but which short-term solutions are long-term nuisances?

The challenges and problems that characterise urban transport systems often bring about enormous creative potential and innovative solutions: new modes of travel are invented, new technologies for managing traffic are created, new forms of revenue are found, new methods to avoid fees are discovered, and so forth. This creativity is one of the fundamental driving forces that bring about innovation—it is an expression of the ability of all sorts of innovators, and adaptors of innovation, to deal with the challenges imposed by local framework conditions. However, creativity can also lead to confusion, because it tends to circumvent regulation and bypass standardisation, leading for example to situations which are impossible to understand for outsiders and even for local planners.

Again, the megacity stories in this book provide ample examples of creativity in the transport system throughout all types of cities, no matter what their level of development. The range of creativity starts on the level of dealing with minor situations in everyday traffic—for example that facing the travellers on the streets of

Ahmedabad, who have to cope with the challenge of a myriad of vehicles ranging in mode from medieval to modern. With travellers used to such flexibility, however, the implementation of a Bus Rapid Transit system has presented a major challenge, as has been the enforcement of the discipline necessary to use it.

Creativity also impacts on short-term mode choice and long-term mode ownership decisions. In São Paulo, motorcyclists circumvent congestion; Londoners opted for the G–Wiz electric vehicle in large part so as to circumvent both taxes and charges imposed on ordinary cars.

Moreover, another megacity story shows how the creativity of entrepreneurs—making use of a loophole in the regulations—filled a gap in South Africa's transport system and created a business which today provides mobility for millions. However, creativity didn't end there: users and providers of the minibus system went on to develop a hand sign language to facilitate communication between each other. This, however, has led to a situation where outsiders are easily lost or even endangered when trying to use the minibuses. Moreover, confusion is not restricted to users unfamiliar with this backbone of the South African transport system: it is not even clear to what degree the minibus taxi business is regulated. On the one hand, the business is based on route-based operating licences given out by the government. On the other hand, the vast revenue created in this business do not form part of the formal economy.

These various examples illustrate what a fundamental driving force creativity is in the development of the transport system. This temperament of mobility culture often collides with the one presented in the next section. How these temperaments interact may be exemplified by the fact that in Beijing there also once existed a paratransit minibus system. However, this was simply eliminated, illustrating that cities differ significantly with regard to their ability to enforce change—the topic of the next section.

Control and Constraint

Leadership and governance are necessary—but can they go too far?

Having control over developments such as land use and provision of public transport is a fundamental pre-requisite for planning the future of local transport systems. Specifically, decision-makers must have authority, and they must also have suitable administrative procedures at their disposal. In short, they need to have basic control over what is going on. However, such control by administrations always has its limits, which are set at different levels depending on the political, cultural and economic context. In some locales, administrators and policymakers have all-embracing authority, taking control over even the details of the transport system, e.g. levels of car ownership or who is allowed to use their car at what times. In other places, however, this level of authority would not be acceptable, and residents would perceive it instead as a profound restriction of individual liberty.

The megacity stories in this book show that there are examples of authorities taking control over development in almost every study city. In many cases, this has to do with constraining vehicle use. Such regulation ranges from the London Congestion Charging scheme to the Peak Hour Operation scheme in São Paulo, in which a fifth of number plates are not allowed to enter the extended city centre during the morning and evening rush hours. Such measures, particularly if the regulation interferes with individual freedom (e.g. to own a car), are usually faced with opposition. Administrations need power to implement the measures. Let us look at Beijing as an example:

To start with, all urban land in China is owned by the government, which gives it substantial control over the development of settlement patterns—in other words, the control of the authorities in this country begins at the very root of travel demand generation. In former times, when the *danwei* system was still widespread, where people lived and where they worked were both heavily influenced by the authorities. This, however, has been abandoned, unleashing enormous demand for travel, which has been further fuelled by economic growth. To control this ever-increasing travel demand and vehicle growth, Beijing has again recently had to introduce several regulatory measures, among them a vehicle plate lottery which significantly constrains car ownership. Even though this lottery has come in for criticism, it is still perceived by the general public as a more equitable way to manage the limited new vehicle quota than market-based approaches, such as the plate auction system adopted by Shanghai. The existence of a difference in approach such as this even within China illustrates how the local *cultural*, economic and *political* context defines which measures for taking control over the development of the transport system are deemed acceptable.

We have seen that the ability and willingness to take control of the development of the transport system is a basic property of cities. The exact nature and extent of this ability is evidently strongly linked to the relationship between a city's authorities and its residents. Hence, in a sense, the following temperament—referring to the influence and power of the citizenry—represents the flipside to the controlling power exercised by the authorities.

Participation and Protest

Citizens participate in shaping the transport system—sometimes much to the dislike of decision-makers

Mobility not only moves people physically—it also moves them emotionally. Citizens often get very animated about transport projects, from the location of metro stations, to the redesigning of streets, to the introduction of new mobility systems. The reasons for such engagement are manifold. As residents, citizens may be affected by negative externalities of transport projects, such as noise or even devaluation of their property. However, contrary to what is often believed to be the case, we saw in the city stories that resistance to transport projects originating from a Nimby ('**n**ot **i**n **m**y **b**ack **y**ard') attitude is by no means the only reason for strong engagement about transport.

The chapter on Berlin describes the case of residents who oppose the upgrading of their neighbourhood streetscape, concerned that the original character of their residential environment could be irrevocably altered. As we saw in the São Paulo chapter, other people's Nimby attitude can ignite an even larger protest: in this city, the citizenry of a well-off neighbourhood opposed the construction of metro station in their vicinity, fearing an influx of unwelcome visitors. This opposition, however, unleashed an even larger protest against the influence of the wealthy on transport planning decisions—forcing authorities to disregard the wishes of these upper-class citizens.

It is not only residents and travellers who are moved to take action about transport projects. In the 1920s it was the shopkeepers in downtown Los Angeles who protested against a parking ban which was intended to provide for the free movement of streetcars through downtown.

However, participation does not necessarily mean resistance to change—it may also take the form of formal or informal support of measures aimed at change. Formal support may, for example, take the form of approval by voters. This was the case with the overwhelming success of a ballot on traffic relief and transport upgrade measures, which were supported by a two-thirds majority of Los Angeles county voters. A form of informal support is the positive attitude with which the citizens of Ahmedabad embraced their new Bus Rapid Transit system. This serves as an example of how publicity and education, by motivating participation, can lead to success in implementing changes to a transport system.

The many examples found in the megacity stories presented in this book illustrate a wide range of participation in—and protest against—transport projects. This popular involvement—and the way in which authorities deal with it—can be decisive for the success or failure of transport projects, in some cases leading to the need to make amendments to them, or even to their complete cancellation. Hence, participation, protest, and how these are handled are all factors that determine where urban transport systems are heading.

Virtuous and Vicious Circles

Mobility interacts with other domains of urban development—with both intended and unintended consequences

The transport system interacts in many ways with other spheres of life, and with the rest of the infrastructural environment. For example, it is a truism that land use influences transport, and vice versa. Planners and policymakers make use of this interaction when aiming to foster integrated land-use/transport planning. The purpose of this integration is to make use of the synergies arising from this interaction between the transport system and other domains of planning. Such synergies, such as that with public transport-orientated residential development, can make certain projects in the transport arena—such as the extension of public transport lines—work, which would not otherwise have succeeded. Synergy can also cause the combined result of individual projects to be of more benefit than the sum total of the advantages brought about by each individual measure. Likewise, planning for transport can be combined with communication, education and many other activities which influence urban development.

The megacity stories have provided us with numerous examples of how the development of a city's transport system is linked to other domains of urban life. This may be intentional, as is the case with public transport-orientated development, or unintended and coincidental, as in the case of the G–Wiz in London: here, public discourses and regulation of car travel in the central city shaped an environment that proved to be fertile ground for the rapid rise of the G–Wiz. Once these favourable conditions disappeared, the success story of the G–Wiz also came to an end.

Some measures in the transport arena are explicitly intended to stimulate positive development by initiating virtuous circles. This is exemplified by the discussion about a revival of the streetcar in order to rejuvenate the downtown of Los Angeles and stimulate local economic growth. Berlin probably provides the most idealistic example of this: here, the authorities convinced themselves that size and social diversity were in themselves drivers of economic growth. Since both can be strengthened through invigorating the city's centrality as regards transport, this has become a transport policy focus with the objective of stimulating economic growth.

However, these kinds of interactions between transport systems and other aspects of city life may also work in the opposite direction, bringing about vicious circles which are almost impossible to break. Segregation is probably the most pertinent example of factors that can lie behind this type of self-perpetuating vicious circle. In the megacity stories we have observed examples of segregation, in places such as Atlanta, where the lack of transport options reinforces poverty, which in turn restricts choice in many areas of life, including transport. Likewise, in São Paulo, the mutual stimulation of private investment and public investment in the infrastructure has created a vicious circle that contributes to the segregation of different segments of the population.

Overall, we have seen that developments in the urban transport system interact with other domains of life with consequences that are sometimes intended, sometimes unintended. Self-evident as this may be, it nevertheless challenges decision-makers to somehow break vicious circles and create virtuous ones. We believe that the emergence of such processes, and the ability to steer them—or even to let them work on one's behalf—is one of the temperaments of urban mobility culture.

Concluding Thoughts on Mobility Culture: Stepping into the Future

This book has tackled two concepts that have proved very difficult to get to grips with: megacities and mobility culture. Although there are definitions of what a megacity is, the problem is that their overwhelming diversity means that it almost impossible to make any generalisations about them. As for mobility culture, the problems start as soon as one seeks to find a workable definition for the term. In the introduction we first derived a simple composite definition based on the words 'mobility' and 'culture': *'mobility culture is the set of values, conventions, or social practices associated with the ability to travel from one point to another and with actual physical travel.'* In dissecting this definition and interpreting it for the urban context, we found that it actually encompasses almost any dimension of urban life which is even faintly related to transport—which again is somewhat unsatisfactory.

So is mobility culture like a bar of wet soap, forever eluding one's grasp? We don't think so—all you have to do is develop the right grip. This is what we attempted to do in the 'Reader's Guide to Mobility Culture': firstly, we identified tangible dimensions of mobility culture that constituted a four-dimensional model. Secondly, we concluded that understanding mobility culture requires the observation of change as it takes place in urban transport systems, and a study of the driving forces of this change. The megacities in Part Two of the book provided food for this type of study.

This takes us back to the other challenge associated with the subject matter of this book: the overwhelming diversity of cities. True, Priester and his colleagues had grouped megacities worldwide into clusters, providing a first step in helping to navigate this diversity. But, when visiting cities around the globe as diverse as London, Ahmedabad and Johannesburg, wouldn't we expect to find that there was little in common between these locales? Of course, in terms of their prevailing mobility culture these places differ significantly—indeed, all cities in the city stories were deliberately chosen to represent different facets of mobility culture. Surprisingly, however, we found significant common denominators in them as regards the dynamics which shape mobility culture. We labelled these dynamics 'mobility culture temperaments'. It appears likely that these temperaments are generic mechanisms which are at work in almost every city of the world, but vary in emphasis from one to another.

How do these temperaments link to our four dimensions of mobility culture? While the dimensions themselves are practically timeless, the temperaments bring in a temporal component, describing the mechanisms of interaction between the dimensions over time. Certain temperaments may thus pertain particularly to certain dimensions. For example, when relating to the built environment, the first temperament, 'heritage and inertia', can describe the forces exerted from the first dimension, 'spatial structure and transport supply', on all the other dimensions of mobility culture. Other temperaments are much more universal—for example, the 'virtuous and vicious circles', which describe in more general terms how positive or negative synergies can build up.

The strength of the mobility culture approach is that it captures these interactions between the dimensions of mobility culture. We have seen in the megacity stories that it is often these mechanisms of interactions, i.e. the temperaments, which are the keys to success or failure of desired changes to the transport system. Hence, for the fate of many single projects—and thus ultimately for entire urban transport systems—it is important for the main players to at the very minimum understand these mobility culture temperaments. Otherwise, they run the risk of finding that their efforts, whilst aimed at improving the transport system, end up gaining no traction or leading to unintended outcomes, simply because of an unfavourable mobility cultural environment.

Finally, there is the fine art of making mobility culture work for you: this takes us to questions such as how fertile ground for mobility culture can be generated, and how virtuous circles can be set in motion. If planners, and others who shape the transport system, were able to master these mechanisms of mobility culture, they would clearly be able to invest their resources much more efficiently.

Given all that has been said above, we are in a position to conclude with a statement that is true in two senses: as we witness the unfolding future of mobility around the world, it will be mobility culture that is our companion—for as this book has demonstrated, it is largely the prevailing mobility culture which determines how cities move forward on their evolutionary paths; so in this sense mobility culture accompanies us on the journey. But we also have set out on an intellectual journey: the adventure of understanding mobility culture. A commitment to deepening our comprehension of this vital concept and continuing to explore it has tremendous potential, giving us the opportunity, to plan our urban futures more efficiently. It is our fervent hope that this book has succeeded in its intention to demonstrate the reality of this potential, and to lead the way in this voyage of discovery.

Index

Institute for Mobility Research (ifmo) (ed.), *Megacity Mobility Culture*,
Lecture Notes in Mobility, DOI: 10.1007/978-3-642-34735-1,
© Springer-Verlag Berlin Heidelberg 2013

C (*cont.*)

D

E

F